DYNAMICS, EXPOSURE
and
HAZARD ASSESSMENT
of
TOXIC CHEMICALS

RIZWANUL HAQUE
Editor

ANN ARBOR SCIENCE
PUBLISHERS INC / THE BUTTERWORTH GROUP

Second Printing, 1980

This book consists of papers presented at the symposium entitled "Dynamics, Exposure and Hazard Assessment of Toxic Chemicals in the Environment." The symposium was held at Miami Beach, Florida, during September 11–13, 1978, and was sponsored by the Division of Environmental Chemistry, American Chemical Society. The content of this book does not necessarily reflect the views and policies of the U.S. Environmental Protection Agency, nor does the mention of commercial products or trade names mean endorsement or recommendation for use.

PREFACE

The Toxic Substances Control Act of 1976 is a historic milestone for the environmental sciences. This legislation has far-reaching consequences, bringing many challenges and opportunities to the chemical profession and industry. The central issue is the evaluation of exposure, hazard and risk of toxic substances to human health and the environment. Because of the complexities involved, scientists possessing expertise in chemistry, biology, ecology, toxicology, engineering, mathematics, computer systems and agriculture must pool their talents to address this issue. Predictive techniques must be developed to assess what happens to a chemical once it is introduced into the environment, in what form it exists, where it goes and when it is finally destroyed. Techniques are also needed to predict exposure concentrations of the chemical in the environment so that this information, in conjunction with "effect" data, may be used to estimate risk.

Chemical technology has played a major role in the discovery, synthesis and trace-level identification of chemicals. In recent years, basic chemical-physical principles have been applied to develop techniques for predicting transport and transformation of chemicals in the environment. The transport and transformation studies in turn provide valuable information on the exposure concentration and potential hazard of chemicals to the environment.

This book describes the use of transport and fate studies in predicting exposure, hazard and risk of toxic substances to the environment. It comprises papers given at the symposium "Dynamics, Exposure and Hazard Assessment of Toxic Chemicals in the Environment" at Miami Beach, Florida, September 11-13, 1978. The symposium was sponsored by the Division of Environmental Chemistry of the American Chemical Society. Representing a broad spectrum of scientists from industry, academia and government, the contributors are experts of national and international reputation.

The first few chapters of this book deal with the historic events pertaining to transport and fate studies, as related to the Toxic Substances Control Act. The addresses by Etcyl Blair (Vice President Dow Chemical Co.), Warren Muir

iii

(Deputy Assistant Administrator, EPA), Courtney Riordan (Associate Deputy Assistant Administrator, EPA) and Virgil Freed (Director, Environmental Health Sciences Center, Oregon State University) provide a good introduction to the role of transport and fate research and its potential in the industrial decision-making process, regulatory postures, federally supported research and impact on academic training.

Stephenson describes how transport and fate data has been used by the Interagency Testing Committee (formed under the Toxic Substances Control Act) in selecting toxic chemicals. His article is followed by introductory papers on analysis of chemicals (Keith), and review of transport and fate studies in exposure estimation and screening of toxic chemicals (Haque et al.). A number of chapters are devoted to a better understanding of the environmental transport and transformation process. They include the topics of photochemistry (Zepp; Dilling and Goersch), volatilization (Mackay et al.; Spencer and Farmer) hydrolysis (Wolfe), biodegradation (Alexander), transport in sediments (Weber et al.), humic substances (Khan), terrestrial microcosm (Gillett) and partition coefficients (Hansch). There are papers in which attempts have been made to integrate transport and fate information in developing models for predicting concentration, movement and persistence of chemicals (Neely; Mill; Eschenroeder et al.). Three papers are devoted to movement, accumulation and transformation in fish and animals (Bungay et al.; Barrows et al.; and Khan and Feroz). The paper by Fine et al. provides background information on the occurrence of toxic impurities in commercial products. Murphy describes toxicodynamics of chemicals and Linjinsky uses transport and fate concepts to predict carcinogenic potential of chemicals. The papers by Kimerle and Stern are concerned with the testing of chemicals for hazard evaluation. The concluding paper by Moghissi et al. is devoted to methodology for risk estimation of toxic chemicals.

It is hoped that this book will be of use to researchers, upper level college students and scientists involved in industrial and regulatory aspects of toxic chemicals. The opinions expressed are those of the authors, and credit must go to them individually. I would like to extend my personal gratitude to them for their contributions. I would also like to thank Virgil H. Freed and Warren Muir for their suggestions, and Courtney Riordan and Thomas Murphy for their support during the planning of the symposium, and editing of this book. I am grateful to the Division of Environmental Chemistry, American Chemical Society, for sponsoring the symposium. Technical assistance from Bill Vaughan and Lisa Yost during the editing of the book is also acknowledged.

Rizwanul Haque

CONTENTS

 **Rizwanul Haque** directs the Toxic Substances Research Program in the Office of Environmental Processes and Effects Research, Office of Research and Development, U.S. EPA, Washington, DC. Previously he worked in the Office of Pesticide Programs and in the Office of Drinking Water. Before coming to EPA he was Associate Professor in the Environmental Health Sciences Center at Oregon State University.

He earned his PhD degree in Physical Chemistry at the University of British Columbia, Vancouver. Dr. Haque has edited two books and has published more than seventy scientific papers in areas of transport, transformation, mathematical modeling, exposure assessment and testing of toxic chemicals and pesticides in the environment.

DYNAMICS, EXPOSURE AND HAZARD ASSESSMENT
OF TOXIC CHEMICALS IN THE ENVIRONMENT:
AN INTRODUCTION

Rizwanul Haque
> Office of Research and Development
> U.S. Environmental Protection Agency
> Washington, D.C. 20460

The development of modern technology has brought a dramatic increase in the production and consumption of chemicals. In a few cases, the benefits of chemical use have been accompanied by unexpected adverse effects. The persistence and bioaccumulation of mercury, polychlorinated biphenyls, kepone and dioxins are classic examples. Such cases have led to public concern that chemicals be fully evaluated in terms of potential risk before being approved for use. In 1976, the U.S. Congress enacted the Toxic Substances Control Act which requires the testing of chemical substances and mixtures for assessment of risk to human health or the environment. The evaluation of chemicals brings a dual challenge to scientists. First, a credible technology of risk assessment must be developed, and secondly, the toxic substances, which run into several thousands in number, must be evaluated rapidly with a cost-effective technology. This book addresses these important issues.

The "dynamics" of a chemical refers to what happens to a chemical, where it goes and when it degrades once it finds its way into the environment. The information of dynamics of chemicals is obtained via "transport and fate" studies. Transport and fate data play a key role in defining the exposure concentration of chemicals in the environment. The main objective here is to explore the full potential of transport and fate studies in the development of predictive techniques for exposure estimation of chemicals, and in defining the potential hazard of chemicals.

The estimation of chemical risk (R) requires a knowledge of its effects and exposure on humans and the environment, and may be expressed as:

$$R = f \text{ (Effects, Exposure)}$$

The data on effects are obtained via toxicological experiments. However, our current knowledge of environmental exposure concentration estimation is inadequate. Current methodology of exposure estimation involves monitoring techniques which are cost-ineffective and suffer from statistical limitations and unextrapolatibility from location to location. There is a need for the development of predictive techniques for exposure assessment. Transport and fate studies present good potential in developing such predictive techniques.

Transport and fate studies also play a role in evaluating hazards of chemicals. For example, the existence of a correlation between the octanol/water partition coefficient and bioaccumulation of chemicals in certain aquatic species may provide a screening method. Similarly, the persistence parameters of chemicals determined from transport and fate studies may also be useful in estimating potential environmental hazard. Finally, transport and fate studies provide a method of predicting in which medium (air, water, soil/sediment) of the environment a chemical is likely to accumulate and persist, thus presenting a red flag signal for that compartment of the environment. This in turn will aid the designing of complex monitoring experiments.

The progress of a chemical from its introduction in the environment to the production of exposure and risk, is depicted in Figure 1. As shown in the figure, the transport and transformation of the chemical in air, water or soil/sediment media is the first step in determining exposure concentration of the chemical. Once this concentration is determined, depending upon the toxicity of the chemical, its effect can be evaluated. This book will address important issues involved in defining the various steps leading to risk assessment. More specifically, it will focus on the following topics:

- the identification of important transport and transformation processes pertaining to air, water, soil/sediment and biota, the measurement of important parameters associated with the processes and investigation of factors influencing such parameters;
- how transport and fate information may be used in defining and building exposure concentration models for toxic chemicals in multimedia environments;

- the role of transport and fate parameters in testing of toxic chemicals, how this information can be utilized in defining hazard signals associated with toxic chemicals, and the utilization of transport and fate data in classification prioritization and selection of toxic chemicals; and
- interface and utilization of transport and fate data in performing ecological and toxicological investigations.

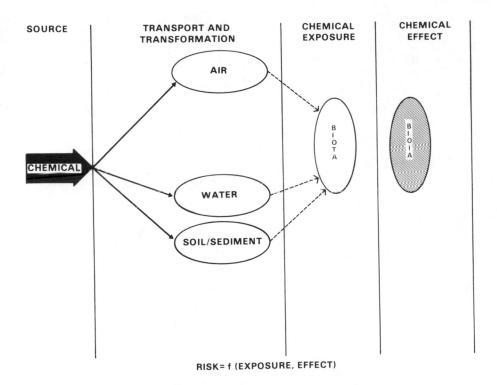

Figure 1. Toxic chemical: transport and fate, exposure, effect and risk.

In addition, the chapters by Muir and Blair delineate the importance of transport and fate research in the implementation of the Toxic Substances Control Act and its impact on the industrial decision-making process. Finally, it is hoped that this book will stimulate novel research ideas capable of solving complex environmental and health problems.

A ROLE OF TRANSPORT AND FATE STUDIES IN INDUSTRIAL DECISION MAKING FOR TOXIC CHEMICALS

Etcyl H. Blair

The Dow Chemical Company
Midland, Michigan   48640

A discussion of the decision-making processes dealing with the role of transport and fate studies cannot be limited to the specific role of these studies. Decision-making, from an industrial point of view, is much more complex and, to a large degree, encompasses a much bigger equation dealing with benefits and risks. This decision equation, of course, is not precisely defined, but it includes the varied benefits derived from the product or technology, counterbalanced against the varied risks—the risk to humans and the environment, the risk in capital investment and the risk of failure in the marketplace.

The role of the American chemical industry is to provide chemical products, intermediates, services and systems for the maintenance and betterment of society. Its origins can be traced from sulfuric acid, bromides and ammonia, to benzene, ethyl alcohol, phenol and formaldehyde; to the high polymers; and to the exotic and tailored structures in drugs, pesticides, photographic chemicals and space components.

Today, the American chemical industry is a high technology industry continuing to look for new products, new processes and innovative ways to support other emerging technologies—be they associated with energy production, the food industry, the housing industry, the clothing industry, the agricultural industry, the metals industry, the transportation industry, or, believe it or not, the U.S. government itself. Regardless of the customer, the product, the processes and the systems must meet a real and lasting need.

In an attempt to portray the role of transport and fate studies in industrial decision-making, it is necessary to understand a few of the events that lead from product development to marketing of an industrial chemical, paying particular attention to the broader aspects of safety to the public, to the manufacturer and to the handling of the products by customers.

The chemical industry generally accepts the mandates for more testing of old products, of careful evaluation of risk potential of new products, and of minimizing chemical incidents and unreasonable risk of injury to health and the environment. The industry is keenly aware of and is participating in the vast changes occurring in societal attitudes toward health and environment. Newly found skills and resources are now intensively channeled toward measuring minute quantities, coping with the uncertainties of government regulatory bureaucracies and explaining industry's actions and attitudes not only to the regulators, but also to the media, the public and even to the scientific community.

The chemical industry is, in reality, a segment of the American business sector with the capability of producing products which satisfy a need as expressed in the marketplace. Only in the western world do products originate by innovation and the supply and demand forces of the marketplace. All other parts of the world—the Communist Bloc, the developing nations and the awakening Third World—make use of western technology as they struggle to leave their primitive ways.

To adequately discuss decision-making on transport and fate, it seems useful to look at the ways that responsible industry has, for many years, paid attention to the concerns of the environment and public health and the systems it has evolved to ensure a minimum impact of its products and processes on the ecosystem.

Sound data are essential to every aspect of decision-making, and coupled with data are dependable technologies—technologies that are reliable and reproducible, and upon which industry can build plants for the present and for the future.

Data and technology, of course, include the research and development phases of commercial development. The manner by which most companies manage research and development is probably similar because of the common background of the scientists and engineers in the American universities, the commonality of business tradition, the rich heritage from those who pioneered the industry, and government regulatory policies.

As research progresses toward a commercial venture—be it in catalysis, in synthesis, in analytical methodologies or in process advancements—frequent assessment must be made of the value or the importance of the chemical substance, the formulation or the fabricated item. Although each company may have its own system for evaluation, it becomes mandatory that more and more individuals become involved. Multidisciplinary resources and talents are needed to deal with procurement, with the biological impact or potential hazards from the material, with the types of formulation or fabrication that may be needed, and even with how the product will be marketed and in what segments of society it will find a use. Thus, there is a need for managing broad ranges of resources—dollars, professional skills (people) and facilities (analytical equipment, pilot plants, etc.).

Industry makes use of planning tools to aid in managing the multidisciplinary resources required. Frequently, a modified critical path network system is employed. The critical path is the shortest time it takes to do the required jobs; it sets the schedule for the many different individuals and functions involved.

Funding and management of each of these factors affect the others. Various stages or phases of these operations have been developed in which there is an attempt to identify and quantify certain jobs that need to be done and the resources—people or dollars—that need to be applied to ensure that the project is conducted properly and on time.

A continuing part of the decision-making, of course, is a constant evaluation of the health and environmental aspects. Consideration is given to the plant site where the material is being made and to the customers who will be utilizing the product. Sophisticated data on transport phenomena and on the ultimate fate of a chemical are critical components of the decision-making process.

In industrial research, R and D activities are frequently segmented into four stages:

Stage I.   exploratory and synthesis
Stage II.   product and use characterization
Stage III. pilot process and field development
Stage IV.   commercialization

It is in Stage I, the exploratory stage, that scientists initiate effort on a problem where the solution will contribute an economic benefit to the company and fill a need that the user is able to identify and for which he is willing to pay. The search is for new concepts, potentially useful compounds and new ways of modifying existing products; the tools are the processes of innovation and invention.

Stage II represents a selection point. More resources are targeted on a given product and/or technology. And Stage III is further targeting of resources—more facilities, more disciplines and involvement of those outside the company.

Several aspects of this four-stage process should be examined. The progression is from many ideas and chemicals to one chemical contained in one or a few formulations for one or a few uses.

Knowledge of the potential use becomes more sophisticated through this progression. For example, a concern for the simple property of tensile strength in a polymer, becomes more sophisticated and turns to brittleness, to adhesive characteristics, to light stability and many more. A dream for a broad use frequently targets on a narrower, but highly unique use.

Toxicological data similarly go through a progression. Newly synthesized compounds may be characterized only as to acute toxicity to the rat, or be screened only for unique drug or pesticidal activity. But as the progression proceeds, and commercial success becomes more likely, the toxicological data base is increased; more species are tested and longer-term tests are conducted.

Environmental data similarly go through a sequence. Considerable insight into transport and fate characteristics can be gained from simple chemical and physical data collected in Stages I and II. Vapor pressure, water solubility, dissociation constants and hydrolysis and oxidation half-lives, along with a simple test for biological oxidation, permit informed judgments on transport and fate.

During the past decade, the proportion of commercial chemicals involved in significant environmental problems has been very low. And if the biologically active and, frequently, persistent pesticides are excluded, the proportion of problem chemicals is quite small indeed. Very few people can name even 10 out of the reported 75-100,000 commercial chemicals. If the total population of chemicals is considered, there is a bell-shaped curve of risk distribution.

This is emphasized here because the intensive and sophisticated study of transport and fate has been and probably will continue to be a highly specialized or targeted area of investigation. The commercial development of most chemicals should encompass an emphasis on simple and essentially predictive tests.

Global monitoring and modeling studies, extension kinetic analysis of degradation, and long-term wildlife studies should be considered for those few large volume chemicals which have large environmental release, are relatively persistent and exhibit relatively high toxicity.

While product development has been broken into four discrete stages, it must be recognized that there is actually a phase-in and a phase-out from each of these stages and that, in reality, many more key steps are involved. This sequence involves many integrated activities with key planning and evaluation, or review points. The process may take up to five years if extensive testing is required.

As mentioned earlier, a new chemical compound is synthesized or a new formulation is developed by an industrial scientist, and the product or formulation is put through a biological activity screening program. Part of the purpose of the screen is to find if the new compound or composition exhibits unusual biological activity for other uses as a pesticide, an herbicide, an antimicrobial agent for fabric or even a drug. These tests frequently can provide qualitative relevance to the health and environmental impact of the chemical.

If the new compound exhibits enough promise to make it seem, even at this early stage, to be a possible product candidate, safety and health are the next immediate concerns, and mammalian toxicological range-finding tests are instituted. These studies are designed to determine the capacity of the material to cause injury from acute exposure if it is inhaled or swallowed, or if it comes into contact with the eyes or skin. These studies must be done to determine if there is any significant degree of danger from incidental exposure to the compound. This information is necessary so that chemists can handle the chemical safely, and it should be made available to engineers if they need to design pilot plants. And this information will become part of the package made available to the customer, to the government and possibly to other segments of society.

Data from these early tests are used to predict the potential effects, if any, from short-term human exposure to the compound. In turn, these tests guide further toxicological testing.

Further into the development process, about the same time that the mammalian toxicological range-finding test is being done, the physical properties of the compound, such as water solubility, vapor pressure, melting and boiling points, and any extreme reactivity, including shock sensitivity, flash point, and differential thermal analysis for decomposition temperature, are generally being determined.

By this time in the life of the new compound or composition, there are usually enough data available on safety and efficacy that a preliminary decision can be made on whether it might be a serious product candidate. If the decision is favorable, environmental testing begins. The environmental analysis includes tests which indicate the compound's tendency to evaporate from water. Its degradability—that is, whether it breaks down readily in soil, air or water—as well as its tendency for possible bioconcentration, are also checked, and an effort is made to determine the new compound's basic toxicity to an indicator aquatic organism.

It is at this point that the first key product review takes place. Up to this point, decisions are generally made by individuals—chemists, biologists, engineers—who are personally involved with the product. A group review takes place and a management decision is made as to what will be done.

The direction taken is decided by the test results that have been developed to date. And now, with TSCA, this review stage becomes more important. This is where decisions on the timing of premarket notification and initial estimates of the scope of the premarket notification data package must be made.

If the compound under scrutiny has shown little potential for environmental insult and has also proved to be very low in toxicity, it might be moved more rapidly toward commercialization without a high degree of further testing, especially if it is site-limited, is useful as an intermediate for the preparation of other types of materials or has only minimal contact with man and the environment.

If, however, projections show that there is opportunity for high exposure, or that the material has a high probability of entering into the environment, it may be necessary to go into subchronic toxicity tests. It should be recognized that these will take up to three to six months and are rather expensive, generally costing from $25,000 to $50,000 each. The tests would be 30- to 90-day studies by ingestion or inhalation in rats, and sometimes other species. In addition, skin sensitization tests may be performed on guinea pigs. Concurrently, product evaluation research is intensified in the laboratories, and frequently potential customers are encouraged to evaluate the product in their selected uses.

The next step is the second major key review stage. Once again the committee of research, development and marketing experts convenes to review the additional data and decide whether to stop or to go either to additional health and environmental testing or to the marketplace. And again, premarket notification planning is evaluated.

If there is opportunity for wide exposure to humans or the environment, and if the initial toxicity and environmental profile data indicate concern for certain effects, an in-depth series of studies, which could take three years or longer to complete, is warranted. These studies would be chosen from a spectrum of animal tests that might include studies of reproductive competency, metabolic pathways, pharmacokinetic parameters, and teratogenic, mutagenic and carcinogenic potential. Normally, the necessary tests are chosen by considering potential use and similarity to already characterized compounds. These decisions require professional judgment by highly experienced experts. In addition, more specialized environmental studies will probably be performed. These might include odor and taste studies and studies of photodegradation, biodegradation and bioaccumulation.

While the additional health and environmental testing is going on, much work in advanced product and process development is being carried out.

Another major key decision review stage is reached at this point. After the expenditure of much time and money, another "go" or "no go" situation presents itself. If the decision is "go", the next step is commercialization, implementing previously planned pilot or manufacturing facilities.

The foregoing is a brief description of industry's methods of coping with the environment and at the same time trying to utilize the technical information produced. This must be incorporated into a program dealing with employee exposure assessment. Consideration must be given to plant-operating personnel, industrial hygiene functions, monitoring systems and personnel functions, and the reported flow of data, along with the characteristics of the data from the plant measurements, from detailed environmental surveys of the industrial hygiene group, to periodic extraction of exposure summaries, to actual work histories of the people. This, then, is brought together and accumulated, and finally assessment is made by physicians and those skilled in epidemiology.

An example to illustrate predictive decision-making uses a well-known benchmark chemical, ethylene oxide (EO). The environmental exposure data are as follows:

1. Low entry to the environment, used primarily as an intermediate.
2. Mode of environmental entry: generally air or water.
3. Movement in the environment: largely air.
4. Estimated lifetime: in air is short, because it is easily degraded by hydroxyl radicals; in water it is short, because it is readily hydrolyzed to ethylene glycol.
5. Bioconcentration potential: nil, because it has a low partition coefficient and is nonpersistent.
6. The 20-day biochemical oxygen demand (BOD) of ethylene glycol is equal to about 89% of theory. The remainder appears to be incorporated as carbon in cells with test organisms.

A 96-hr static test shows that EO, at 40 g/l, has no adverse effects on fish. Its hydrolysis product, ethylene glycol, at 100 g/l, likewise has no adverse effects. Terrestrial organisms will have nil exposure. It can be concluded from these data that neither environmental exposure nor environmental effects appear to be unreasonable. Thus, very little further testing need be conducted.

A hypothetical chemical with a profile different from that of ethylene oxide would have to be treated differently. If the chemical is stable in water, it could be concluded that large-scale testing is not necessary because the rate of release to the environment is low, and that volatility will result in fairly *rapid transport from water to air* where it will be degraded. On the other hand, if the chemical is stable in water and there is much environmental release, it might be necessary to develop test data to determine more precisely how rapidly the chemical transports to air.

If the data show that the transport is not rapid enough and that significant water concentrations might exist, then additional aquatic toxicity tests might be in order.

Four fundamental questions must be addressed:

1. What is the spectrum of biological activity?
2. How does the material partition between water and fatty substances, that is, does it bioconcentrate?
3. What is the stability or persistence of the material?
4. How much enters the environment and how does it move after it enters?

These factors must be considered in terms of use, frequency and time of exposure, and the quantities involved in various environmental compartments and populations. Test information must be fully utilized and integrated into day-to-day operations, and on a periodic basis the accumulation of data in respect to these four parameters must be brought together so that management at certain levels within a company can make the decision whether to move ahead or to seek more data. The important thing is that technical information and sound data must be in a position to be utilized by all members of the development and decision-making teams.

Clearly, obtaining sound data is essential, but it must be emphasized that the future of chemistry and the chemical industry also rests in the arenas of politics and law.

In *Harper's Magazine* recently, William Tucker articulated that while scientists are seeking answers in the laboratory, the environmental movement is seeking answers in Washington. According to Tucker, the legal, bureaucratic "solution" to our problems has killed the incentive for innovation and scientific investigation, which is our best hope for solving problems. And this movement has many energetic disciples.

The role of transport and fate studies is an important one today. Good data, sound interpretation and, perhaps more importantly, rational perspective on relative risks, are essential for the future.

It is important that scientists in industry become more involved with scientists in universities and in government. Discussions concerning health and environmental affairs must become more candid, with focus much more on scientific interpretation than on political expediency.

As indicated earlier, sound data and technology are needed for appropriately minimizing health and environmental risks. To accomplish this we need specialized skills, interdisciplinary interpretation and judgment.

TSCA and the authority vested in EPA are now superimposed on the chemical development scheme. Legalistic and bureaucratic processes have a tendency to place burdens of proof foreign to the normal processes of scientific interpretation and the tests of consistency of the data as a whole. EPA has articulated scenarios that involve requirements for testing and criteria for decision-making. Many of these scenarios, when carefully examined, raise specters of testing ourselves to extinction.

Industry today is part of a great debate and decision process which is much more transcendental than the role of transport and fate studies. Our arena requires perspective on the complexities of unreasonable risks and the balancing of important human, economic and natural resources.

3

CHEMICAL SELECTION AND EVALUATION:
IMPLEMENTING THE TOXIC SUBSTANCES CONTROL ACT

Warren R. Muir*

Office of Toxic Substances
U.S. Environmental Protection Agency
Washington, D.C. 20460

INTRODUCTION

Until January 1977, Chemical Hazards were covered by the federal statutes shown in Table I. At that time a new, comprehensive law called the Toxic Substances Control Act (TSCA) was added to the list. Its enactment culminated years of debate and consideration by Congress, the White House, the chemical industry, scientists and environmentalists, and it is perhaps the most important of all the statutes. Under this law, the U.S. Environmental Protection Agency has broad, generally discretionary powers that include three key features: regulation, reporting and testing.

EPA AUTHORITY UNDER TSCA

The first of the agency's authorities is the ability to regulate the production, distribution or use of chemicals that cannot be addressed under other statutes. Despite the seeming plethora of preexisting laws concerning chemicals, there were important gaps in statutory coverage and instances where front-end controls were more appropriate. For example, the production and use of industrial chemicals such as PCBs could not be controlled adequately under any of these laws. (TSCA provided direct statutory language banning PCBs.) Similarly, while laws such as the federal Food, Drug

---

*Deputy Assistant Administrator for Testing and Evaluation.

Table I. Federal Regulations Supplemented by TSCA (1977)

Clean Air Act	1972
Clean Water Act	1970
Consumer Products Safety Act (CPSA)	1972
Federal Dangerous Cargo Act	1952
Federal Food, Drug and Cosmetic Act	1939/1976
Federal Hazardous Substances Act	1960/1976
Federal Insecticide, Fungicide and Rodenticide Act (FIFRA)	1972
Federal Ports and Waterways Safety Act	1972
Federal Railroad Safety Act	1976
Hazardous Materials Transportation Act	1975
Lead-Based Paint Poison Prevention Act	1976
Marine Protection Research and Sanctuaries Act	1972
Occupational Safety and Health Act (OSHA)	1970
Resource Conservation and Recovery Act	1976
Safe Drinking Water Act	1974
Wholesome Meat Act	1967
Wholesome Poultry Act	1968

and Cosmetic Act could be employed to protect people from contaminated food, often authority was lacking to prevent the food from being contaminated in the first place. TSCA provides such authority.

The second of the agency's authorities is the ability to require manufacturers or processors to report information such as production, use, properties or effects of commercially produced chemicals. The first action by EPA under this provision of the law has been to compile a list of all the chemicals produced in or imported into the United States, including production location and amount. This will be the first definitive list of what chemicals are produced and where. In the future, EPA will selectively require reporting by manufacturers and others participating in the commercial marketing chain of a chemical to fill gaps in our knowledge about specific chemicals and to provide the basis for developing priorities for commercial chemicals.

The third feature is the agency's authority to require testing of substances in an effort to determine possible health or environmental effects. Testing may be required of manufacturers and processors where a chemical or group of chemicals may pose an unreasonable risk or where a chemical is produced in amounts large enough to give rise to high exposures.

Both the reporting and testing features of TSCA support other acts as well, for example, OSHA, under which there is no authority to require testing. In implementing TSCA, the goal of the agency's program is twofold: to identify chemical risks that must have priority attention under

TSCA and to support control options with responsible risk assessment.

DETERMINING PRIORITIES

There are an estimated 43,000 chemicals being produced commercially in the United States. Between 200 and 1,000 new substances are introduced into commerce each year, with an infinite number of possible structures. However, because only a relatively small proportion of these chemicals can be reviewed in depth, tested or regulated, EPA must develop approaches for setting priorities for assessing chemicals; ultimately only a few will be regulated. This is the purpose of the EPA assessment process, which has several stages.

Initial Chemical Selection

Chemicals are fed into the assessment process via several routes. To date, selection has been made *ad hoc*, primarily based on data submitted to the agency in the form of nominations from within EPA and other agencies, substantial risk notices reported under section 8(E) of TSCA, petitions and notifications from the public. However, in the future, it is expected that half or more of the selections will be made systematically through studies of major trends in the chemical industry and through a scoring and evaluating process. The scoring system to be used initially is that used by the Interagency Testing Committee (ITC), which is charged with recommending chemicals and classes of chemicals for test rules under TSCA.

The multistep screening process the ITC followed in order to limit the number of chemicals likely to need priority for health and environmental effects testing will be used initially by EPA for chemical selection. The committee began its scoring by compiling lists of chemicals that were of environmental concern to various agencies and other groups.

The ITC's second step in the screening process was to abbreviate its initial list by eliminating those chemicals and categories that were not in commercial production or that were used primarily as food additives, pesticides or drugs, that is, chemicals already being regulated. This recognition of legislation already in place for certain chemicals also has its counterpart in the assessment process: EPA is committed to turning over the investigation—and perhaps ultimately the regulation—of a chemical to the most appropriate agency. The responsibility for determining where and by whom chemicals are best evaluated belongs to EPA under TSCA as an offshoot of its assigned role as an information gatherer.

The third step in selection taken by the ITC was to eliminate chemicals on the basis of production volume, environmental release, occupational exposure and general human exposure. In the assessment process, a parallel of this evaluation is the Chemical Source/Effects Report or Phase I Report, our initial hazard assessment.

In its fourth step in selection, having eliminated thousands of chemicals, the committee further reduced the number by considering potential biological activity and the need for health and environmental effects testing. Final selections for the list of chemicals the Committee would recommend for testing to EPA were arrived at after the committee reviewed dossiers on the chemicals (the equivalent of our CHIP), public comments and information from other agencies.

In the assessment process, however, this level of evaluation is supplemented by a determination of what the economic impact of regulation will be.

In sum, scoring in assessment process serves as a tool to focus limited analytical resources on probable high-priority problems.

Preliminary Evaluation: CHIP

Once selected for assessment, a chemical is generally the subject of a chemical hazard information profile (CHIP). The profile is put together by the Assessment Division (AD) after a preliminary review has been made of the available literature, of existing TSCA files and of computer information systems. The profile is essentially a condensation of known effects and exposure information on a chemical. The review and the preparation of a profile take 1-2 weeks of a staff member's time. Several of these profiles are distributed at a time for discussion at an EPA-wide meeting.

It has been our experience that at this stage in any chemicals are either accorded a low priority or referred to other offices. However, a good number of them require more information before any conclusions can be drawn; a few are found to be sufficiently well characterized as being of concern to warrant a place in the next stage of the assessment process.

Initial Hazard Assessment

In this stage an expanded literature search and an interagency information-gathering of available effects and exposure information are made as part of the initial hazard assessment. The new data uncovered are incorporated in the Chemical Source/Effects Report, known to AD staff in the vernacular as the Phase I report.

Essentially, the Phase I report provides a source analysis of reported effects. It summarizes and analyzes the major sources of exposure, available statutory authority, regulatory status, information gaps still to be filled and the key issues raised by discussion of the chemical. In the course of its preparation, the report undergoes EPA-wide review, as did its predecessor, the CHIP.

Upon completion of the report, the Program Integration and Control Action Division in the Office of Toxic Substances prepares an analysis of statutory options relevant to the sources of exposure described in the Phase I report. The Toxic Substances Priorities Committee, after studying the Phase I chemical disposition, will select an option from those recommended and propose either regulation or further evaluation. Having come this far, it is likely that a chemical will be kept in the assessment process. However, a decision will be made here about who should carry out the evaluation.

If regulation seems likely, technical information about the compound needs careful evaluation. This extremely labor-intensive task is done only at the last stage of assessment.

In the Chemical Risk Evaluation Document or Phase II report, the reported effects are validated and estimates of risk and of the risk-reduction potential of the control options are made.

THE PROGRAM'S PRESENT STATUS

An overview of how TSCA is being implemented, at least so far as chemical selection and evaluation are concerned has been presented above. The assessment program is currently unfolding as follows.

- The envisaged system of chemical selection under TSCA described above is used by the ITC and will soon be adopted by OTS for assessment purposes.
- The approximate number of section 8(E) notices reviewed is at 240, and about 20 of those are being tracked in the assessment process as subjects for CHIPs.
- Approximately 35 CHIPs have been reviewed this year, and 5 were recommended for continued assessment in the Office of Toxic Substances.
- Sixteen chemicals are currently being reviewed at the initial hazard assessment stage, six more are awaiting consideration, and other candidates are being sought.

As yet, no chemicals have reached the stage of a detailed risk assessment through the formal chemical selection and evaluation assessment process outlined here, but three assessments, on asbestos, NTA and chlorofluorocarbons are contemplated, and a few of the sixteen chemicals currently undergoing chemical source/effects analysis are likely candidates.

PARTICIPATION OF THE SCIENTIFIC COMMUNITY IN TSCA

OTS welcomes the input of the scientific community throughout the stages of the assessment process. Indeed, public review is written into the TSCA. The participation of scientists is sought not only in the review phase but also during the preparation of documents. Those willing to review documents for EPA are encouraged to notify the agency.

4

TOXIC CHEMICAL TRANSPORT AND FATE RESEARCH
IN U.S. ENVIRONMENTAL PROTECTION AGENCY

Courtney Riordan*
 Office of Research and Development
 U.S. Environmental Protection Agency
 Washington, D.C. 20460

INTRODUCTION

Transport and fate research in the Environmental Protection Agency
(EPA) is a fairly open-ended topic. It is possible to struggle over the substantive problem of defining the area of transport and fate, or to take a
very narrow bureaucratic view and simply describe transport and fate research as that which happens to be funded under the EPA research categories titled transport and fate. Neither approach seems appropriate here.
Instead, both substantive and programmatic concerns of the EPA research
program in the area of transport and fate research will be discussed selectively—no detailed topography, just a brief overflight identifying patterns
and peaks.

The Office of Research and Development (ORD) has direct responsibility
for three programs that explicitly address the problem of transport and
fate: (1) transport and fate of pollutants in air, (2) transport and fate of
pollutants in water and (3) transport and fate of toxic substances. The
resources to support this research activity derive from three respective
authorities: The Clean Air Act (CAA), The Clean Water Act (CWA) and
The Toxic Substances Control Act (TSCA). It is anticipated that beginning
in 1980, a small program in research transport and fate of pesticides in
direct support of the Federal Insecticide, Fungicide and Rodenticide Act

*Acting Deputy Assistant Administrator.

(FIFRA) will be initiated. The EPA also administers a transport and fate of energy pollutants research program as part of the multimillion dollar inter-agency environment/energy research program. A portion of the research funded under the interagency program is implemented in the EPA laboratories, but only that research is funded which prevails in priority over other agency proposals in competition for the limited pool of resources available.

In 1979, the total resources devoted to research in these areas are planned to be:

	Amount	Positions
Air Transport and Fate	$11.0M	44
Water Transport and Fate	1.3	7
Toxics Transport and Fate	2.2	11

These efforts are relatively well funded in terms of extramural research funding. Some EPA research is primarily in-house with the ratio of in-house expenditures to total funding being 80% or more. In the transport and fate area, the ratio of extramural expenditures to total resources is 50% or more.

MISSION ORIENTATION

There is a definite reason for tying the funding of the research program to congressional authorities. For many years now the EPA research organization has been the object of a great deal of criticism from inside and outside the agency. For a while it looked as though the ORD couldn't do anything right. It was criticized because its research did not address the long-term and more fundamental problems associated with the maintenance and protection of environmental quality. At the same time, it was being criticized by the program offices, i.e., the regulatory arm of the agency, for being unresponsive. They often failed to see, and ORD was not able to demonstrate to them, the value of much of the research that was being conducted. Finally, it had a bad image in terms of the research community. There was substantial opinion that the EPA laboratories were not cultivating adequately the expertise outside the agency in ways to best utilize those resources in both the short and long run.

Largely as a result of this criticism, a congressional mandate to reassess the planning and management of research in EPA, and the leadership of a new assistant administrator for ORD, significant studies have been undertaken in an effort to solve these problems. Changes have been and are being made that will have an important impact on the planning and conduct of the transport and fate research program.

As part of the overall planning and management scheme, research has been ordered along a scale ranging from that in direct support of regulatory activities to that which the program or regulatory offices are willing to support as exploratory, whether to improve our understanding of existing problems, to identify new problems, or to represent what has been called anticipatory research. This latter class of activities is viewed as either long-term or basic or high-risk research that the operating side of the agency would not typically support in preference to alternative mission-oriented studies. The anticipatory research decision represents an important recognition by the agency that such freedom of activity is indispensable to maintain interest of researchers and allow new discovery.

This scale of relevance is applied to all major categories of research, *e.g.,* health effects, ecological effects and transport and fate. Although there is no strict formula, it is the intention of the agency that in the course of time, research in support of regulation (operationally defined as that amount of research actively supported by the client program office) will comprise a major share of the total research activity, *i.e.,* 50-60% or more.

Research relevancy is achieved through the zero-based budgeting planning and management system. ORD has to defend its research program in what are called media-ranking exercises. For example, the air transport research has to be ranked in priority in competition for resources with the regulatory and enforcement activities dealing with air pollution. There is a great burden to demonstrate the contribution of the research to the mission of air pollution regulation. Similarly, the water transport and fate research must be ranked by a water media team and toxics by a toxics team.

Experience has shown that he survives best who has given the most effort and thought to designing his proposed research program in such a way that its value to the operating arm of the agency is clear. Contrary to dire predictions, the horizons for research have not been radically truncated as a result of this process. The rest of the agency has generally been receptive to research on the merits of its overall importance and has not demanded lock-step adherence to projects with very narrow focus and quick turn-around times. This always involves effort in the form of discussions, meetings, agreements, etc.

From the point of view of the extramural researcher seeking support for a study in which he is interested, however, the EPA system does present a serious problem. Unless his particular research happens, by coincidence, to address one of the priority problems sanctioned by the research plan via the ZBB process, his chances for support by EPA are slim. The FY'79 program which started on October 1, 1978, went through the ZBB process one year ago, and work plans which specify individual projects are due for completion this month. Preparation of requests for proposals and negotiations for grants for specific research activities are already underway.

Given this kind of system, the researcher interested in receiving support from EPA typically cannot anticipate much success unless he becomes somewhat of a student of the agency. Ideally, he should be familiar with the mission and current directions within a particular program office where he thinks his research would be of some value. Further, he should become acquainted with the particular laboratories within ORD which try to fulfill those program needs through their combined in-house and extramural efforts. Where possible, he should extend this interest to correspondence with laboratory personnel who deal directly with his particular area of study, either at professional meetings or by letter or telephone.

WHO DOES WHAT?

The research on transport and fate of pollutants in air is implemented by the Environmental Sciences Research Laboratory (ESRL) at Research Triangle Park, North Carolina. In addition, ESRL is responsible for implementing the air media portion of the research dealing with transport and fate of toxics and energy pollutants. In general, this area of research has two major objectives: (1) the determination of atmospheric and chemical processes that affect the formation and removal of gaseous and aerosol air pollutants and (2) the development, evaluation and validation of models for predicting impacts of emissions of air pollutants on ambient quality air. The kinds of research that are supported range all the way from bench-scale laboratory investigation of chemical and physical processes to elaborate chamber studies to intensive large-scale field characterization and measurement programs. The latter can involve studies of several weeks' duration with elaborate combinations of airborne and ground-level measurements of meteorological, chemical and physical parameters.

The water transport and fate research program is administered by the Environmental Research Laboratory (ERL) at Athens, Georgia. Compared to the program in air, the water program has always been a poor relation in terms of funding and manpower resources. One significant reason for this apparent discrepancy has been the differences in scale and complexity between the air regulatory activities and those in water. The problems of oxidant formation in urban atmospheres, sulfate formation and interregional transport and deterioration of air quality in pristine airs from remote sources have demanded sophisticated studies involving very large quantities of resources. The transport and fate problems associated with conventional water pollutants, on the other hand, have not presented the same challenge. One-dimensional models incorporating first-order linear-process descriptions have often proved adequate to predict the transport of BOD and nutrients such as phosphorus and nitrogen. The concern with toxic materials in the water programs and the emerging priorities of the toxic substances control program suggest that the relative amounts of resources in these areas may shift toward toxic substances research over the next few years.

At the present time, the transport and fate research at the Athens Laboratory includes the study of environmental processes such as photolysis, hydrolysis, adsorption and microbial degradation, singularly or in combination in laboratory, field and real world environments and the incorporation of these process parameters into models of varying levels of complexity and sophistication, from qualitative, evaluative type models to quantitative predictive models.

NEAR-TERM PROGRAM DIRECTIONS

Air—CAA

In the study of air, concern continues to focus upon problems of ozone/ oxidants and the particulates, especially sulfates and nitrates. Some of the specific kinds of activities that will be given highest priority are:

- studies in the laboratory and in the field on the formation, growth, and removal of fine particulates with an emphasis on inhalable particulate matter; part of the reevaluation of total suspended particulate standards;
- continuation of field studies on long-range transport and transformation of sulfates and nitrates emphasizing winter and night physical and chemical behavior of plumes;
- field studies on interregional oxidant transport, rural oxidant chemistry and sources, and the role of aromatic chemistry in oxidant formation;
- development of regional scale models for sulfates and oxidants; and
- studies to assess the role of specific aerosols on visibility.

Water—CWA

Current efforts in this area are to:

- expand existing water quality models to handle nonconservative chemical pollutants;
- develop methods for estimating mass balances for priority chemicals;
- continue research on environmental processes to provide test data for decision making;
- develop evaluative models to be used in early stages of decision making; and
- develop exposure assessment techniques for pesticides in food chain and freshwater and estuarine environments.

Toxics—TSCA

The main focus here is to develop test methods and protocols and exposure assessment models that can be used in making before-the-fact decisions with respect to production, use and disposal of chemicals.

Uncertainties

The transport and fate research program owes its support to the need to be able to estimate exposures to chemical substances released to the environment. Depending upon the chemical, its mode of production, use or disposal and the characteristics of the receiving environment, the transport, transformation and fate of a particular chemical may vary considerably. On the transformation side are extremely complex processes involving chemodynamics, biotransformation and bioaccumulation. On the transport side are problems of extremely complex meteorological and hydrological processes at work.

The area has the potential for adsorbing tremendous amounts of research dollars, providing many advances, and yet still leaving much to be done. The accuracy of prediction could probably always be measurably improved with the expenditure of more time and effort.

FUTURE PLANNING

Planning for the future transport and fate research program will require two things above all else: a clear sense of purpose and appropriate emphasis. The purpose will be reflected in the degree of resolution that we attempt to achieve in understanding the effects of environmental processes on substances and in the construction of predictive systems to assess potential exposures to substances. Most agree that methods for prediction should be designed that are hierarchical in their demands for data and in their degree of complexity. Ultimately if the system is to work it should be supported by judicious monitoring to discover errors that will be inherent in any practicable decision-making system. The program emphasis is directly related to purpose. This implies more difficult problems for EPA in that it suggests shift at times among studies in support of different media. These shifts require painful redirection and reprogramming decisions.

5

THE UNIVERSITY CHEMIST—
ROLE IN TRANSPORT AND FATE OF
CHEMICAL PROBLEMS

Virgil H. Freed

Oregon State University
Corvallis, Oregon 97331

INTRODUCTION

Since the dawn of time, man has sought chemicals to alleviate his ills, improve comfort and provide an escape from the misfortunes of life. However, the materials available were of limited and dubious quality until the rise of the chemical profession. Thereafter, the enthusiastic endeavors of this group of scientists led to the ushering in of a chemical age some four to five decades ago. Through the clever and ingenious efforts of the chemists, vast new areas of knowledge of the chemical aspects of the physical world and life were opened up. This new knowledge resulted in development of more effective medicines, textiles of great utility and a veritable avalanche of other chemical products to add to the comfort and security of man's life.

The chemicals discovered and the processes developed for their synthesis provided a basis for the growth of a huge chemical industry. The industry supplied products for better buildings, more durable clothing, increased food production and protection and preservation of life. Depending on the figures used, it is estimated that there are some four million compounds now known to the chemical profession with new ones being added everyday. Of these four million compounds, about 100,000 are believed to be in daily use and approximately 7,000 produced in quantity.

CHEMICALS—ADVERSE EFFECTS

Those who are old enough to have a memory of life during the 1930s and even on into the 1950s perhaps best appreciate the contributions of chemistry to mankind. The slogan of one of the big chemical firms, "better things for better living through chemistry," was being proven true almost daily during those years. Value of these contributions to better medicine, more abundant, stable and nutritious food supply and comfort of life would seem to be without question. But, as it turns out, it was not without cost.

With the increase in chemical production and a longer experience with the exposure of man not only to the ultimate chemical but also to some of the by-products and intermediates, the adverse effects or costs became more apparent. Water and air pollution accompanied the manufacture and use of many such products. However, as the skills and tools of the analytical chemist became more sophisticated, it was possible to detect chemicals at much lower concentrations and consequently at a greater distance from either the source or place of use. Soon, effects of this pollution were to be observed on lower organisms and subsequently, after longer periods of exposure, on man. As increasing quantities of chemicals were manufactured and used, evidence began to mount that many of these substances were being transported long distances. Thus, we became aware that some of these substances had the stability to persist in the environment and physical characteristics that permitted mobility.

The health and ecological impacts of chemicals in the environment gave rise to considerable concern not only in this country but also all over the world. Obviously, even though the chemicals were of great utility, their usefulness must be considered of decreasing value if the consequence of manufacture and use was an ever increasing health and ecological cost. The response, therefore, was to develop more sophisticated pollution control devices, laws and regulations that put constraints on the availability and use of the chemicals.

REGULATION OF CHEMICALS

Congress and the various state legislatures have enacted numerous laws governing the manufacture, sale and use of chemicals. They have established agencies that promulgate regulations having to do with handling, transport and use of chemicals as well as occupational exposure and environmental residues to which the general populace may be exposed. A statute recently enacted is the Toxic Substances Control Act which, when combined with the laws governing use of pesticides, drugs and food additives, comprises a package for regulation of the commerce and use of most chemicals. Added to this are the various laws dealing with air and water pollution. From these laws, agencies derive their authority to promulgate their seemingly endless stream of regulations.

The laws and regulations are social instruments to deal with a perceived problem. In part, they are based on science, but all too often the scientific base has not been adequate for the job. Those who passed the laws and promulgated the regulations often were ignorant of the more sophisticated aspects of the chemistry of the compounds with which they had to deal and certainly evidenced little appreciation for the complex problems of transport and fate of the substances in the environment. Yet, such information, together with some elementary knowledge of the toxicology of the compounds, would seem to be fundamental requirement for rational regulation.

Surely, no one would argue that chemists and their colleagues in the other sciences have the final answer to problems in transport and fate of chemicals in the environment. Nonetheless, even at the time of the writing of the laws, substantial progress had been made on the elucidation of the principles involved in these phenomena. The laws and regulations, however, have given great impetus to further development of our knowledge and understanding of transport and fate. The state of this knowledge has reached the point where it is now reasonable to expect to see it applied in the solution of many of the problems being experienced. The symposium recorded in this book will undoubtedly serve as a vehicle for dissemination of this knowledge and catalyze its acceptance and application. However, further development and systematic application of knowledge regarding transport and fate of chemicals in the environment will require the efforts of a large number of additional chemists. Only the chemist with the unique skills and instruments of this profession can develop and apply the fundamental knowledge that is required in this area. It is the chemist, after all, who has the training and the tools for the elaborate analyses needed to identify and quantitate a specific substance within a mixture of thousands of others. Again, it is only with these skills that a water solubility in the low parts per million or parts per billion can be determined. Similarly, who but one trained as a chemist could evaluate photochemical stability or make predictions regarding the probably environmental behavior of a new compound? Thus, in the administration of the Toxic Substances Control Act as well as similar legislation, the need for many well-trained chemists is apparent.

For study of the behavior and fate of the chemicals in the environment, a specialized training and perspective is needed. In other words, like other specialties in chemistry, this one has its particular perspective and skill requirement. Whereas in many specialties the focus is on the mechanistic and microscopic, in this one the focus must be more on the macroscopic. In other words, the chemist is continually trying to simulate the vastly larger and more complex chemistry of the ecosystem. This viewpoint is not difficult to master, yet it can hardly be said to be prevalent throughout the entire profession. It is, after all, relatively new in development and of more common currency in regulatory laboratories than in the university.

ROLE OF UNIVERSITY CHEMISTRY IN TRANSPORT AND FATE— TRAINING AND RESEARCH

In view of what will probably become an increasing demand for chemists to work on environmental behavior and fate of chemicals, it is appropriate to ask what role the universities can play in meeting this need. Certainly to respond to the need is well in accord with the universities' traditional roles of teaching, research and public service. The question, therefore, is not one of whether the universities should attempt to respond to such a need, but rather how they might go about it.

One of the traditional roles of the university is that of research. The chemistry departments in almost all universities have pursued this role to the limit of their funding and teaching commitments. University research, in large measure because of the structure and funding of the universities, has been occupied with basic disciplinary-oriented research. Problems of mission orientation, though equally basic, have often been left to specialized institutes and government agencies. In the field of environmental behavior of chemicals, however, the university chemistry research program has an opportunity for expansion and demonstration of the relevance of its research to societal needs. This is not to advocate that the entire research program be turned to such efforts. Rather it is an attempt to encourage those who have an interest in such problems to pursue it.

One can hardly think of any field of chemistry that is not relevant to the particular problem of transport and fate. Analytical chemistry, of course, is an obvious one. So also, is physical chemistry. Equally germane are the studies that might be conducted in organic chemistry and inorganic chemistry. To illustrate the point, consider the matter of the behavior and fate of a given chemical in the soil matrix. Here, the knowledge of colloidal chemistry, solution chemistry, surface chemistry and analytical chemistry all come to bear. Similarly, the organic chemist and the biochemist will find of interest the reactions and mechanisms of reactions by which the compound may be degraded in this matrix.

There are many problems involving environmental behavior and fate of compounds to challenge the chemists interested in doing research. To illustrate, one starts with the observation that the environmental behavior of the compound is related to its physicochemical properties. Thus, the mobility of the compound, whether it will be accumulated in a given sink of the environment or taken up by organisms, is related to its properties as well as to the processes to which it is subjected. Much more needs to be done to determine the properties of many of the compounds now or potentially to be in commerce and, on the basis of the relationship of these properties to behavior, to develop a system of predicting the behavior of new introductions. There is also the matter of the photochemistry of many of the commercially important compounds. It is important to know the mechanism of the reactions by which the photochemical breakdown occurs, what products are formed, the rate of this breakdown and whether the compound is less photostable as a gas, in solution or in the adsorbed phase.

Numerous other research problems exist, such as those of gaseous behavior, rates of breakdown in different media, and so on, that await the attention of chemists. There is also the matter of putting together all the facets of information, such as adsorption, rate of degradation, photodegradation, vapor loss, etc., into a single model to get material balances, flow and points of accumulation. All such studies are appropriate to a university chemistry department. Not only are they appropriate, but if such a chemistry department were hoping to train chemists to deal with transport and fate problems, such research would prove a stimulus and benefit to their teaching.

Training

In order to understand the dynamics of chemicals in the environment and assess the various hazards that may be encountered, much more information regarding the chemistry and properties of the compound will be required. Since the behavior of the chemical in the environment and, to some extent, its fate are related to the physicochemical properties, determination of these properties will be essential. Further, rather than risk adverse effects by waiting to observe the behavior of large amounts of chemical in the environment, laboratory tests to predict the probable behavior will be essential. To do all this work will require a much larger number of chemists devoted to these types of studies.

Undoubtedly, many practicing chemists have gravitated or will gravitate to this type of work, but it is reasonable to assume that many more will be required. To meet this demand for additional professionals, appropriate university training would seem to be desirable.

It would seem a relevant role for the university to train chemists to work on transport and fate problems for industry or regulatory laboratories or to conduct research on such problems.

The training of chemists, either undergraduate or graduate, to enable them to work on the problem of transport and fate of chemicals in the environment or hazard assessment should not result in diluting their basic chemical training with strictly applied courses. Their basic education in the fundamentals of chemistry should continue to be just as important as the specialized courses preparing them for this type of activity. However, while maintaining the rigor of fundamental courses, there would be many opportunities to introduce pedagogic material and examples of transport, fate and hazard assessment. Additionally, there is always room in a curriculum for seminars and special problems to further augment training in areas of interest.

It would be difficult to think of any basic course in chemistry that did not offer an opportunity to introduce examples and laboratory experiments relating to transport, fate and hazard assessment. A most obvious course, perhaps, is that of analytical chemistry, where the student could learn the principles of analytical methods while working on chemicals of environmental significance. Almost any type of analytical method being taught could be applied to one or more of the multitude of substances of concern in the environment. Thus, gas chromatography, mass spectrometry, polarography and other techniques of electrochemistry could be taught as well using chemicals of environmental interest as any others. The addition to the pedagogic program at this point would be instruction of the student on the importance of the chemical being analyzed. Other techniques of analytical chemistry such as separation, purification and statistics of the analytical procedure would be applicable to samples of water, air or biota.

There would be many rich examples of environmental chemistry to be taught in physical chemistry without sacrificing any of the principles to be conveyed to the student. For example, study of water solubility, partition coefficients, adsorption, kinetics of reactions in degradation and photochemical studies are of immediate applicability to environmental problems.

The teaching of organic and biochemistry also would afford many opportunities to relate these subjects to the study of environmental behavior and fate of compounds. Of special interest to the organic chemist would be reaction, photochemical and other mechanisms in alteration of the chemical. The biochemist would find fascinating problems related to the action of the chemical on the living system and enzymatic reactions.

Since in almost any curriculum there are electives, that is, a choice of optional courses, it would seem possible to introduce some special courses for training chemists having an interest in this area. One such course which should be of limited credit and duration would deal with the matter of chemicals in the environment and assessment of the hazard being posed. Such a course might deal not only with the dynamics, fate and effects of the chemicals but with the various types of sources, sinks and, of course, use.

Finally, it would seem that chemists as a whole, not just those concerned with environmental behavior, should be better prepared to evaluate the cost-benefit-risk of chemicals. It seems that it is altogether too easy to become preoccupied with the risk and the dynamics of behavior and totally overlook the benefits that accrue from the use of chemicals. By seminars, special courses or teaching units in regular courses, the chemist should be made aware of the broad aspects of the profession. Most chemicals now used by society have not been accepted for trivial reasons. They fill some economic, social or physical need but, as pointed out in the beginning, they are attendant with cost in terms of health or ecological impact. The professional chemist, particularly one working in the area of toxic substances, should have a grasp on both the benefits and the costs in order to develop the scientific information that would permit a rational decision. In the same vein, in order to assess both the benefits and risk from manufacture and use of a chemical, the chemist should have a basic knowledge of toxicology. This, again, is a subject that could be introduced by a special elective course and would be of benefit to all chemists, not just those working in a specialized area.

It would appear obvious that the chemist plays a crucial role in regulation of chemicals and understanding the dynamics, fate and effects of chemicals. What has been offered here are a few suggestions to enable the chemist, particularly the university chemist, to assume the proper role in these activities. Surely more can be added to the list of suggestions and innovative and imaginative ways of implementing them can be offered by many others in the field. It is hoped that the various agencies and the public in general will give support and encouragement to a profession eager to fulfill its important role.

6

SETTING TOXICOLOGICAL AND ENVIRONMENTAL
TESTING PRIORITIES FOR COMMERCIAL CHEMICALS

Marvin E. Stephenson

National Science Foundation
Washington, D.C. 20005

The setting of testing priorities for toxic chemicals is one of the most important and complex steps in regulation under the Toxic Substances Control Act (TSCA). This is particularly important since the number of chemical substances and mixtures subject to the provisions of the act range up to 70,000. Section 4(e) of TSCA established an Interagency Testing Committee which has the continuing responsibility to identify and recommend to the Administrator of the Environmental Protection Agency chemical substances or mixtures which should be tested to determine their hazard to human health or the environment. The committee is composed of representatives from eight member agencies, which are the Council on Environmental Quality, the Department of Commerce, the Environmental Protection Agency, the National Cancer Institute, the National Institute of Environmental Health Sciences, the National Institute for Occupational Safety and Health, the National Science Foundation and the Occupational Safety and Health Administration. Liaison membership includes the Consumer Product Safety Commission, the Department of Defense, the Department of Energy, the Department of the Interior and the Food and Drug Administration.

In April and October of each year the committee reports on its recommended revisions to the priority list and the reasons for these revisions.

In arriving at the decision to designate a particular chemical substance or mixture or category of materials as a likely candidate for further testing, the committee attempts to consider all relevant factors including:

    a. the quantities in which the substance or mixture is or will
       be manufactured;

    b. the quantities in which the substances or mixture enters or
       will enter the environment;

    c. the number of individuals who are or will be exposed to the
       substance or mixture in their places of employment and the
       duration of such exposure;

    d. the extent to which human beings are or will be exposed to
       the substance or mixture;

    e. the extent to which the substance or mixture is closely related
       to a chemical substance or mixture which is known to present
       an unreasonable risk of injury to health or the environment;

    f. the existence of data concerning the effects of the substance
       or mixture on health or the environment;

    g. the extent to which testing of the substance or mixture may
       result in the development of data upon which the effects of
       the substance or mixture on health or the environment can
       reasonably be determined or predicted; and

    h. the reasonably foreseeable availability of facilities and person-
       nel for performing testing on the substance or mixture.

Clearly, the intent of these considerations is to provide at least a qualitative decision matrix where the multiple of the probability of the occurrence of the event (the exposure potential) and the magnitude of the event (the adverse toxicological effects) provides an estimate of the expected risk for a given chemical.

The practical need for priority rating of toxicological and environmental testing requirements for chemical substances in commercial production today, which number over 70,000, is obvious. These practicalities involve not only the limitations of available financial resources for this purpose but, more importantly, the shortage of trained toxicologists and pathologists to conduct the tests—a dearth of human resources which will take years to remedy.

At the outset, the implementation of an interagency responsibility for recommending specific testing of chemicals seemed to require that the review must be as inclusive and comprehensive as possible in order to provide adequate overseeing of the entire set of commercial chemicals which are clearly in current use in substantial amounts. It was also recognized that a final recommendation is to be made on only a few compounds of highest concern. Furthermore, a strategy must be developed so that the review procedure can be documentable and iterative such that all undesignated chemicals may be reexamined at timely intervals.

The process which has been developed is a multistep approach as shown in Figure 1, in which a very large array of chemical substances is successively reduced by statutory limits; by circumstances such as the inability to define the substance or mixture—materials generally regarded as inert and natural products; by profiles of exposure; and lastly, by the assessment of biological hazard potential.

PHASE I

**MERGING OF
SOURCE LISTS**

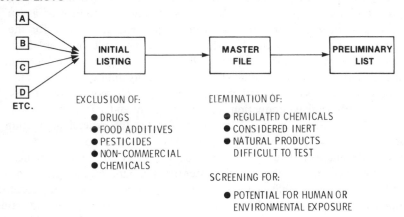

PHASE II

**FEDERAL REGISTER
NOTICE**

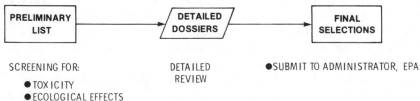

Figure 1 Process of prioritization.

In order to direct attention to the chemical substances most likely to require toxicological or environmental effects testing, the initial list was assembled from previously developed arrays of chemical substances reflecting priorities for regulation, research and hazard potential evaluation. This collation resulted in a list of approximately 3,650 entries. Those substances believed to be used principally or exclusively as drugs, food additives and/ or pesticides were then removed from the list, as well as those chemicals which were judged not likely to be in current commercial production. This first screen resulted in a list of approximately 1,700 items which was termed the Master File.

While a concurrent judgment of exposure and biological hazard potential is the evaluation protocol of choice, the practical feasibility of assembling basic toxicological information from current sources needed for the evaluation required that a first priority of review be made only on the basis of exposure to reduce the master list to a range where the biological hazard potential could be evaluated on a reasonably limited number of materials. The exposure evaluation was carried out by scoring four factors: (1) the current annual production, (2) the environmental release rate and persistence, (3) the occupational exposure, and (4) the general population exposure. The score for the general population exposure combined estimates of the number of people having direct contact with the substance, the frequency of their exposure, the intensity (amount of substance) involved and the penetrability of the substance (the amount inhaled, absorbed or otherwise taken into the body). Using a combination of published data and personal knowledge it was possible to provide numerical scores for each of the four exposure factors on about 700 items in the master file. Normalized values of each exposure factor were summed and used to develop the linearly weighted rank-ordered list. Items already under strict regulation, of well-known hazard potential or ill-defined for testing purposes were then struck from the list of 700. Others of the scored substances were rejected on the basis of relative ranking or were retained on the argument that further review was needed. This screening then resulted in a sufficiently narrowed list which could be scored for biological purposes. Special nominations were also made of specific chemicals which were not on the initial list to form a total of about 450 entries which was termed Preliminary List.

The preliminary list was then evaluated for biological activity factors of carcinogenicity, mutagenicity, teratogenicity, acute toxicity, other toxic effects (such as reproductive effects or organ-specific toxicity) bioaccumulation and ecological effects—a total of seven factors. Scores assigned for each of the seven factors were represented by either a number from the set 0, 1, 2, 3 or a letter score (generally x, xx or xxx). The assignment

of a numerical score by an evaluator indicated a judgment that further
testing of the substance is not needed for the factor being considered,
while the magnitude of that numerical score indicated the degree to which
the effect had been confirmed or the dose level at which it had been found.
Conversely, the assignment of a letter score indicated a decision that further testing should be carried out, with the number of x's indicating the
monotonic increase in concern. For example, in scoring a substance for
carcinogenicity, a score of 3 meant that the substance is well established
as a carcinogen in humans or experimental animals, while a score of xxx
meant that the substance is strongly suspected of carcinogenic activity but
has not been adequately tested. The numerical score 0 had a special property in that it represented a judgment that the substance did not possess
the indicated effect whether or not it had been adequately tested. For
each chemical the average score for a single effect was either a numerical
or a letter score. All participating evaluators were required to reach con-
sensus on which scale was appropriate but not on the magnitude of the
agreed scale.

From the exposure and biological activity scoring procedures a number
of analyses were obtained, including rank-ordered lists by a single biological effect, cumulative totals of all effects and the multiple of exposure
and biological effect(s). This information then served as a guide to select
from the preliminary list approximately 80 entries which would be examined in detail for possible designation on the priority list.

From these deliberations, the priority list shown in Figure 2 has been
developed which currently contains 18 designated chemical compounds
and categories of chemical compounds which are recommended for further
testing and evaluation for specific toxicological and effects. By considering
10 categories of chemical compounds in the current list, the total number
of individual chemicals in substantial current production recommended for
testing is approximately 45; however, the potential number of individual
compounds which could be represented in the current list is several hundred.

Because of the large number of compounds which have to be considered,
methods of dealing with them in blocks rather than individually seem desirable. There are a number of ways to define these blocks, but the most
useful to the committee seems to be a definition based on similarities of
chemical structure.

The principal bias which the method of exposure evaluation involved
favored those chemicals exhibiting multiple routes of exposure. A chemical of intermediate exposure through occupational and general population
scores normally would be ranked above a chemical having an extremely
high value for only one type of exposure. Of course the single limiting
factor for developing improved estimates of exposure is the restricted

1. Acrylamide
2. Alkyl Epoxides
3. Alkyl Phthalates
4. Aryl Phosphates
5. Chlorinated Benzenes, Mono- and Di-
6. Chlorinated Naphthalenes
7. Chlorinated Paraffins
8. Chloromethane
9. Cresols

10. Dichloromethane
11. Halogenated Alkyl Epoxides
12. Hexachloro-1, 3-Butadiene
13. Nitrobenzene
14. Polychlorinated Terphenyls
15. Pyridine
16. Toluene
17. 1, 1, 1-Trichloroethane
18. Xylenes

Figure 2. The ESCA Section 4(e) Priority List, by Alphabetical Arrangement

access to production and use information which would enable the construction of elementary material balances and exposure profiles. This feature, the exposure profile, is currently far less satisfactorily estimated than the biological activity term and needs a level of effort similar to that currently being expended on the development of biological testing protocols.

Setting priorities for testing still involves almost entirely subjective judgments. Property-effect relationships, models of toxicity and indexing systems are of value when used at the level of their specificity but, at this point, do not substitute for the thoughtful judgment of experienced and expert people dedicated to this very complex problem.

7

ANALYSIS OF TOXIC CHEMICALS
IN THE ENVIRONMENT

L. H. Keith

Radian Corporation
Austin, Texas 78766

Environmental analysis of toxic chemicals offers great analytical challenge because these compounds usually appear at trace levels, often in the presence of hundreds of other, more concentrated chemicals. Advances over the last ten years have increased the sensitivity and speed of analyses of environmental samples. In addition, a wide variety of techniques have been developed for sample collection, concentration and analysis.

The overall objectives of an environmental analytical program must be the satisfaction of analytical goals in a cost-effective and technically sound manner. Thus, two decision frameworks constrain the development of an environmental analytical program: (1) a cost-effectiveness decision framework, and (2) a technical evaluation decision framework.

In the cost-effectiveness decision framework illustrated in Figure 1, the cost increases from left to right. Cost-effectiveness is influenced by the intended use of the analytical results. The first consideration in the cost-effectiveness decision framework involves the selection of low-cost collective parameter measurement, high-cost individual compound identification, or surrogate analyses. The choice must be based on the intended use of the resulting data. Collective parameter measurements, *e.g.,* BOD, TOC and color, often help determine the magnitude of a toxic chemical hazard. Surrogate analyses for parameters such as total organic chlorine (TOCl) or total polynuclear aromatics are sometimes valuable as screening methods. In many cases, however, it is only through precise identification of the

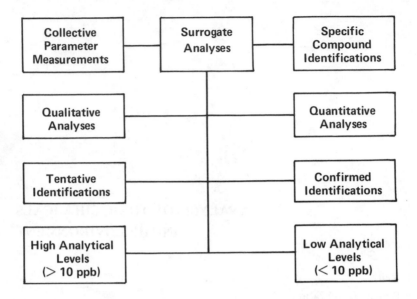

Figure 1. Cost-effectiveness decision framework.

source of a problem, *i.e.,* through individual compound identification that
a cost-effective effort can be made to eliminate or correct the problem.
If a cheaper method does not give the information needed, it is not very
cost-effective.

The second consideration involves the choice between qualitative and
quantitative analyses. There are few situations where quantitative data are
not useful, especially since toxicity is almost always related to concentration. Quantitative data, however, are always more expensive, and the differential increases at low detection levels where analytical interference can
be more significant.

The need for tentative or confirmed identification is the third point to
be considered in the cost-effectiveness decision framework. Tentative identification often suffices for screening studies or for monitoring needs, but
confirmed identification is required for regulatory purposes. Confirmed
identification is also necessary if the samples are being analyzed for the
first time. Confirmation may be achieved by comparing spectral or physical properties of a standard with those of the sample analyzed under identical conditions. Another valid technique for confirmation involves comparison of mass spectral data from a computerized or tabular library with
mass spectral data on the compound that was first identified tentatively
by other methods, such as gas chromatography (GC), using specific

detectors. Once specific compounds have been confirmed, their continued presence can be monitored, at reduced cost, using tentative identification methods. For example, trihalomethanes in drinking water may be monitored using GC retention times, if identification has been confirmed with standards and the same peaks are observed repeatedly, with no interferences.

Consideration of the analytical concentration levels required is the fourth point in the cost-effectiveness decision framework. The decision must be based on the environmental problem under investigation. For example, analysis of drinking water requires lower levels of detection than are usually required in analysis of water for recreational purposes. Analyses near the detection limit of any technique are subject to more interference with reduced precision and accuracy and, as a result, acceptable precision and accuracy may be more expensive to achieve.

The same types of logic may be considered with the technical decision framework shown in Figure 2. The first and often the most important consideration is how the organic pollutants in the sample are to be accumulated. In most instances toxic organic compounds appear at trace levels in environmental samples. Therefore, concentration is the first step in an analytical scheme. Common accumulators in use today are columns of polymeric resins such as XAD, Tenax, or Ambersorb; carbon; polyurethane foams; purge and trap columns of Tenax in combination with silica gel; and liquid-liquid extractors.

The column accumulators work by adsorbing organic compounds on the surface or in the pores of the packing material. Large volumes of sample are passed through the column and the organics present are accumulated on or in the packing material. The accumulated organic pollutants are eluted from the column with a small volume of an appropriate solvent. Problems with column accumulation techniques include variations in breakthrough volumes with different chemicals; variations in adsorption efficiency with contact time, flow rate, temperature and pH; and variations in desorption effeciency with solvent polarity, temperature, pH and contact time.

Purge and trap accumulators work by stripping volatile organic compounds out of the water and onto a resin trap with an inert gas. At the end of the accumulation period the organic pollutants are desorbed thermally onto the head of a cool gas chromatographic column. Purging efficiency varies with the breakthrough volume of each compound, gas flow and temperature. The traps are usually operated at ambient temperature and sample volumes are relatively small, usually 5 to 200 ml.

Liquid-liquid extractors likewise are restricted to relatively small samples, 0.5 to 2 liters. Extraction efficiencies vary with the partition coefficients of each compound, the solvent polarity and contact time with the solvent.

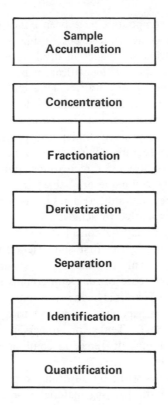

Figure 2. Technical decision framework.

Extractions are usually relatively lengthy and may take from several hours to several days.

Samples accumulated on adsorption columns or with a liquid-liquid extractor must be further concentrated. Kuderna-Danish and micro-Snyder concentrators and vacuum rotary evaporators are commonly used for concentration of organic extracts. The first two concentrations are generally considered to be most appropriate for compounds with boiling points between 50° and 300°C. Vacuum concentration, freeze drying and reverse osmosis have been used for concentrating nonvolatile or poorly extractable compounds from water.

Environmental samples are often so complex that some sort of fractionation scheme must be employed prior to instrumental analyses. In practice, a mixture containing hundreds of compounds is fractionated into several mixtures containing substantially fewer compounds, thus simplifying the ultimate separation of each compound. Common fractionation techniques include vacuum distillation; steam distillation; acid, base and neutral fractionation; column chromatography on silica gel, alumina or Florisil; and partition extractions with acetonitrile, nitromethane and other solvents.

Usually, when analyzing highly polar compounds such as carboxylic acids, phenols and amines, more volatile derivatives must be prepared. Methyl and trimethylsilyl derivatives are the two types most commonly prepared. Diazomethane is one of the most popular reagents for methylating carboxylic acids, and dimethylsulfoxide is sometimes used for methyating phenols. On-column methylating reagents, such as methElute® and Methyl-8®, are convenient for "screening analyses" using GC. These reagents are mixed with the sample and co-injected into the gas chromatograph where heat drives the methylation reaction.

Final separation is usually achieved by GC or high pressure liquid chromatography (HPLC). GC separations can be carried out using packed columns, support coated open tubular (SCOT) columns or capillary columns. There are many choices of stationary phases and solid supports. The most promising techniques of the future are high pressure liquid chromatography and capillary gas chromatography.

The identification and quantification of trace organic compounds is often easier than preparing the sample for analysis. Computerized low resolution gas chromatography-mass spectrometry (GC/MS) is the most commonly used analytical method. Often 60 to 80% of the volatile pollutants can be identified by low resolution GC/MS if they can be sufficiently concentrated and separated from one another to provide useful mass spectra. Fourier transform GC/IR is another technique that has future potential. It is not yet widely used because infrared bands in the gaseous phase are shifted sufficiently to make impractical the computerized matching of spectra from liquid or solid samples. GC with electron capture, thermionic, electroconductivity, flame photometric and photoionization detectors is often used for tentative identifications. These detectors produce peaks that are quantifiable by comparing peak heights or areas to those of standards.

Errors at any of these decision points will result in higher analytical costs, misrepresentative or incomplete data or perhaps no usable information at all. The analyst must recognize the advantages and disadvantages of each variable at each decision point if he is to make sound technical judgments and keep analytical costs at a minimum. Improvements in the techniques of sample accumulation, fractionation, derivatization and separation are continuing at a rapid pace. As a result an even larger number of options which must be considered can be expected in the future.

8

ROLE OF TRANSPORT AND FATE STUDIES IN THE EXPOSURE, ASSESSMENT AND SCREENING OF TOXIC CHEMICALS

Rizwanul Haque, James Falco*, Stuart Cohen†
and Courtney Riordan

> Office of Research and Development
> U.S. Environmental Protection Agency
> Washington, D.C. 20460

INTRODUCTION

The basic discipline of transport and fate has grown significantly during the last decades, as far as its potential use in evaluating complex environmental problems is concerned.[1,2] Transport and fate studies on chemicals provide information on the extent of partitioning and lifetime of chemicals in air, water and soils, and **on their** potential bioaccumulation in the biota.[2] An integrated evaluation of this information provides guidance as to the potential impact of chemicals on the environment. Transport and fate information plays a key role in the process of development and regulation of chemicals.[3-5]

The estimation of exposure concentration of chemicals which are used in risk analysis requires information on the transport and transformation of chemicals. This chapter reviews the basic principles of transport and fate studies as they relate to the development of exposure techniques capable of predicting the concentration of chemicals in the environment. Attempts have also been made to describe the potential use of transport and fate data in the screening of toxic chemicals for movement, bioaccumulation and persistence.

*Environmental Research Laboratories, Athens, Georgia.
†Office of Toxic Substances.

47

TRANSPORT AND FATE

The basic elements of transport and fate studies are the knowledge of
(1) physicochemical properties of the chemical, (2) transport processes in
the environment and (3) transformation processes parameters. The introduction of a chemical into the environment will result in the transport
of the chemical in the air, water, soil/sediment and biota (Figure 1).
The intrinsic physical-chemical properties of the chemical will influence
the intracompartmental transport of the chemical, whereas intercompart-
mental transport processes will contribute to the net concentration of
the chemical in a particular compartment of the environment. Finally,
there are several transformation processes which will operate at one time
or simultaneously to convert a chemical from one form to another or
to degrade the chemical. The net concentration of the chemical in the
environment may be expressed by the following equation:

$$\frac{dc}{dt} = \Sigma \frac{k_i \, C}{1 + KS} \tag{1}$$

where C is the original concentration, K is the partition coefficient between two phases of the environment; k_i is the rate constant for various
processes and S is the mass of the sorbing site in a particular medium.
Therefore, a knowledge of the partition coefficients and rate constants
is important in estimating exposure concentrations of chemicals in the
environment. We now consider various properties and transport and
transformation processes pertaining to the four compartments of the
environment: water, air, soil/sediment and biota.

WATER

The water solubility is an intrinsic property of a chemical which is
important for determination of the transport of the chemical in the aquatic
environment. The water solubilities of several toxic chemicals are given in
Table I. Toxic chemicals include compounds ranging from highly soluble
(e.g., aniline, cresols, benzene, methylene chloride) to sparingly soluble
(e.g., polychlorinated biphenyls). Many toxic chemicals contain one or more
chlorine atoms in the molecule. In general, an increase in the number of
chlorine atoms in the molecule decreases the water solubility.[6] Although
water solubility is one of the simple properties, it is useful in predicting
certain environmental transport parameters. For example, as shown in the
latter section of this chapter, the water solubility of chemicals may serve
as an indicator for partitioning,[7] adsorption[8,10] and bioaccumulation of
chemicals.

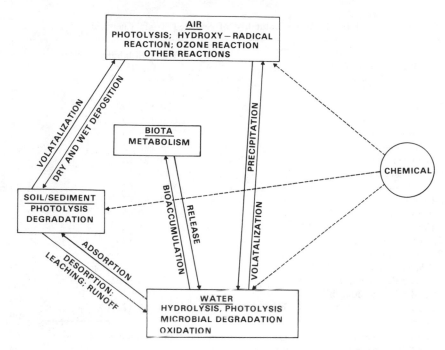

Figure 1.   Diagram showing transport and transformation process for toxic chemicals in the environment.

Transport processes pertaining to aquatic environments include partitioning to biota, adsorption to soils and sediments, volatilization, precipitation from air, runoff, desorption and leaching. Adsorption, partitioning and volatilization decrease the concentration of chemicals in water, whereas leaching, desorption, runoff and precipitation from air increase the concentration.

The volatilization of chemicals from an aqueous system may be represented by the equation

$$P_i = C_i \ P_i s/C_i s \qquad (2)$$

where P_i is partial pressure in air, $P_i s$ and $C_i s$ are vapor pressure and solubility, respectively, and C_i is the concentration in water and may be expressed as

$$C_i = C_{io} \ exp \ (-k_{iL} \ t/L) \qquad (3)$$

where C_i and ki are constants and L is the depth of the water. The volatilization of half-lives of many toxic chemicals have been estimated by several workers,[11,12] and typical values are given in Table II.

Table I. Water Solubility of Selected Toxic Chemicals

Chemical	Water Solubility (ppm)
Nitrobenzene[a]	2,000 (25°C)
Xylene[a]	130 (25°C)
Aniline[a]	37,000 (30°C)
Benzene[a]	1,790 (25°C)
p-Dichlorobenzene[a]	79 (25°C)
o-Dichlorobenzene[a]	145 (25°C)
Methylene Chloride[a]	13,000 (25°C)
Ethyl Chloride[a]	5,740 (20 C)
o-Cresol[b]	1,300 (50°C)
m-Cresol[b]	2,700 (50°C)
p-Cresol[c]	17,000 (35°C)
2,4 dichlorobiphenyl[c]	0.637 (25°C)
2,2,5-Trichlorobiphenyl[c]	0.248 (25°C)
2,2,5,5-Tetrachlorobiphenyl[c]	0.0265 (25°C)
2,2,4,5,5-Pentachlorobiphenyl[c]	0.0103 (25°C)
2,2,4,4,5,5-Henzchlorobiphenyl[c]	0.000953 (25°C)
Polychloroinated Biphenyl Arochlor 1254[a]	0.056 (25°C)

[a]Research Program on Hazard Priority Ranking of Manufactured Chemicals. Phase II, Final Report, Stanford Research Institute.
[b]A study of Industrial Data on Consolidated Chemicals for Testing, OTS, EPA, Final Report, (June 1978).
[c]Bull. Environ. Cont. Toxicol. 14:13-18 (1975).

Table II. Volatilization of Selected Chemicals from Water[a]

Chemical	Half-Life (hr) For L = 1 m
n-Octane	5.55
Benzene	4.81
Toluene	5.18
O-Xylene	5.61
Napthalene	7.15
Biphenyl	7.52
DDT	73.9
Arochlor 1254	10.3
Mercury	7.53

[a]From Environ. Sci. Technol. 19.1178 (1975)

The transformation processes pertaining to aquatic environments include hydrolysis, photolysis, microbial degradation and oxidation. However, oxidation is not as important as the other three. Hydrolysis of chemicals is strongly dependent on pH and temperature. The hydrolytic half-lives of toxic chemicals (Table III) span over a wide range.[13] The structure of the molecule and the substituents are sometimes useful in estimating hydrolysis rates.[14] The photolysis of chemicals depends upon the energy of the sunlight, the absorption spectrum of the molecule and the presence of sensitizers in the environment. Recent studies of Zepp and Cline[15] provide an expression for the calculation of photolytic half-lives ($t_{1/2}$) of chemicals in sunlight from laboratory data,

$$t_{1/2} = \frac{0.693}{ka\phi} \tag{4}$$

where ϕ is the quantum yield and ka is the term depending on the amount of light of a certain wavelength absorbed by the molecule. The work of Zepp and Cline also takes into consideration seasonal variation in sunlight intensity, latitude, time of day, depth in water bodies and ozone layer thickness. Humic substances and organic material present in water may affect the photolytic rates. The photolytic half-lives of selected toxic chemicals are given in Table IV. Microbial reactions in aquatic environments are most important in defining degradation of chemicals. The concentration of chemicals, the nature of the microbes, environmental conditions and temperature are important in defining microbial degradation.[16] Microbial-mediated reactions with certain metals result in methylation of metals and thus in the synthesis of other toxic chemicals.[17] The biomethylation of mercury into methyl mercury is a good example.[17] Compared to hydrolysis and photolysis, relatively little is known about expressing the microbial degradation quantitatively.

Air

The vapor pressure (Table V) of a chemical is an important physical property and is useful in defining the extent to which a chemical will be transported into the atmosphere. Although vapor pressure data alone can provide valuable information on the potential transport of the chemical into air, the adsorption of chemicals to surfaces may alter the vapor transport. It has been observed that the vapor loss of the polychlorinated biphenyl Arochlor 1254 is appreciable from poorly adsorbing sand surfaces, whereas it is negligible when it is bound to strongly adsorbing soil surfaces.[18,19] The presence of chlorine atoms in molecules also affects the vapor loss, and the vapor loss decreases with an increase in the

Table III. Hydrolytic Half-Life of Selected Organic Chemicals[a]

Chemical	Half-Life (298°K and pH = 7)
C_2H_5Cl	38 days
$CH_2=CHCH_2Cl$	69 days
$CHCl_3$	3500 yr
$CH_3CH(0)CHCH_3$	15.3 days
$CH_3C(0)OC_2H_5$	2.0 yr
CH_3CONH	3,950 yr
$C_2H_5OC(0)N(C_6H_5)$ (CH_3)	44,000 yr
$CH_3P(0)$ $(OCH_3)_2$	88 yr
$(NP)_3$ PO	11 min
$CH_3P(0)(F)CH_3$	2.9 min
$C_6H_5C(0)Cl$	16 sec

[a]From *J. Phys. Chem. Ref. Data* 7:383 (1978).

Table IV. Photolytic Half-Life of Selected Chemicals in Aquatics[a]

Compound	Half-Life in Hours
p-Cresol	5.8×10^3
Benz(a)anthracene	5.8×10^{-1}
Benzo(a)pyrene	5.3×10^{-1}
Quinoline	5.5×10^2
Benzo(f)quinoline	5.2×10^{-1}
9H Carbazole	1.0
7H Dibenzo(c,g)Carbazole	3.5×10^{-1}
Benzo(b)thiophene	3.4×10
Debenzothiophene	1.3×10^2
Methyl Parathion	2.0×10^2
Mirex	3.9×10^3

[a]From EPA Report 600/7-78-074.

number of chlorine atoms in the molecule as observed from the volatilization studies of several chlorinated biphenyls from metal surfaces.[20] The volatilization of chemicals from soil surfaces can best be studied by the method developed by Spencer and co-workers.[21] The volatilization of chemicals from the aquatic phase may increase the concentration in

Table V. Vapor Pressure of Selected Chemicals[a]

Chemical	Vapor Pressure (mm Hg)
Nitrobenzene	1 at (44.4°C)
o-Dichlorobenzine	1 (20°C)
p-Dichlorobenzine	0.4 (25°C)
Toluene	40 (31.8°C)
Hexachloro-1,3-butadiene	0.15 (20°C)
m-Cresol	1 (52.0°C)
o-Cresol	1 (38.2°C)
p-Cresol	1 (53.0°C)
Chlorinated Paraffins (various)	10 (65°C)
Methyl Chloride	5 (22°C)

[a]Reported from initial report of the TSCA Interagency Testing Committee to the Administrator, Environmental Protection Agency.

air. The presence of particulate matter, such as dust in air, may also increase the chemical concentration in the air.

Other important transport factors pertaining to air are air-water and air-land equilibria (dry deposition, wet deposition and gaseous diffusion), horizontal mixing in the troposphere (both on the localized point source or areal source scale, and the hemispheric or global scale) and vertical mixing between the troposphere and the stratosphere. The particle size and the manner in which the chemical is released (aerosol, solid, etc.) may also affect the transport in air. The transformation processes occurring in the atmosphere are complex and may be classified into major categories. These are (1) photolysis and (2) reaction of the chemical with other species.

1. The photolytic process is important for predicting degradation of many organic chemicals.[22,23] Since many molecular species have bond energies that are roughly equivalent to the energy of the sea level solar cutoff, approximately 95 kcal/mole photons, photolysis may be an important degradation process for some chemicals that absorb at solar wavelengths. At higher altitudes, the total solar flux is greater, resulting in an increased amount of photolytic and excited state reactions. The destruction of ozone by halocarbons in the stratosphere is a result of such reactions.[24]

2. Reaction with other chemical species includes the reaction of toxic chemicals with hydroxyl radicals (OH) and ozone. These two are the most important reactant species for the degradation of hydrocarbons in

the troposphere. The three major sources of hydroxyl radicals in the lower atmosphere are:

$$H_2O + O\,(^1D) \longrightarrow 2\,OH \tag{5}$$

$$NO + HO_2 \longrightarrow OH + NO_2 \tag{6}$$

$$H_2O_2 + \xrightarrow{h\nu(\lambda < 370\ nm)} 2\,OH \tag{7}$$

Altshuller[25] has recently reviewed several estimates of average global and hemispheric OH concentration in ambient air and found them to vary from 2.9 to 6.3 x 10^5 radicals/cm^3. The hydroxyl radical reactivity with chemicals may be represented with a second-order rate mechanism. For hydrocarbons the rate constant averages[27] in the range of 10^9–10^{10} m^{-1} sec^{-1}. The rate constant reduces to 10^5–10^8 m^{-1} sec^{-1} for halocarbons[28] containing labile hydrogen atoms or double bonds.

The major source of ozone in the polluted atmosphere is due to the following pathways[22]:

$$NO_2 \xrightarrow{h\nu(\lambda < 430\ nm)} NO + O(P) \tag{8}$$

$$O(^3P) + O_2 + M \longrightarrow O_3 + M \tag{9}$$

where M is a nonreactive third body such as N_2 or H_2O. The ozone concentration in urban polluted atmospheres may range in the tenths of parts per million, whereas in clean air the concentration is in the 20-80 parts per billion range. Alkenes react much faster with ozone than do most other organic compounds, and the rate constants generally range[30] between 10^2 and 10^5 m^{-1} sec^{-1}. Therefore, based on concentrations of the reacting species and the appropriate second-order rate constants, organic atmospheric reactions with ozone will compete with OH-organic reactions only if the organics are alkenes. These reactions do not always transform the chemicals into nontoxic products, and, in some cases, the tropospheric transformation products are more toxic than the original compounds. For example, it has recently been demonstrated[31] that two different polynuclear aromatic hydrocarbons were transformed to direct bacterial mutagens upon exposure to ambient or simulated smog.

Soil/Sediment

Once chemicals find their way into the environment, a major portion reaches the soil and sediment, and in some cases soil and sediment serve as sinks.[19] The important transport pertaining to soils and sediments are adsorption and leaching. Factors influencing the adsorption[32,33] process are (1) structural characteristics of the chemical, (2) the organic content of the soil, (3) the pH of the medium, (4) particle size, (5) ion exchange capacity and (6) temperature. Most adsorption processes take a shorter time to attain equilibrium and the desorption rate is much slower.[34] The adsorption data can best be represented by the Freundlich isotherm as follows:

$$\frac{x}{m} = KC^n \tag{10}$$

where   x = the amount of chemical sorbed,
        m = the mass of soil,
        C = the equilibrium concentration of the chemical
        K = a constant describing the extent of adsorption and
        n = a constant describing the nature of the adsorption.

Studies on adsorption of several polychlorinated biphenyl-type compounds on several surfaces indicate that the adsorption increases with a decrease in the water solubility of the adsorbate, and also increases with the organic content of the adsorbent.[9] The organic content of the soil/sediment has been emphasized by several workers for predicting adsorption. It has been suggested that the organic content of the soil is the controling factor in determining adsorption.[35] A good correlation between the octanol/water partition coefficient and the adsorption coefficient has also been demonstrated.[36,37] It should be pointed out that the above concept for predicting adsorption does not apply to ionic-type compounds or neutral organic chemicals capable of becoming charged species in solution.[38,39] For example, the organic cations diquat and paraquat, which are highly water soluble, adsorbed to the clay portion of the soil with a different mechanism involving intercalation of the molecule between clay layers. Similarly, triazine-type compounds form charged species on clay surfaces.

Leaching of chemicals may reduce chemical concentration in soil/sediment and may pose ground-water contamination problems.[40] The adsorption coefficient of a chemical may provide an indicator of leaching potential. Volatilization of toxic chemicals from soils and sediments may also reduce the concentration in soils.

The two major degradation pathways for chemicals in soil/sediment are biodegradation and photolysis. However, at present the contribution of photolysis to the total degradation of chemicals is not very well known, and biodegradation represents the major route. Degradation[41] in soil depends upon such factors as concentration of the chemical, temperature, moisture, anaerobic conditions and organic content of the soil. The mechanism of degradation in soil/sediment is poorly understood, and most data on half-lives have been calculated using a first-order kinetic expression.

Biota

The biota compartment of the environment contains living organisms, including animals, plants and humans. Transport and fate analysis can provide only limited but useful information in predicting the behavior of chemicals in the biota. Information on the exposure concentration of a chemical, the bioaccumulative capacity of the biota for the chemical, the capacity of the biota to metabolize the chemical and the rate of release of the chemical from the biota can be obtained from transport and fate studies. These studies play a key role in evaluating chemical hazard and risk. The octanol/water partition coefficient of toxic chemicals, defined as the ratio of the concentration of the chemical in octanol and water phase at equilibrium, as given in Equation 11, has been found to be very useful in predicting the bioaccumulation ratio of chemicals in aquatic organisms.

$$P = \frac{[C] \text{ octanol}}{[C] \text{ water}} \qquad (11)$$

Neely et al.,[42] first reported a semiempirical correlation between the octanol/water partition coefficient and the bioconcentration factor for many organic chemicals:

$$\log (\text{Bioconcentration Factor}) = 0.524 \log [P] + 0.124 \qquad (12)$$

This correlation was based on the assumption that most organic chemicals, once partitioned into the biotic phase, are transported into the lipid phase. The correlation was good, in spite of the fact that the P values used in the correlation were calculated theoretically. Laboratory studies have now shown the existence of the following correlation between P values and ecological magnification of many chemicals in fish:[43]

$$\log \text{E.M} = 0.7285 + 0.6335 \log P \qquad (13)$$

However, such correlations are qualitative or semiquantitative at best. In both the studies there was considerable scatter. Such a correlation may not hold for low P values or for chemicals where ionization or a binding with protein or other biological polymers may occur.

A correlation between the partition coefficient and the water solubility of many chemicals also exists.[7,8] Caution should be exercised in quantitative treatment because of the use of the log-log relationship. The log of Equation 11 gives the following:

$$\log P = \log [C] \text{ octanol-log } [C] \text{ water} \qquad (14)$$

It may be possible that at equilibrium, for most hydrophobic chemicals, [C] water and [C] octanol approach values of the solubility of the chemical in water and in octanol, respectively. It may be possible that the solubilities of most hydrophobic organic chemicals in octanol are in the same range. Since most chemicals possess much higher solubility in octanol than in water, Equation 14 simply gives a linear correlation between water solubility and P.

Since there is a correlation between the partition coefficient and biomagnification, a relationship between water solubility and the bioconcentration factor might be expected. Metcalf and Sanborn[44] plotted ecological magnification (bioconcentration factor) against water solubility on a log-log scale, and there was indeed a trend between biomagnification and water solubility. This trend was also reported for other organic chemicals.[45]

The P values have also been found to be very useful in predicting toxicity of drugs. Hansch[46] has developed empirical correlations involving toxicity, partition coefficient and electronic factors in a series of chemicals having the same basic skeleton with varying substituents. However, the application of this method in predicting environmental toxicity has not been fully explored.

EXPOSURE ASSESSMENT

The exposure concentration of chemicals, defined here as the concentration to which humans or the environment are exposed at a particular time, can be estimated from transport and fate information. The concentration of a chemical actually reaching a living target will be different. The three basic elements in estimating exposure concentration are (1) source term, (2) transport and transformation characteristics, including physical-chemical properties, and (3) population at risk.

1. *Source Term.* The relevant source information on chemicals may be obtained by considering such factors as production data, release rate during use (discharge, emission), rate of disposal, use pattern and contribution due to natural sources. Source term data are needed in a mass-balance analysis of the chemical.

2. *Transport and Transformation Characteristics.* The transport and transformation characteristics are needed to define the major environmental pathways, and are the key to estimating exposure concentration of chemicals.

3. *Population at Risk.* This information is needed in determining the number of population groups which are exposed to chemicals at a certain time. Occupational characteristics, medical surveillance data and socioeconomic use habits of population groups are used in estimating population at risk.

Models for Exposure Assessment

The information of input data, physical-chemical properties and transport and transformation characteristics are used to build mathematical models for exposure assessment. Neely[47] has applied this concept to develop models capable of predicting concentrations of toxic chemicals and pesticides in soil, air, water and fish. This model uses transport (volatilization, adsorption and bioaccumulation or partition coefficient) parameters, physical-chemical properties, degradation rate constants and release rate by fish or biota. Smith *et al.*[48] have utilized the transport and fate information in developing a model for toxic chemicals in aquatic environments. This model uses compartments which are arbitrary in size and which represent segments of the water column or sediments. By adjusting various constants, this model may be applied to aquatic systems, such as streams, ponds or lakes. It is based on the following equation:

$$M''_{li} = I\,(\Delta t) + \underset{in}{\Sigma Ml ji} + M_{li} - \underset{out}{\Sigma M_{lij}} \tag{15}$$

where

$I\,(\Delta t)$ = the input of pollutant to the ith compartment,

M_{li} = the mass of the component,

$Mlji$ = the mass of the pollutant in the aqueous phase, which is added from jth the phase, and

$Mlij$ = the mass of pollutant in the aqueous phase that flowed out of compartment i to compartment j.

The concentration of the chemical in the sediment phase may be represented as:

$$M''_{si} = IS\,(\Delta t) + \underset{in}{\Sigma M_{sji}}\ 3\ M_{si} - \underset{out}{\Sigma Msij} \tag{16}$$

where

$IS\,(\Delta t)_i$ = the external input to the soil/sediment phase,

$MSji$ = the mass added to the solid from adjacent site i, and

M_{Sji} = the suspended solid mass that flowed from compartments i to j.

To account for transformation processes, a function, f, is defined as:

$$f_{li} = \frac{dM'li}{dt} = (rp + rh + ro + rv + rb)_i\ V_i \tag{17}$$

where r represents the rate constants for photolysis, hydrolysis, oxidation, volatilization and biodegradation and V_i the volume. This concept has been found to be useful in defining the major environmental pathways for several toxic pollutants. Baughman and Lassiter[49] have used similar concepts in predicting concentration of chemicals in aquatic environments.

Exposure assessment models are only at the beginning stages. Such models must be validated in model ecosystems, microcosms and field conditions.

Chemical Runoff Model

Transport and transformation information has been found useful to predict the concentration profile of pesticides in water sediments of rivers, where the concentration of pesticide results from usage and runoff. The model for this prediction is commonly known as the Agricultural Runoff Management (ARM) model, as developed by Donigian et al.[50]

The model requires (1) rainfall records, (2) data on rates and amounts of pesticide applications, (3) transport and transformation data, including physical-chemical properties of pesticides, and (4) data on the parameters necessary to define the physical characteristics of the basin. The practical application of the ARM first requires loading of sediments and pesticides into the basin. Data on the average size of the field are needed to estimate this loading. The concentration of a pesticide at the time of entry into a river can be estimated with the following equation:

$$C^T(t^+) = C^T(t^-) + \frac{L}{Q \times \tau}$$

(18)

where: C (t) is the concentration after loading has occurred, $C^T(t^-)$ is the concentration before loading occurred, Q is the volume flow rate of the river at the time of loading, τ = the average time of travel down the length of the river, and L is the load of pesticide input into the river. The average daily flow rate for most major rivers can be obtained from U.S. Geological Survey data.

The concentrations predicted using the above equation are the amount of dissolved and sorbed pesticides per unit volume of river water. It is easy to show that the concentration of dissolved material is related to the total concentration by the following equation:

$$C(t^+) = C^T(t^+) * \frac{1}{1 + K \cdot S \times 10^{-6}}$$

(19)

where K is the sediment water partition coefficient and S is the concentration of suspended sediments.

For most major rivers, suspended sediment concentration can be obtained from U.S. Geological Survey data published annually for each state. Partition coefficients can be obtained from laboratory data. The above treatment is based on assumptions that the equilibrium partition between adsorbed and dissolved pesticide is attained instantaneously and that the pesticide load is evenly distributed along the river segment analyzed.

Once the dissolved concentration of a pesticide in a river is estimated at the time of loading, degradation processes must be determined to estimate the concentration of the pesticide as a function of time at a position downstream of the point of entry. The processes of hydrolysis, volatilization, oxidation, bacterial degradation and photolysis may be used to make such estimates. In general, for reasonable lengths of river segments, environmental conditions remain relatively constant. Thus, most degradation processes can be represented by pseudo-first-order reactions. Knowing the rate of degradation and velocity of the water, a mass balance can be written which describes the behavior of pesticides in water as follows:

$$\frac{v \, dc}{dx} = \frac{-kc}{1 + K.S.10^{-6}} \qquad (20)$$

where v = the average velocity of the river,
          x = the distance down the length of the river, and
          k = the rate constant for degradation.

Using Equations 18 and 19 as initial conditions, the above differential equation can be solved to give the following result:

$$C = C(t^+) \, e^{-\left\{\frac{k \cdot x}{v(1 + K.S.10^{-6})}\right\}} \qquad (21)$$

The assumption implicit in using Equation 20 to describe the mass balance is that the rate of transport by turbulent diffusion is small compared to the rate of transport by bulk flow. Note that by replacing the term x by time t in Equation 21, one can determine the concentration at the mouth of a river from the time of a loading to either the average time of travel down the length of the river or the occurrence of a new loading event. If a new loading event occurs, equations are used to recalculate the initial conditions, and the time is set to zero.

Once the concentration vs time graph has been established over the period of rainfall record, a concentration-frequency-duration table can be developed by counting the number of times given concentrations have been exceeded and recording the length of time each occurrence counted lasted. These data can in turn be compared with toxicity data for important species residing in receiving waters to estimate potential environmental impacts of use of the pesticide.

SCREENING OF TOXIC CHEMICALS

Transport and fate concepts may be utilized to screen toxic chemicals as to their fate and behavior in the environment. The two approaches, (1) the Benchmark[51,52] concept and (2) the Structure Activity concept, may be used to screen chemicals.

Benchmark

This concept involves the selection of one or more benchmark chemicals from important classes of toxic chemicals, and measurement of key environmental parameters and physical-chemical properties. This information is then integrated to build an environmental profile[53] of the class of the chemical. In order to predict the behavior of a new chemical, the first step is to find out from structural similarity to which benchmark the chemical corresponds. By then comparing it with the benchmark, and by pattern recognition methods, it is possible to predict the behavior of the new chemical. The benchmark method can be described by the following schematic representation:

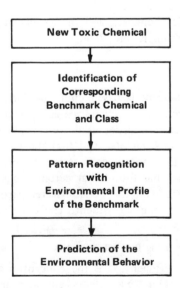

The key environmental parameters and physical-chemical properties included in the Benchmark concept are water solubility, vapor pressure, hydrolysis, soil degradation, adsorption, volatilization, photodegradation and partition coefficient. This concept also specifies that the measurement of the above parameters be carried out under standard conditions. These parameters may also provide certain indicators for the behavior of toxic chemicals. For example, a high octanol/water partition coefficient will indicate the potential bioaccumulation, and a high persistence value may indicate that the chemical will remain in the environment for a relatively long period of time, and thus the probability of its exposure to humans and the environment will increase. In other words, the benchmark concept may give certain environmental flags for toxic chemicals which may be used in the screening.

Information from the benchmark concept may be used to build an environmental profile of chemicals.[53] The concept of environmental profile involves the estimation of concentrations and half-lives of chemicals in major environmental compartments. A hypothetical environmental profile of a chemical is shown in Figure 2. These profiles may be used to predict the behavior of any chemical via the pattern recognition methods with the corresponding benchmark chemical. Although the Benchmark concept presents a potential predictive tool for toxic chemicals, much research is needed to develop this system. Validity of the concept, validation of results from field studies and collection of key parameters are some of the work that must be carried out before this system could be developed. Furthermore, it appears that this concept may be applicable only in predicting the transport and fate of toxic chemicals, and the extension of the concept to effects and toxicity prediction is uncertain.

Structure-Activity

The structure-activity concept is based upon the variation in certain properties and biological activity of chemicals as a function of substituents.[46] The most common chemical properties used are octanol/water partition coefficient and toxicity. It has been observed that the octanol/water partition coefficient, as a function in a homologous series, correlates with certain toxic characteristics. Hansch has found that certain toxicity of a series of chemicals may be correlated with the octanol/water partition coefficient and electronic parameters of the substituent. The correlation may be expressed in general terms as

$$\log C = a \log P + b \log P^2 + K \tag{22}$$

where P is the partition coefficient, C is the concentration of the chemical required to produce certain toxic effects, and a, b and K are constants. The constant K represents the electronic parameters such as the σ constant. For many systems, b approaches zero.

At this point, a distinction must be made between the Benchmark and the Structure-Activity concepts. The structure-activity-type concept is

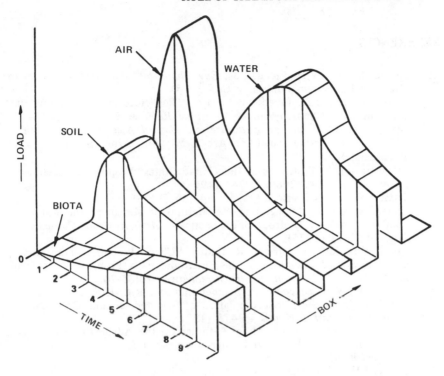

Figure 2. Diagram showing a hypothetical environmental profile of a chemical.

applicable to a series of chemicals where the basic skeleton of the molecule is the same, but variations are due to changes in the functional groups and substituents. The Benchmark concept applies to broader classes of chemicals where major changes in the structure of the chemical may take place, as long as chemicals can be grouped in general classes.

Relatively little work has been done on correlating the structure of chemicals with environmental degradation parameters. The recent studies of Darnell et al.[54] describe a relationship whereby hydroxyl radical reactivity with alkanes may be predicted based solely on the number of primary, secondary and tertiary hydrogens. The work of Wolfe et al.[14] provides a correlation whereby hydrolysis rates can be predicted from free energy data of carbamate pesticides.

NOTE: The contents of this paper do not necessarily reflect the views and policies of the U.S. Environmental Protection Agency, nor does mention of trade names or commercial products constitute endorsement or recommendation for use.

REFERENCES

1. Haque, R., and V.H. Freed, Eds. *Environmental Dynamics of Pesticides* (New York: Plenum Publishing Co., 1975).
2. Haque, R. and V.H. Freed. "Behavior of Pesticides in the Environment: Environmental Chemodynamics," *Residue Reviews* 52:89-116 (1974).
3. Blair, E. H. In *Dynamics, Exposure and Hazard Assessment of Toxic Chemicals*, R. Haque, Ed. (Ann Arbor, MI: Ann Arbor Science Publishers, Inc., 1979). p. 5
4. "Registration of Pesticides in the United States: Proposed Guidelines," *Federal Register* 43:29696-29741 (1978).
5. Toxic Substances Control Act, PL 94-469 (1976).
6. Haque, R. and D. Schmedding. "A Method of Measuring the Water Solubility of Very Hydrophobic Chemicals," *Bull. Environ. Contam. Tox.* 14:13-18 (1975).
7. Freed, V.H., C.T. Chiou and R. Haque. "Chemodynamics: Transport and Behavior of Chemicals in the Environment—A Problem in Environmental Health," *Environ. Health Perspectives* 20:55-70 (1977).
8. Chiu, C.T., V.H. Freed, D.W. Schmedding and R.L. Kohnert. "Partition Coefficient and Bioaccumulation of Selected Organic Chemicals," *Environ. Sci. and Technol.* 11:475 (1977).
9. Haque, R., and D. Schmedding. "Studies on the Adsorption of Selected Polychlorinated Biphenyl Isomers on Several Surfaces," *Environ. Sci. Health,* Bll 2:129-138 (1976).
10. Smith, J.H., W.R. Mabey, N. Bohonus, B.R. Holt, S.S. Lee, T.W. Chou, D.C. Venberger and T. Mill. "Environmental Pathways of Selected Chemicals in Freshwater Systems: Part II. Laboratory Studies," EPA Report 600/7-78-074 (1978).
11. Mackay, D., and P.J. Leinonen. "Rate of Evaporation of Low-Solubility-Contaminants from Water Bodies to Atmsophere," *Environ. Sci. Technol.* 9:1178-1180 (1975).
12. Dilling, W.L., N.B. Tetertiller and G.J. Kallos. "Evaporation Rates and Reactivities of Methylene Chloride, Chloroform, 1,1,1-Trichloroethane, Trichloroethylene, Tetrachloroethylene and Other Chlorinated Compounds in Dilute Aqueous Solutions," *Environ. Sci. Technol.* 9:833-838 (1975).
13. Mabey, W., and T. Mill. "Critical Review of Hydrolysis of Organic Compounds in Water Under Environmental Conditions," *J. Phys. Chem. Ref. Data* 7:383 (1978).
14. Wolfe, N.L., R.G. Zepp and D.F. Paris. "Use of Structure Reactivity Relationships to Estimate Hydrolytic Persistence of Carbamate Pesticides," *Water Res.* 12:561-563 (1978).
15. Zepp, R.G., and D.M. Kline, Rates of Direct Photolysis in Aquatic Environment," *Environ. Sci. Technol.* 11:359-366 (1977).
16. Paris, D.F., D.L. Lewis, J.T. Barnett and G.L. Baughman. "Microbial Degradation and Accumulation of Pesticides in Aquatic Systems," EPA Report 660-3-75-007 (1975).
17. Wood, J.M. "Biological Cycles for Toxic Elements in the Environment," *Science* 183.1049 1052 (1974).

18. Haque, R., D. Schmedding and V.H. Freed. "Aqueous Solubility
     Adsorption and Vapor Behavior of Phlychlorinated Biphenyl Aroclor
     1254," *Environ. Sci. Technol.* 8:139 (1974).
19. Haque, R. In: *Environmental Dynamics of Pesticides,* R. Haque and
     V.H. Freed, Eds. (New York: Plenum Press, 1975), p. 97.
20. Haque, R., and R. Kohnert. "Studies on the Vapor Behavior of
     Selected Polychlorinated Biphenyls," *J. Environ. Sci. Health* Bll:
     253-254 (1976).
21. Spencer, W.F., and M.M. Clieth. In: *Environmental Dynamics of
     Pesticides,* R. Haque and V.H. Freed, Eds. (New York: Plenum Press,
     1975), p. 61.
22. Finlayson, B.J., and J.N. Pitts, Jr. "Photochemistry of the Polluted
     Troposphere," *Science* 192:111-119 (1976).
23. Graedel, T.F., L.A. Farrow and T.A. Weber. "Chemical Calculations
     with Altered Source Conditions," *Atmos. Environ.* 12:1403-1412
     (1978).
24. Chou, C.C., R.J. Milstein, W.S. Smith, H.V. Ruiz, M.J. Molina and
     F.S. Rowland. "Stratospheric Photodissociation of Several Saturated
     Perhalo Chlorofluorocarbon Compounds in Current Technological
     Use," *J. Phys. Chem.* 82:1-7 (1978).
25. Altshuller, A.P. "Lifetimes of Organic Molecules in the Stratosphere
     and Lower Troposphere," In: *Advances in Environmental Science and
     Technology* (in press).
26. Darnall, K.R., A.C. Lloyd, A.M. Winer and J.N. Pitts, Jr. "Reactivity
     Scale for Atmospheric Hydrocarbons Based on Reaction with Hydroxyl
     Radical," *Environ. Sci. Technol.* 10:662-669 (1976).
27. Lloyd, A.C., K.R. Darnall, A.M. Winer and J.N. Pitts, Jr. "Relative Rate
     Constants for Reactions of Hydroxyl Radical with a Series of Alkanes,
     Alkenes and Aromatic Hydrocarbons," *J. Phys. Chem.* 80:789-794
     (1976).
28. Cox, R.A., R.G. Derwent, A.E.J. Eggleton and J.E. Lovelock. "Photo-
     chemical Oxidation of Halocarbons in the Troposphere," *Atmos. En-
     viron.* 10:305-308 (1976).
29. Singh, H.B., F.L. Ludwig and W.B. Johnson. "Tropospheric Ozone:
     Concentration and Variabilities in Clean Remote Atmospheres,"
     Atmos. Environ. 12:2185-2196 (1978).
30. Japar, S.M., C.H. Wu and H. Niki. "Rate Constants for the Reaction
     of Ozone with Olefins in the Gas Phase," *J. Phys. Chem.* 78:2318-
     2320 (1974).
31. Pitts, J.N., Jr., K.A.V. Cauwenberghe, D. Gorsjean, J.P. Schmid, D.R.
     Fitz, W.L. Belser, Jr., G.B. Knudson and P.M. Hynds. "Atmospheric
     Reactions of Polycyclic Aromatic Hydrocarbons: Facile Formation
     of Mutagenic Nitro Derivatives," *Science* 202:515 (1978).
32. Hamaker, J.W., and J.M. Thompson. *Organic Chemicals in the Soil
     Environments,* C.A.I. Goring and J.W. Hamaker, Eds. (New York:
     Marcel Dekker, 1974) p. 139.
33. Bailey, G.W., and J.L. White. "Review of Adsorption and Desorption
     of Organic Pesticides by Soil Colloids with Implication concerning
     Pesticide Bioactivity," *J. Agric. Food Chem.* 12:324 (1964)

34. Lindstrom, F.T., R. Haque and W.R. Coshow. "Adsorption from Solution in a New Model for the Kinetics of Adsorption-Desorption Process," *J. Phys. Chem.* 74:495 (1970).
35. Lambert, S.M. "Omega, A Useful Index of Soil Sorption Equilibrium," *J. Agric. Food Chem.* 16:340 (1968).
36. Briggs, G.G. "A Simple Relationship Between Soil Adsorption of Organic Chemicals and their Octanol/Value Partition Coefficients," *Proc. 7th British Insecticide and Fungicide Conference* (1973), pp. 73-75.
37. Karickhoff, S.W., D.S. Brown and T.A. Scott. "Sorption of Hydrophobic Pollutants on National Sediments," *Water Research* 13, in press (1979).
38. Haque, R., S. Lilley and W.R. Coshow. "Mechanism of Adsorption of Diquat and Paraquat," *J. Coll. Interface Sci.* 33:195 (1970).
39. Weber, J.B. "Interaction of Organic Pesticides with Particulate Matter in Aquatic Soil System," *Adv. Chem. Series* 55 (1972).
40. Hamaker, J.W. *Diffusion and Volatilization in Organic Chemicals in Soil Environments,* C.A.I. Goring and J.W. Hamaker, Eds. (New York: Marcel Dekker, 1972), p. 341.
41. Goring, C.A.I., D.A. Laskowski, J.W. Hamaker and R.W. Meikle. *The Environmental Dynamics of Pesticides,* R. Haque and V.H. Freed, Eds. (New York: Plenum Publishing Co., 1975) p. 135.
42. Neely, W.B., D.R. Branson and G.E. Blau. "Partition Coefficient to Measure Bioconcentration Potential of Organic Chemicals in Fish," *Environ. Sci. Technol.* 8:1113-1115 (1974).
43. Lu, P., and R.L. Metcalf. "Environmental Fate and Biodegradability of Benzene Derivatives as Studied in a Model Aquatic Ecosystems," *Environ. Health Perspective* 10:269-284 (1975).
44. Metcalf, R.L., and J.R. Sanborn. In: *Bull. Illinois Natural History Survey* 31:381-436 (1975).
45. Haque, R., P.C. Kearney and V.H. Freed. In: *Pesticides in Aquatic Environments,* M.A.Q. Khan, Ed. (New York: Plenum Publishing Co., 1977) p. 39.
46. Hansch, C. *Biological Activity and Chemical Structure,* (Amsterdam: Elsevier Press, 1977), p. 47.
47. Neely, W.B. "A Preliminary Assessment of the Environmental Exposure to be expected from the addition of a Chemical to a Simulated Environment," *J. Environ. Studies,* in press (1978).
48. Smith, J.H., W.R. Mabey, N. Bohonos, B.R. Holt, S.S. Lee, T.W. Chou, D.C. Bemberger and T. Mill. "Environmental Pathways of Selected Chemicals in Freshwater Systems, Part I. Background and Experimental Procedures," **EPA Report** 600/7-77-113 (1977).
49. Baughman, G.L., and R. Lassiter. In: *Estimating the Hazards of Chemical Substances to Aquatic Life,* J. Cairn, K.L. Dixon and A.W. Maki, Eds. ASTM Publication STP 657 (1978), p. 35.
50. Donigian, A.S., D.C. Beyerlein, H.H. Davis and N.H. Crawford, "Agricultural Runoff Management (ARM) Model, Version II Refinement and Testing," EPA Report 600/3-77-098. (1977).
51. Goring, C.A.I. In: *Organic Chemicals in the Soil Environment,* C.A.I. Goring and J.W. Hamaker, Eds. (New York: Marcel Dekker, 1972) pp. 79-86 (1975).

52. Haque, R. "Benchmark Chemistry Program," In: *Proceedings of a Symposium on Substitute Chemical Programs,* Vol. III, pp. 79-86 (1975).
53. Haque, R., and A. Eschenroeder. Paper presented at National American Chemical Society Meeting, San Francisco, CA, September 1976.
54. Darnall, K.R., R. Atkinson and J.N. Pitts Jr. "Rate Constants for the Reaction of OH Radical with Selected Alkanes at 300 K," *J. Phys. Chem.* 82:1581-1584 (1978).

9

ASSESSING THE PHOTOCHEMISTRY OF ORGANIC
POLLUTANTS IN AQUATIC ENVIRONMENTS

Richard G. Zepp

Environmental Research Laboratory
Office of Research and Development
U.S. Environmental Protection Agency
Athens, Georgia   30605

INTRODUCTION

During the past 30 years, increasing amounts of chemicals have been
added to the environment by human activities. More than 20 years ago it
was realized that the gases from factory and automobile emissions were
somehow transformed to highly visible, noxious photochemical smog.
Early concern over the smog problem prompted research into how gases
are transformed and transported in the atmosphere.[1] More recently, computer models, based on rate data for transformation of atmospheric chemicals, have been employed to assess photochemical smog formation[2] and
other atmospheric problems.[3,4]

Unlike photochemical smog, organic pollution in water cannot be seen
by the human eye. The widespread contamination of water bodies by
certain toxic chemicals such as polychlorinated biphenyls (PCBs) did not
become apparent until the past 10 years as analytical techniques improved.
Thus, efforts to understand and quantify the processes that influence the
behavior of organic water pollutants have not kept pace with corresponding efforts in the air pollution field. Only recently, for example, have computer models been developed that utilize quantitative rate and equilibrium
data to predict the behavior of chemicals in natural waters.[5,6]

In this chapter, various processes that influence rates of photochemical transformations of aquatic pollutants are discussed. Two general types of photochemical processes are shown in Figure 1. Direct photolysis involves light absorption by the chemical itself; indirect or sensitized photolysis is initiated through light absorption by other substances in the system. To assess the role of direct or indirect photolysis in the aquatic environment, procedures are being developed that employ data measured under controlled conditions in the laboratory to predict how rapidly photolysis occurs under the wide range of conditions extant in the environment. The discussion of this approach will first center on rate expressions that describe photochemical transformation. This will be followed by sections concerning simulation of solar spectral irradiance reaching the surface of water bodies, light attenuation and mixing in natural waters, and the effects of sorption on photolysis in natural water. Finally, atmospheric photolysis of volatile water pollutants and computer simulations that assess the role of photolysis in various types of aquatic environments are briefly examined.

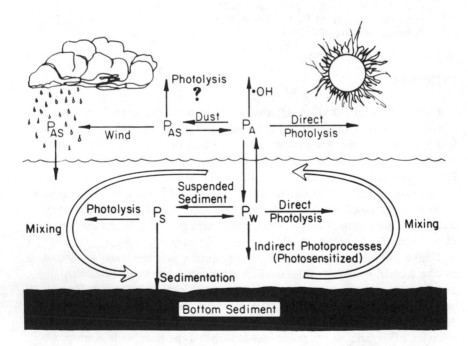

Figure 1. Processes influencing photochemical transformation in the environment.

PHOTOLYSIS RATE EXPRESSIONS

The rate of a photochemical process is determined by the rate of light absorption I_a and the quantum yield for the process, ϕ (Equation 1). The disappearance quantum yield is an important parameter that describes the fraction of absorbed light resulting in photoreaction.

$$\text{Rate} = I_a \phi \tag{1}$$

The net rate of photochemical transformation is the sum of the rates of all direct and indirect processes. Under most environmental conditions, pollutant concentrations are very low, and direct or indirect photoprocesses can be described by first-order rate expressions, that is, the rate is directly proportional to pollutant concentration, [P] (Equation 2). The rate constant for photolysis, k_p, is then the sum of the rate constants for direct and indirect photolysis, k_d and k_s, respectively:

$$(\text{Rate})_{net} = (k_d + k_s) \, [P] = k_p[P] \tag{2}$$

The rate data are sometimes expressed in terms of half-lives, $t_{1/2}$ (Equation 3):

$$t_{1/2} = \frac{0.69}{(k_d + k_s)} \tag{3}$$

Direct Photolysis Rates

Direct photolysis is described by a first-order rate expression because at low concentrations the light absorption rate I_a is proportional to [P]. Thus, the direct photolysis rate constant at wavelength λ equals $\phi_d k_{a,\lambda}$ where $k_{a,\lambda}$, the specific light absorption rate, is expressed in units of reciprocal time, and ϕ_d, the quantum yield, is unitless.[7]

The disappearance quantum yield, ϕ_d, is an inherent property of a chemical that can be readily computed from kinetic data obtained in the laboratory.[8] Direct photolysis is the only pathway for photochemical transformation in distilled water. For many chemicals it is likely that ϕ_d obtained in distilled water is nearly the same as that observed in natural water. Zepp and Baughman[9] have argued that this is usually true because concentrations of natural substances such as nucleophiles that could undergo direct photoreactions with or quench photolysis of aquatic pollutants are generally very low. Molecular oxygen can act as a quencher but its effects can be studied in distilled water. According to Kasha-Vavilov's law quantum yields in solution are generally wavelength-independent, although there are some exceptions.[10] However, the lower ϕ_d is in distilled water, the greater the probability that it may exhibit wavelength dependence or may be enhanced in natural waters.

The specific light absorption rate, $k_{a,\lambda}$ is essentially a measure of the overlap between the spectrum of sunlight or other light sources, and another inherent property of the photoreactive chemical, its electronic absorption spectrum (Equation 4).

$$k_{a,\lambda} = 2.303 \, E_0(\lambda)\epsilon_\lambda \qquad (4)$$

where: $k_{a,\lambda}$ is the specific absorption rate at wavelength λ,
$E_0(\lambda)$ is the scalar irradiance (discussed later in the chapter),
and ϵ_λ is the molar absorptivity at λ.

The molar absorptivity reflects the probability that the chemical will absorb light at λ, and it is defined by Equation 5.

$$\epsilon_\lambda = \frac{A\lambda}{\ell[P]} \qquad (5)$$

where $A\lambda$ is the absorbance measured in a spectrophotometer and ℓ is the cell pathlength in centimeters. If the irradiance is held constant and [P] is very low, variations in the direct photolysis rate constant with wavelength directly follow changes in ϵ_λ, that is, the "action spectrum" for direct photolysis is the same as the electronic absorption spectrum of the chemical.

The value of k_d is assumed to equal $2.3 \, \phi_d \int E_0(\lambda)\epsilon_\lambda d\lambda$ where the integration is over the range of $E_0(\lambda)$ that is absorbed by the compound. The maximum value of ϕ_d is 1, so it follows that the maximum direct photolysis rate constant can be computed with knowledge of the spectral irradiance from the light source and the absorption spectrum of the compound. Some recent data (Table I) indicate that certain chemicals are extensively transformed after only brief full exposure to sunlight. Note that rapid direct photolysis can occur even with compounds that react with very low quantum efficiencies.

Indirect Photolysis Rates

Several recent studies have indicated that some chemicals photolyze more rapidly in natural water samples than in distilled water.[11-17] In certain cases, compounds that are completely nonphotoreactive in distilled water have been shown to react rapidly in natural water samples, e.g. 2,5-dimethylfuran (DMF)[16] and the insecticides disulfoton[14] and aldrin.[13] Possible explanations for these phenomena include enhanced efficiencies for direct photolysis (see above), formation of complexes and occurrence of various indirect photoprocesses.[9,18] The latter are discussed in detail here because they are believed to be of general importance, whereas the other effects are significant only with certain types of chemicals.

Assessment of indirect photolysis is complicated by the fact that the molecular structures of the natural substances or photosensitizers that

Table 1. Disappearance Quantum Efficiencies and Near-Surface Half-Lives for Direct Photolysis of Selected Chemicals in Water, Latitude $40°N$, Summer

Chemical	ϕ_d	Midday $(t_{1/2}, min)$	Reference
Naphthalene	0.0015	4300	22
Anthracene	0.0030	45	22
Benz[a]anthracene	0.0033	36	6
Benzo[a]pyrene	0.00089	32	6
Pyrene	0.0022	41	22
Naphthacene	0.013	1.8	22
3,3'-Dichlorobenzidine	——	1.5	83
Ferrocyanide complex	0.14	20	36
Trifluralin	0.0020	22	7
N-Nitrosoatrazine	0.30	4.8	84

absorb the light and mediate indirect photolysis have not been identified. This lack of information makes it more difficult to define the action spectra for photosensitized reactions. Another complication is the fact that a variety of photochemical processes could result in indirect photolysis.[9,18] Thus, conceivably, more than one process could be involved in the photosensitized reaction of a given chemical. Despite these complexities, however, general rate expressions for sensitized photolysis can be defined. These expressions are discussed here and applied to analyze kinetic data obtained for several types of photosensitized reactions.

Unlike direct photolysis, indirect photolysis involves light absorption by the photosensitizer, and the rate of light absorption is

$$I_{a,\lambda}^s = 2.3 \ \epsilon_{\lambda,s} \ E_0(\lambda) \ [S] \qquad (6)$$

where $\epsilon_{\lambda,s}$ is the molar absorptivity of the sensitizer and $[S]$ is its concentration. This expression predicts that the rate constants for indirect photoreactions should increase with increasing $[S]$ if $E_0(\lambda)$ is held constant.

The expression for the quantum yield for photosensitized reaction varies from one type of indirect process to another. Generally, when the concentration of the pollutant is very dilute, the quantum yield is directly proportional to $[P]$. (This proportionality constant will be referred to as C_λ.) The total rate expression is, therefore:

$$(Rate)_\lambda = 2.3 \ E_0(\lambda) \ \epsilon_{\lambda,s} \ C_\lambda \ [S] \ [P]$$

With constant sensitizer concentration, as is usually the case, the rate expression is in the first-order form that various experiments have indicated.[9]

Light attenuation effects must be taken into account to correctly interpret the data obtained in experiments with natural waters. These effects will be discussed in more detail in a later section, but an introductory discussion is necessary here. The scalar irradiance $E_0(\lambda,Z)$ at depth Z approximately equals $E_0(\lambda,0)e^{-K\lambda Z}$, where $E_0(\lambda,0)$ is scalar irradiance just below the water surface and K_λ is the attenuation coefficient of the water at λ. The average irradiance in a well-mixed system $E_0^{av}(\lambda,Z)$, is[7,19]

$$E_0^{av}(\lambda,Z) = E_0(\lambda,0) \frac{F_\lambda}{K_\lambda Z} \qquad (7)$$

where F_λ, the fraction of light attenuated by the system at depth Z, equals $1 - e^{-K\lambda Z}$. The average rate constant for sensitized photolysis is expressed as

$$k_{s,\lambda} = E_0^{av}(\lambda,Z) X_{s,\lambda} \qquad (8)$$

where $X_{s,\lambda} = 2.3 \, \epsilon_{\lambda,s} \, C\lambda[S]$.

When $K_\lambda Z$ is less than 0.2, the average irradiance approximately equals the surface irradiance, $E_0(\lambda,0)$. In the rest of this chapter, rate constants or half-lives that pertain to this condition are referred to as "near-surface." To correct for light attenuation, the observed rate constant is multiplied by $K_\lambda Z F_\lambda^{-1}$.

Having defined general equations for photosensitized reactions, it is of interest to discuss efforts to predict the variations in near-surface k_s from one natural water to another. These variations can be very large. For example, the half-life for the photosensitized oxygenation of DMF in 13 natural waters in winter sunlight ranged from 13 min in swamp water to 150 min in clear spring water.[16] To learn more about predicting these variations, the DMF reaction was studied in detail in several natural water samples. Because light scattering complicates correction for light attenuation, the water samples were all centrifuged to remove suspended particles. Plots of $X_s K_\lambda^{-1}$ vs wavelength, that is, action spectra, were very similar for two natural waters, both showing decreases with increasing wavelength.[9] Ultraviolet and blue radiation are effective, but longer wavelength light is not. Maximum spectral overlap between these action spectra and near-surface solar irradiance occurs at about 410 nm. The action spectra are important pieces of information because they can be used in conjunction with solar spectral irradiance to compute rate constants for photosensitized reactions as described above, i.e., $k_s = \int E_0(\lambda) X_{s,\lambda} d\lambda$. Also, as discussed in the next section, action spectra are essential to accurately compute the depth dependence for photosensitized reactions. Because a great deal of effort is required to obtain these spectra, a simple method has been sought to approximately compute near-surface rate constants for photosensitized reactions. As a result of these efforts, it was found that the rate constants for reaction in sunlight correlated to a_{366}, the absorption coefficient of the water expressed as 2.303 times absorbance per centimeter at 366 nm (Figure 2). An empirical relation (Equation 9) was derived by applying a linear least-squares fit to the data where k_o, the rate constant, expressed in cm^2/microeinstein, is calibrated in terms of photosynthetically active light (total radiation from 350 to 700 nm).

$$\log k_o = 0.67 \log a_{366} - 2.11 \ (r^2 = 0.93) \tag{9}$$

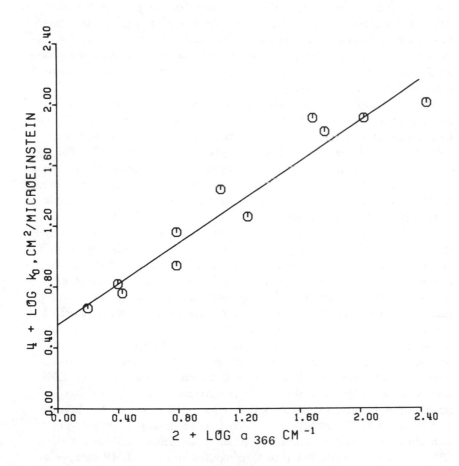

Figure 2. Relation between rate constants for photosensitized oxygenation of 2,5-dimethylfuran, k_o, and the (napierian) absorption coefficients of water samples from the United States, a_{366}. Each data point represents a different water body

Justification for such a calibration is discussed in the following section. To compute the near-surface first-order rate constant, k_o is multiplied by photosynthetically active light (PAL) expressed as microeinsteins cm^{-2} sec^{-1}. To convert from PAL expressed as watts to microeinsteins sec^{-1} the former is multiplied by a conversion factor of 4.48 microeinsteins sec^{-1} watt^{-1}.[20] For latitude 40°N, mean annual near-surface PAL is about 0.11 microeinsteins cm^{-2} sec^{-1} on clear days at sea level. Because the water samples were centrifuged, the dissolved organic matter, part of which was humic or "yellow" substances, was responsible for the photosensitization. Additional studies are required to define the influence of suspended particles upon photosensitized reactions. It is worth noting, however, that most of the water samples collected for the study[16] were clear, and photosensitized reaction for those samples occurred at the same rate in centrifuged and uncentrifuged water.

The rate constants for the DMF reaction can be employed to assess photosensitized reactions of other chemicals in natural waters. A skeletonized mechanism for these reactions is shown in Figure 3. Absorption of light promotes the sensitizer, S, into its excited singlet state, $^1S^*$. This state is usually so short-lived that it does not play an important role when the concentration of pollutant is low, as is often the case in natural water. Intersystem crossing from $^1S^*$ puts a molecule into its longer-lived triplet state, $^3S^*$. "Triplet sensitizers" are usually mainly responsible for photosensitized reactions,[21] particularly when the pollutant concentration is low. Exceptions could be reactions involving free radicals generated by photolysis of naturally occurring substances in the water,[18] which are discussed later in the section. A variety of processes can deactivate $^3S^*$. The most important are radiationless decay back to ground state and quenching by energy transfer to molecular oxygen or natural substances in the water. Most of the transferred energy goes to oxygen, the most concentrated energy acceptor in the environment. It has previously been shown that the product of energy transfer to oxygen, singlet oxygen $^1O_2^*$, mediates the photosensitized oxygenation of DMF in natural waters.[16] In this reaction DMF acts as the acceptor, A in the figure. The concentration of DMF was sufficiently low in these experiments that it had little effect on the lifetime of $^3S^*$ or $^1O_2^*$. The steady-state concentrations of singlet oxygen generated photochemically by sunlight were computed from half-lives for the DMF reaction.[16] Employing $[^1O_2^*]$ computed from the DMF data, near-surface rate constants have been determined for various photosensitized oxidations.[9,16,22] Although singlet oxygen does react rapidly with certain types of chemicals,[21] e.g., cis-dienes, sulfides, phenols, certain amino acids and polycyclic aromatics,[22] it is a highly selective agent and thus may be

$$S \xrightarrow{h\nu} {}^1S^* \longrightarrow {}^3S^*$$

$${}^3S^* \xrightarrow{k_d} S$$

$${}^3S^* + O_2 \xrightarrow{k_{ox}} S + {}^1O_2^*$$

$${}^3S^* + Q \xrightarrow{k_q} S$$

$${}^3S^* + PH \xrightarrow{k_+} I \left\langle \begin{array}{l} \text{Products} \\ \text{pH} \end{array} \right.$$

$$I = \left\{ \begin{array}{l} S + {}^3PH^* \\ SH\cdot + P\cdot \\ S^- + PH^+ \end{array} \right.$$

$${}^1O_2^* \xrightarrow{1/\tau_0} O_2$$

$${}^1O_2^* \xrightarrow{k_A} AO_2$$

Figure 3. Skeletonized mechanism for photosensitized reactions.

ignored in assessing the likelihood of photosensitized reaction of most chemicals.

Concentrations of ${}^1O_2^*$ can also be used to compute the steady-state concentration of excited triplet sensitizers, $[{}^3S^*]$, capable of transferring energy to oxygen

$$[{}^3S^*] = \frac{[{}^1O_2^*]}{k_{ox}\, \tau_0\, [O_2]} \tag{10}$$

where: k_{ox} is the rate constant for energy transfer from ${}^3S^*$ to O_2,
   τ_0 is the lifetime of singlet oxygen.
   $[O_2]$ is the concentration of molecular oxygen.

At 25°C, k_{ox} is 3 x 10^9 m sec^{-1}, τ_0 is 2 x 10^{-6} sec, and $[O_2]$ in air-saturated water is 2.5 x 10^{-4} m; $[{}^3S^*]$ is thus equal to 0.67 $[{}^1O_2^*]$. It is reasonable to assume that oxygen, a very low energy acceptor, can accept energy from essentially all the $^3S^*$ that are generated by absorption of blue and ultraviolet light. Thus $[^3S^*]$ computed by Equation 10 can be viewed as an upper limit for the steady-state concentration of excited triplet sensitizers available for other photosensitized reactions. The maximum rate constant for such reactions equals $k_+\phi_I[^*S^3]$, where k_+ is the rate constant for interaction between $^*S^3$ and pollutant PH and ϕ_I is the probability that the product of such interactions, I, goes on to products (Figure 3). Some illustrative kinetic data computed for photosensitization by energy transfer, hydrogen abstraction and electron transfer are shown in Table II. The computations are crude because few kinetic data are available concerning such processes in water. Although the processes are generally slower than some of the direct photolyses that have been studied (Table I), it does appear that energy transfer and electron transfer may be rapid enough to be important. Hydrogen abstraction from saturated hydrocarbons is slow; however, certain compounds such as phenols react with k_+ of $> 10^9$ m sec^{-1},[23] although ϕ_I is low for this reaction, it still may be much more rapid than the example in the table. A discussion of the nature of the products derived from these reactions is beyond the scope of this discussion; for a review and references, the interested reader is referred to the recent book by Turro.[23b]

Recently, interest has developed regarding the possible role of photochemically generated free radicals in the transformation of aquatic pollutants or naturally occurring species.[6,14,17,24,25] Draper and Crosby[14] and Zika[25] have established that peroxides form when fresh water or marine waters are exposed to sunlight. These peroxides may form via reactions of singlet oxygen and naturally occurring substances, or perhaps via other pathways involving hydrated electrons or superoxide radicals. Thermal or photochemical homolysis of such peroxides yields free radicals that can initiate oxidation of organic compounds in water. Mill, for example, has presented evidence that cumene is oxidized slowly by a free radical mechanism in a freshwater sample exposed to sunlight.[24] In marine systems, Zafiriou and True[17] have recently presented evidence that photolysis of nitrite in sea water from the upwelling Peruvian area can generate free radicals at near-surface rates of up to 9 x 10^{-8} m day^{-1}.[17b] Swallow has suggested that hydrated electrons generated by photolysis of aromatic substances in the sea can lead to formation of superoxide radicals or carbonate-derived radicals.[26] These radicals could play an as yet undetermined role in the indirect photolysis of chemicals in the sea.

SIMULATION OF SOLAR SPECTRAL IRRADIANCE

Methods have been developed to compute $E_0(\lambda)$ for sunlight as a function of season, latitude and depth in a water body. The scalar irradiance

Table II. Illustrative Near-Surface Rate Constants for Various Photosensitized Reactions

Water Body	10^{13} x Mean[a] $[^3S^*]$, M	Energy Transfer[b] (k_s, hr^{-1})	H Abstraction[c] (k_s, hr^{-1})	Electron[d] Transfer (k_s, hr^{-1})
Aucilla River Lamont, FL	12	0.13	3.0×10^{-4}	3.95
Mississippi River Baton Rouge, LA	3.9	0.042	9.8×10^{-5}	0.99
Gulf of Mexico Shell Point, FL	2.9	0.031	7.3×10^{-5}	0.74
Columbia River Portland, OR	1.8	0.020	4.5×10^{-5}	0.46

[a]Mean annual steady-state concentration of excited triplet sensitizers at latitude 40°N (Equation 10).

[b]Computed assuming k_+ is 3×10^9 M sec^{-1} and ϕ_I is 0.01.

[c]Computed assuming data for H-abstraction from p-xylene by acetophenone, $k_+ = 7 \times 10^5$, $\phi_I = 0.10$.[85]

[d]Computed for photoreduction of benzophenone by benzylamine, $k_+ = 1.5 \times 10^9$, $\phi_I = 0.47$.[86]

in a water body is related to the global (sun and sky) irradiance at the surface G_λ as described below. Several computer codes have been employed to calculate G_λ, given standard atmospheric conditions, the most detailed computations being those of Braslau and Dave.[27] Solar ultraviolet radiation is of primary interest here because it is mainly responsible for inducing direct and sensitized photoreactions of aquatic pollutants. Mo and Green[28] have developed a program for solar ultraviolet radiation that has been employed to assess various important photobiological phenomena such as UV-induced skin cancer. The code developed by Cline and Zepp[7] to compute photolysis rate constants utilizes empirical relations described by Bener[29] to calculate intensity values of UV radiation for clear days. Bener's relations are based on his extensive measurements of solar UV radiation in Switzerland. At present, aside from Bener's work, few extensive measurements of UV radiation have been made. Data describing UV radiation in natural waters are virtually nonexistent, with the exception of some broadband measurements by Calkins[30] and recent studies by Smith and Baker[31] of UV radiation in the sea.

Values of G_λ computed for clear days can be adjusted to take clouds into account. Mo and Green[28] provide approximate assessments of the effects of clouds on solar UV light assuming that the reduction in light intensity F_c can be described by the empirical relation of Büttner[32]

$$F_c = 1 - 0.056\ C \tag{11}$$

where C is the fraction (in tenths) of the sky covered by clouds. Employing this equation and the voluminous data available on average cloud cover, Mo and Green[28] determined that the average reduction in UV intensity caused by clouds over several large U.S. cities was less than a factor of two on an annual basis. The main problem with employing relations such as Equation 11 to predict cloud effects, however, is that the degree of light attenuation strongly depends on cloud thickness. An alternative approach is to apply the extensive data concerning reduction of *total* solar radiation by clouds to assess their effect on solar UV radiation. Clouds transmit UV radiation to a greater extent than total radiation, but the difference in fraction transmitted is small with the exception of thick clouds at large solar zenith angles.[34] Spinhurne and Green[34] have presented equations that can be used to estimate reduction of solar UV radiation by clouds using data concerning total solar radiation. In agreement with theory, measurements of solar radiation in Germany have shown that the ratio of UV-A (320-400 nm) radiation to total solar radiation between 300 and 2000 nm is approximately constant.[33]

In urban areas, smog is also an important determinant of G_λ. Comparison of data concerning smog and cloud effects indicates that more solar UV radiation is transmitted through heavy smog than through clouds on an overcast day.[35] Nader's studies also indicate that light attenuation by the smog in Los Angeles exhibits little wavelength dependence.

The above discussion implies that the decrease in photolysis rate constants on cloudy days should be related to the decrease in total (or visible) solar radiation expressed as a percentage of the total radiation available under clear skies at the same time of year. This relationship was recently tested by Broderius[36] in detailed studies of the direct photolysis by sunlight of iron cyanide complexes in water. Based on the data shown in Figure 4, Broderius concluded that the photodecomposition of the cyanide complexes is a direct function of visible sunlight intensity, regardless of meteorological conditions. It should be noted thet $Fe(CN)_6^{3-}$ absorbs visible (440 nm) sunlight most strongly and that $Fe(CN)_6^{4-}$ absorbs UV (330 nm) most strongly.

Perry and co-workers[37] and Mancini[38] have suggested that direct photolysis rate constants can be calibrated in terms of total or visible solar radiation, and this procedure was also used above in the case of rate constants for indirect photolysis. This technique is valid for many, but not all chemicals. Employing the computer program of Cline and Zepp,[7] ratios of $E_0(\lambda,0)$ to near-surface PAL were computed for several wavelengths

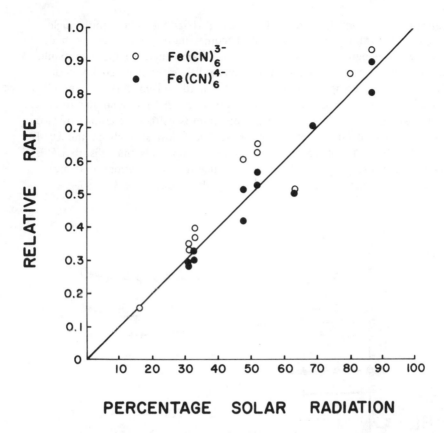

Figure 4. Photolysis rate constants for iron cyanide complexes vs. integrated visible radiation (400-700 nm), both normalized to clear sky values (from Reference 36).

(Figure 5). The ratios for irradiance in the UV-A region and visible region are approximately constant for most solar zenith angles. Diurnal and seasonal variations of visible solar radiation and of near-surface photolysis rate constants of compounds that absorb sunlight most strongly at wavelengths > 320 nm are thus similar. N-Nitrosamines, polycyclic aromatic hydrocarbons and dyes are classes of compounds that fall into this category. Also, UV-A and blue radiation are mainly responsible for photosensitized oxygenation, and it is thus valid to calibrate k_o in terms of PAL. Because the spectral distribution of sunlight is a function of depth, it is not correct to assume that the depth dependence of photolysis rates and PAL are the same. The diurnal and seasonal variations of G_λ in the UV-B (280-320 nm) spectral region, moreover, are strongly influenced by stratospheric ozone absorption.[7,34,35] As indicated in Figure 5, the irradiance in the UV-B region drops off much more rapidly than PAL as the solar zenith angle increases.[34,35] Moreover, the dropoff becomes more pronounced with decreasing wavelength. As a consequence, UV-B-absorbing compounds exhibit considerably larger seasonal swings in photolysis rates than do UV-A or visible-absorbing compounds or PAL.[7] Burkhard and co-workers[39] noted this effect when comparing seasonal fluctuations in total solar radiation with changes in half-lives for the acetone-photosensitized reactions of two pesticides in sunlight. Acetone absorbs sunlight most strongly in the UV-B spectral region. Photolysis rate constants of compounds that absorb UV-B radiation are also subject to considerable day-to-day variations that are caused by fluctuations in the thickness of the ozone layer.[7]

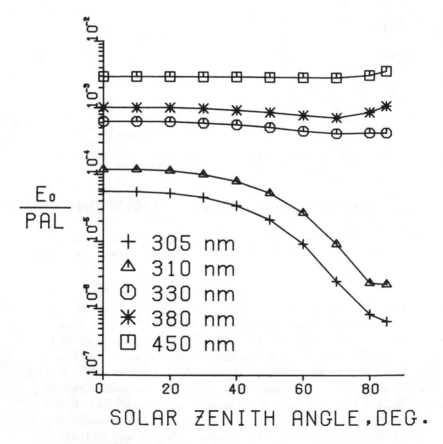

Figure 5. Computed ratio of scalar irradiance at various wavelengths E_0 to integrated scalar irradiance in the 350-700 nm interval PAL. The ratios are computed for just below the surface of a water body, assuming an ozone layer thickness of 0.32 cm.

The computer codes discussed above are used to calculate the solar spectral irradiance on a horizontal surface. Because molar absorptivities reflect the probabilities of photon absorption, the irradiance values used for computing photolysis rates should be expressed as photons (or einsteins) per unit area and time. Horizontal irradiance just beneath the surface of a water body is slightly lower than $G\lambda$ because of reflection.[7] To transform horizontal irradiance to photochemically relevant scalar irradiance, E_0, it is necessary to multiply the horizontal irradiance by a term called the distribution function, $D\lambda$.[40,41] The distribution function near the surface of a water body is approximately the ratio of the mean light pathlength to depth for downwelling radiation and is described by Equation 12.

$$D\lambda = (1 + R\lambda)^{-1} (\sec \theta + 1.2\ R\lambda) \qquad (12)$$

where $R\lambda$ is the ratio of diffuse (skylight) to direct radiation and θ is the angle of refraction of direct radiation.[7] The value of $R\lambda$ increases with decreasing wavelength and solar altitude. $D\lambda$ is generally somewhat lower in water bodies than in air because of the collimating effect of refraction of sunlight at the water surface. Theoretical values of $D\lambda$ for near-surface UV and blue light in clear water bodies are typically 1.3 or less (Figure 6). Studies by Miller and Zepp[42] and Jerlov[43] indicate that $D\lambda$ values

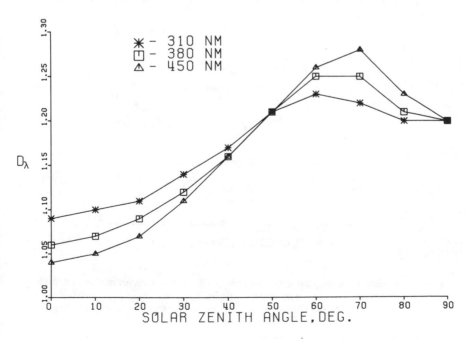

Figure 6. Computed values for near surface distribution functions $D\lambda$ as a function of solar zenith angle.

for UV light, as expected, are somewhat larger in turbid natural waters having high concentrations of suspended matter. Scattering by suspended matter makes the light more diffuse, thus enhancing the distribution function. If the suspended particles do not strongly absorb UV light, photolysis rates can actually be enhanced in the suspensions compared to distilled water solutions (Figure 7).

Employing simulated $E_0(\lambda)$ values, photolysis rate constants can be calculated for various locations and times, that is, a "climatalogy" of photolysis rate constants can be computed.[7,12] The computed rate constants or half-lives have been found to closely predict photolysis rates observed under sunlight at several locations in the United States. The most extensive comparison of computed and experimental rate constants was conducted by Broderius and Smith[36] in their studies of the sunlight-induced conversion of iron cyanides to hydrogen cyanide. Broderius employed the computer program described by Zepp and Cline[7] in these studies. Diurnal and seasonal variations observed for photolysis rates of the cyanide complexes were in agreement with those computed for St. Paul, Minnesota, the site of the study (Figures 8 and 9). In addition to Broderius' study, other investigators at Palo Alto, California,[6] and Athens, Georgia,[7,9] have found that rate constants observed in sunlight agree well with computed values.

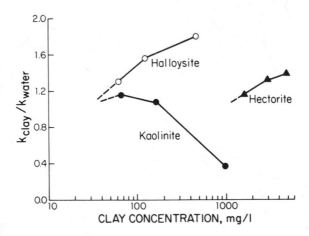

Figure 7. Effects of suspended clays on photolysis of γ-methoxy-m-trifluoromethyl-butyrophenone in water under sunlight. k_{clay} and k_{water} are first-order rate constants for photolysis in clay suspensions and in distilled water respectively.[42]

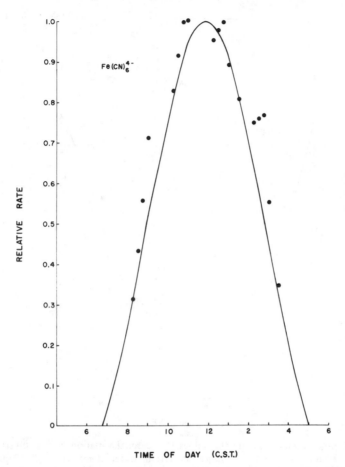

Figure 8. Comparison of computed and experimental rate constants for diurnal variation of photolysis of hexacyanoferrate (II) in water on October 20, 1977, at St. Paul, MN (from Reference 36).

LIGHT ATTENUATION IN NATURAL WATERS

As mentioned in the preceding section, not much information is available concerning the transmission into natural waters of solar UV radiation of wavelengths measuring less than 350 nm. Available data have been reviewed by Jerlov,[43] Smith and Tyler,[41] and Zaneveld.[44] The symbols and terminology used for radiometric quantities in this chapter are discussed in these reviews. The data have usually been presented in terms of beam attenuation coefficients, c, or diffuse attenuation coefficients, K_λ. The former is a measure of the transmittance of a collimated beam of light through

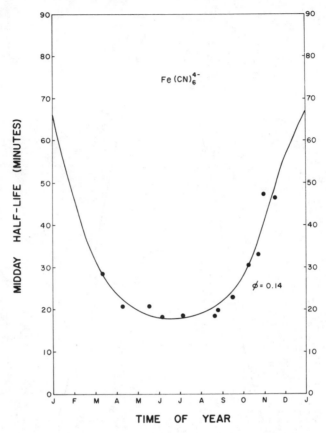

Figure 9. Computed and experimental values for seasonal variation of photolysis rate constants for hexacyanoferrate (II) in water at midday at St. Paul, MN (from Reference 36).

water. Beam attenuation occurs through absorption or scattering of the light. In a natural water, most scattering is in the forward direction. The diffuse attenuation coefficient is computed from measurements of solar irradiance at various depths in a water body. It is the slope of a plot of the natural logarithm of solar irradiance vs depth. Irradiance at depth Z, $E_0(\lambda,Z)$, is thus approximately $E_0(\lambda,0)e^{-K_\lambda Z}$, where $E_0(\lambda,0)$ is surface irradiance. For a nonhomogeneous medium, K_λ is usually smaller than c because the irradiance measured in a water body includes forward-scattered light as well as light that has been neither absorbed nor scattered. For a clear water body, c is theoretically equal to or slightly smaller than K_λ. Comparisons of c and K_λ values for clear ocean water in the UV spectral region have revealed that the c values, contrary to expectation, are consistently larger.[41,44] Apparently, even careful purification has not produced distilled water that is as transparent as deep ocean waters. The diffuse attenuation coefficient, therefore, provides the best measure of UV penetration into turbid as well as the clearest natural waters. It is easier to visualize K_λ in terms of a photic zone that equals 4.6 K_λ^{-1}, the depth at which 99% attenuation of the surface irradiance occurs. The photic zone for open ocean water at various wavelengths is shown in Figure 10. It is considerably deeper for UV radiation than is commonly believed, ranging from about 30 m for UV-B radiation to close to 100 m for wavelengths in the UV-A spectral region. Thus the UV photic zone is 20 to 60% as deep as the depth of maximum penetration of solar radiation into the open ocean. Ryther[45] has estimated that the open ocean, a region of the sea with very low productivity, comprises about 90% of the ocean's area. Thus, the data in Figure 11 provide approximations that pertain to most of the natural water on earth. Coastal and upwelling regions of the ocean are considerably more productive, and because microorganisms absorb light, sunlight does not penetrate as deeply into these waters. For such areas, the magnitude of K_λ in the 350- to 700-nm range can be calculated approximately by employing empirical relations described recently by Smith and Baker.[46] Typically, the photic zone for 350- to 500-nm light is 5 to 10 times shallower in coastal than in open sea water.[41] Figure 11 demonstrates the simulated depth dependence for direct photolysis of three aromatic compounds in two types of sea water. Photolysis half-lives, averaged over time and depth (the top 35 m), are in parentheses. The diffuse attenuation coefficients used for these calculations were recently reported by Smith and Baker.[31] These data indicate that large differences in depth dependence can occur from one pollutant to another and that although

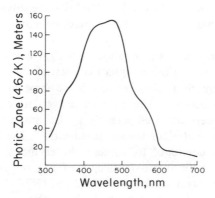

Figure 10. Approximate photic zone for the open ocean as a function of wavelength.[41]

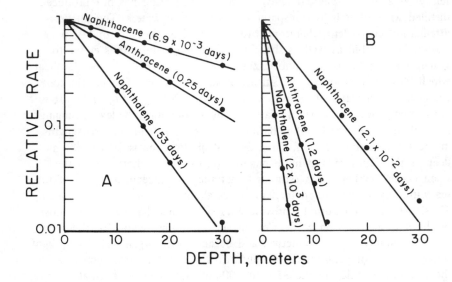

Figure 11. Simulated depth dependence for direct photolysis of several aromatic compounds during summer in the Gulf of Mexico: A—Mid-Gulf; B—coastal water near Tampa, FL. Average half-lives in top 35 m are in parentheses.[22]

light attenuation does slow photolysis, it is still very rapid for some compounds at great depths. For example, the half-life for naphthacene at a point 150 m deep in the Gulf of Mexico is estimated to be only about 4 hr, about 1% of its near-surface value. Naphthacene absorbs sunlight most strongly in the blue spectral region where sea water is most transparent.

Unfortunately, even less is known about UV light attenuation in inland water than in the ocean. Diffuse attenuation coefficients for Crater Lake and other exceptionally clear lakes are nearly the same in the UV-A region as those for the open ocean.[47] The most extensive data concerning shorter wavelength UV light were obtained recently by Calkins, who employed a submersible Robertson meter for his measurements.[30] The Robertson meter, which was specifically developed to simulate the erythema action spectrum, has been assigned a specific mean wavelength of 312 nm for a solar zenith angle of 0° and ozone layer thickness of 0.?2 cm (NTP).[48] In most natural waters, light attenuation increases with decreasing wavelength in the 300- to 500-nm range, and the photic zone for UV-B radiation thus represents the *minimum* that would pertain to aquatic pollutants. Some of Calkins data are summarized in Table III. The table also includes concurrent Secchi disk depths, Z_{sd}, and the ratio of the Robertson meter photic zone, 4.6 K_R^{-1} to Z_{sd}. It is apparent that the penetration of light into inland waters is, as expected, shallower than into the ocean and quite variable. Because few data exist concerning UV light in natural water, it is of interest to relate UV to visible light penetration. Holmes[49] has reported that for visible light the median value for KZ_{sd} in turbid marine water is 1.44; Poole and Atkins[50] had earlier suggested that KZ_{sd} is about 1.7 for visible radiation. The median value of $K_R Z_{sd}$ for Calkins' data is 9.15. It may be tentatively concluded, then, that typically the *maximum* difference in photic zones for photolysis of aquatic pollutants and photosynthesis is roughly a factor of 5 or 6. Calkins data indicate that this ratio may be as high as 25 in waters in which light attenuation is dominated by absorption by humic substances, or as low as 2 to 3 in turbid waters containing high concentrations of sand, silt and clay.

Clearly these conclusions are based on a very limited data set, and additional data, especially concerning spectral irradiance at wavelengths less than 350 nm, are badly needed. In particular, relations such as those defined by Smith and Baker[46] between pigment concentration and K_λ values

Table III. Data Concerning Penetration of UV-B Radiation into Natural Waters Measured with Robertson Meter (about 312 nm)[a]

Site	Z_{sd},[b] (m)	$4.6/K_R$,[c] (m)	$K_R Z_{sd}$	$\dfrac{4.6}{K_R Z_{sd}}$[d]
Puerto Rico	34.4	29	5.46	0.84
	4.6	18	3.83	1.2 (max.)
Delaware Bay	6.6	3.4	8.99	0.51
Chesapeake Bay	2.64	1.3	8.8	0.52
Patuxent River	2.72	1.3	9.3	0.49
Lake Superior	14.2	3.2	20.2	0.23
Lake Michigan	6.1	3.4	8.26	0.56
Lake Huron	6.88	3.6	8.73	0.53
Lake Erie	2.51	1.8	6.49	0.71
Douglas Lake, MI	3.46	0.38	41.7	0.11 (min.)
Lake Herrington, KY	3.6	0.61	27.1	0.17

[a]Partial summary of data obtained by Calkins.[30]
[b]Secchi disk depth.
[c]Depth at which 99% attenuation of surface UV-B radiation occurred.
[d]Ratio of UV-B photic zone to Secchi disk depth.

need to be defined for inland waters. Such relations would permit assessment of photolysis of aquatic pollutants in water bodies in which solar radiation has not been measured.

Characterization of the penetration of UV light into water bodies is only part of the problem of assessing the effects of light attenuation. Another problem involves estimating the influence of vertical mixing on the amount of light received by the pollutant. At present, it is often assumed that a water body is completely mixed with respect to photolysis and volatilization.[5-7,51] According to this assumption, photolysis and volatilization occur more slowly than mixing and, consequently, the pollutant concentration should be uniform throughout the water column. Mackay[51] has suggested that this assumption is valid for most rivers. If the water is completely mixed, the average irradiance is described by Equation 7. In many water bodies, essentially all the light is absorbed in the upper part, F_λ is nearly 1, and the average rate constant at depth Z_{av} equals

$$k_p = \frac{1}{Z_{av}} \int \frac{k^o_{p,\lambda}}{K_\lambda} \, d\lambda \tag{13}$$

where $k^o_{p,\lambda}$ is the near-surface rate constant.

Vertical mixing is slower in lakes and the ocean than in rivers. During warm periods, lakes and the sea develop diffusion floors called thermoclines through which mixing is extremely slow. In lakes, the part of the water column below the thermocline, the hypolimnion, is much more poorly mixed than the region above it, the epilimnion. Pollutants usually enter lakes and the ocean in the mixed layer where they reside until, with the advent of cold weather and fall turnover, they are mixed more deeply into the water body. Extremely hydrophobic pollutants may also be removed from the photic zone by sorption to dead microorganisms or suspended sediments that are settling downward. The data in Table III and Figure 11 indicate that the photic zone for aquatic pollutants is usually shallower than the thermocline. Once the pollutant is transported below the thermocline, it undergoes photolysis slowly if at all, and, as long as the water body remains stratified, upward mixing back into the photic zone is extremely slow. Thus, a situation may develop in which the concentration of a photoreactive chemical is considerably lower in the epilimnion of a lake than in its hypolimnion, even though the chemical is introduced into the epilimnion. Volatilization can also cause a depletion in concentration in the epilimnion relative to deeper water. Near-surface depletions have been noted in the vertical profiles of chlorinated hydrocarbons in Lake Zurich.[52]

Vertical mixing in the mixed layer of a lake or the ocean depends mainly on wind speed at the surface. According to Csanady,[53] above a wind speed of 5 m/sec Langmuir circulations cause rapid vertical mixing down to the diffusion floor, *i.e.*, the thermocline. Broecker and Peng[54] state that the average scalar wind for the world's oceans is about 15 knots (7.7 m/sec), indicating that the upper layer of the ocean sometimes is very well mixed. That radon profiles measured in the mixed layer of the ocean are sometimes uniform further supports the contention that mixing can be rapid in the upper layer. Because chlorophyll a vs depth typically exhibits a nonconstant profile in the ocean,[20] mixing may be slower than the growth rate of algae. Nonetheless, because the deep ocean is very clear, the average rate constant computed for the mixed layer assuming complete mixing is probably close to correct. For coastal regions of the ocean and for inland lakes, the water is less clear, and photolysis rates may be limited by vertical transport of compounds such as those included in Table I. Adequate assessment of photolysis in such systems requires knowledge of their vertical mixing rates. For an interesting discussion of mixing limitations to volatilization, the reader is referred to the chapter by Mackay and co-workers[51b] in this book.

Finally, it is of interest to consider the effects of light attenuation on the average rate constant for a photosensitized reaction. The data obtained for the photosensitized reaction of DMF in several natural waters indicated that the fraction of attenuated light, $X_s K_\lambda^{-1}$, that resulted in reaction was roughly constant.[16] In Figure 12 are plotted computed values of average k_s (Equation 13) for the photosensitized reaction of DMF vs depth in water bodies with differing attenuation coefficients; the computer program developed by Cline and Zepp[7] was used for the calculations. For illustrative purposes, near-surface rate constants were computed using Equation 9; action spectra obtained for the Okefenokee Swamp (Georgia) were assumed throughout, and the spectral distribution of the K_λ values was assumed to be the same for the water bodies. At shallow depths, the photolysis rate constant is largest in the water body with the largest attenuation coefficients, or least light penetration. Light attenuation effects, however, are also greatest in this water body, and with increasing depth the *average* k_s become about the same in both water bodies. Although the average rate constant converges, it should be kept in mind that the sensitized photoreaction is compressed into a zone of high reactivity near the surface in the opaque water but is spread out more throughout the water column in a weakly absorbing natural water.

PARTITIONING TO SEDIMENTS AND BIOTA

Most natural waters contain varying amounts of suspended particles of both terrigenous and biogenous origin. The particles in rivers and some

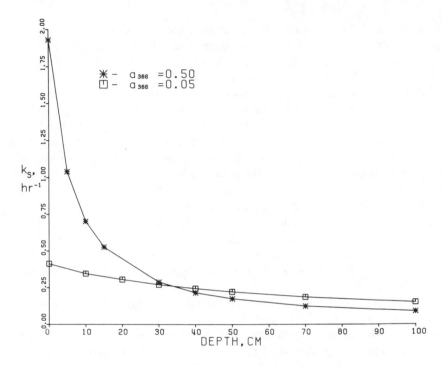

Figure 12. Average rate constants k_S for photosensitized oxygenation of 2,5-dimethylfuran in water columns of various depths. The curves are computed for water bodies having absorption coefficients a_{366} in cm^{-1} indicated on the figure. The near-surface k_S values were computed using Equation 9 and the depth dependence was computed as described by Zepp and Cline[7] assuming the same spectral distribution of attenuation coefficients and X_S values observed for the Okefenokee Swamp.

lakes are mainly terrigenous, that is, the sands, silts and clays are derived from soil runoff. In the ocean, suspended particles are mainly living microorganisms or detritus derived from dead organisms. The concentrations of particles in natural water range from less than 10^{-2} mg/l in the open ocean to greater than 10^4 mg/l in some rivers. These particles are expected to influence photolysis of aquatic pollutants in several ways. First, in the ocean and turbid inland water bodies, suspended phytoplankton and sediments, respectively, are primarily responsible for the light attenuation that was discussed in the previous section. Second, light scattering by the particles renders downwelling radiation more diffuse, enhancing the distribution function. Average photolysis rate constants are thus larger in a turbid water body than in a nonturbid water body that has the same attenuation coefficients. Third, partitioning to suspended sediments or microorganisms is likely to alter the photoreactivity and photoproducts of aquatic pollutants. Finally, in inland waters, partitioning into bottom sediments partially removes pollutants from underwater light, thus obviously influencing photolysis rates. The latter two effects are considered in more detail in this section.

Photoreactivity on Suspended Particles

Sediments

It is well recognized that chemicals in water bodies exhibit varying tendencies to partition into the sediments and biota. Partitioning to suspended sediments is a reversible phenomenon that can be quantified in terms of an equilibrium constant known as the partition coefficient, K_p.[55] K_p is the ratio of the concentration of a chemical in the sediment to its concentration in the water. It can be shown that the fraction of a chemical dissolved in water in a sediment-water mixture equals $(1 + K_p\rho)^{-1}$, where ρ is the ratio of sediment to water. The magnitude of K_p depends upon both the compound and the nature of the sediment. For a given sediment and various nonionic chemicals, K_p correlates inversely with water solubility of the chemicals and directly with the octanol-water partition coefficient, P.[55] For a given nonionic chemical, the K_p for various clay and silt-sized sediments is proportional to the organic content of the sediment. Initial studies by Karickhoff et al.[55] indicated that sorption and desorption are very rapid processes and, thus, that equilibrium would be maintained even when chemicals were being transformed. Assuming sediment-water exchange is rapid, Wolfe et al.,[56] employing known partition coefficients and sediment concentrations, computed that even extremely hydrophobic chemicals are mainly dissolved and not sorbed in the water columns of most rivers and lakes. Miller and Zepp[9,57] similarly concluded that appreciable fractions of hydrophobic nonionic organics are sorbed only at very high sediment concentrations. Because light is completely attenuated within the top few centimeters of such concentrated sediment suspensions,[42] it was suggested that photolysis on sediments probably is unimportant.[9] More recent studies by Karickhoff have shown that a portion of the very hydrophobic chemicals that partition to sediments desorb more slowly than was originally believed.[58] These results indicate that significant fractions of hydrophobic pollutants may be sorbed to suspended sediments even in aquatic environments that permit substantial light penetration. Miller and Zepp[57] investigated the products and the kinetics of the photolysis of several hydrophobic chemicals predominantly sorbed onto natural sediments in water and concluded that the sorbed chemicals were in a microenvironment that was less polar and a better hydrogen donor than water. Additional studies are required to better elucidate the photoreactivity of extremely hydrophobic chemicals on sediments.

Biota

In the ocean and in many lakes, algae and bacteria make up the bulk of the suspended particles. Nonpolar organic chemicals are known to accumulate in bacteria via a rapidly reversible equilibrium process,[56,59,60] but nothing is known about the photoreactivity of chemicals sorbed on bacteria. Algae absorb sunlight much more strongly than bacteria, and thus it is probable that indirect photochemical transformations of sorbed pollutants in the sea, if important, would occur predominantly on the algae. Some scattered data indicate that algae are capable of mediating photobiological transformations of chemicals.[9,61-63] Most of these studies, however, were conducted in such a way that the results could not be extrapolated to environmental conditions. With two pesticides, malathion[61] and methoxychlor,[63] controlled studies established that the transformations did not involve reactions photosensitized by the culture medium or dissolved substances released by the algae. Presumably the reactions involve chemicals sorbed to the algae. Unfortunately, there are very few reliable quantitative data concerning the partitioning of chemicals to algae. Baughman and Paris,[60] after reviewing the literature pertaining to sorption of chemicals by algae, have suggested that uptake by algae may be viewed as a passive process that, like sorption to sediments, can be quantitated in terms of partition coefficients. Studies of Ellgehausen and co-workers[64] further indicate that sorption of pesticides to algae is reversible and that K_p for algae may correlate with the octanol-water coefficient. Employing available partition coefficients for algae and known algae concentrations in natural waters, it is estimated that typically only a small fraction of the chemical in a water body is sorbed to the algae.[56] Thus, rate constants for phototransformation on the algae must be considerably larger than for photolysis processes in the water in order for the algae to appreciably influence the net rate of photolysis in the system.

Partitioning to Bottom Sediments

Certain pollutants, such as polycyclic aromatic hydrocarbons and PCBs, have been found in the bottom sediments of lakes and rivers. It has been suggested that the bottom sediments are essentially a permanent sink for pollutants.[65] Much evidence, however, shows that sorption to the bottom sediments is reversible, at least with nonionic organic pollutants. It is well known that biota in the water column can become contaminated with a pollutant long after external input of the pollutant to the water has ceased. The main pathway for such contamination is now believed not to be biomagnification up the food chain but rather direct uptake of pollutant that has been released from the bottom sediments into the overlying water.[66] Moreover, experiments in model ecosystems such as those by Kearney et al.[67] have clearly demonstrated that pesticides can be released from the bottom sediments to the water column.

Factors that influence bottom sediment-water column exchange are not well understood at present. Vertical profiles of radionuclides in bottom sediments of various water bodies indicate that mixing of the upper layer[68] occurs mainly by movements of the macrobenthos in the case of lakes.[69] Such mixing can move pollutants from deeper in the sediments to the surface where exchange with overlying water occurs. The mixed layer of the sediment can extend as deep as 12 cm, but mixed layer depths of 5 cm or less are most commonly observed in lakes.[68] In parts of some lakes or rivers, the mixed layer may be only a few millimeters deep. Presumably, barring some major perturbation of the sediments by nature or man such as dredging, pollutants buried deeper than the mixed layer are cut off from exchange with the water column.

To assess the influence of partitioning to bottom sediments on photolysis, Zepp and Baughman[9] assumed that exchange between sediment and water column was rapid compared to photolysis. Under these conditions, the ratio of concentrations in sediment to water column is always equal to K_p (see preceding section), and the apparent photolysis rate constant is reduced to $k_p(1 + K_p\rho)^{-1}$, where ρ is the weight ratio of bottom sediment in the mixed layer to water in the water column.

A more realistic assessment is afforded by use of a computer model[70] that was designed to estimate the behavior of pollutants under a variety of environmental conditions. Assumptions of this model are described in detail by Burns.[70] To illustrate the effects of sorption on behavior of photoreactive chemicals in a hypothetical 2-m deep pond, system and water column "half-lives" were computed as a function of K_p for cases in which photolysis does or does not occur in the water column (Figure 13). In the latter case, water-borne export was assumed to be the only process that resulted in loss of the chemical from the system. The term "half-life" refers to the approximate period of time required for half of the pollutant to be lost either from the system, including both water column and bottom sediments, or from the water column alone. Important assumptions made for these calculations are:
(1) that the mean near-surface k_p is 0.69 hr^{-1} in all cases of "with

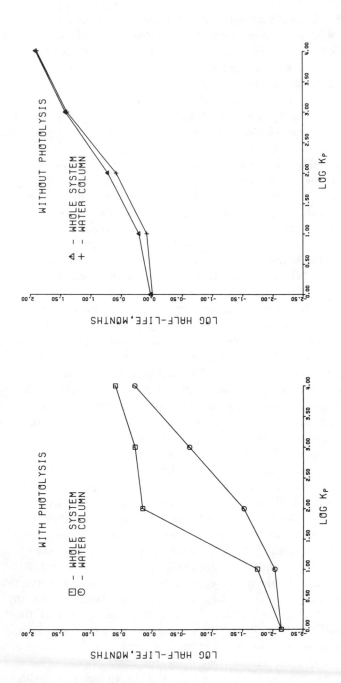

Figure 13. Computed half-lives for photolysis in a pond of a compound having a near-surface half-life of 1 hr and various assumed values of the partition coefficient for sorption to the sediment, K_p. Assumptions: pond depth 2 m; mixed layer of bottom sediment, 5 cm; suspended sediment concentration, 300 mg/l; no photolysis on suspended sediments.[70]

photolysis" and that light attenuation in the pond decreases photolysis to 0.2 of its near-surface rate; (2) that photolysis of pollutant sorbed to to suspended sediments does not occur (probably a generally invalid assumption in view of other studies[57]); and (3) that the top 5 cm of the bottom sediment is mixed and thus a part of the system being considered. The half-lives pertain to a pond in which cessation of pollutant input was preceded by a continuous discharge of pollutant into the pond for a period of time that was sufficiently long for the system to approach a steady state.[70] The following conclusions are derived from the computations.

1. The system half-life and water column half-life, both with or without photolysis, increase with increasing K_p. In the "with photolysis" case, the increase in water column half-life with increasing K_p primarily results from the assumption that photolysis does not occur on the suspended sediment. The system half-life for the with photolysis case sharply increases between a K_p of 10 and 100. This phenomenon is attributable to a change in the distribution of the pollutants in the system. Pollutants that are hydrophilic, i.e., K_p less than 10, are predominantly in the water column at steady state and thus are, for the most part, rapidly degraded by photolysis. More hydrophobic pollutants, with K_p values greater than 100, are predominantly in the bottom sediment protected from photolysis. For such chemicals the system half-life is much longer because the slow flux out of the bottom sediment is the rate-limiting step for photochemical loss from the system.

2. The system and water column half-lives are considerably shorter with photolysis than without it even for hydrophobic chemicals that are predominantly in the bottom sediment.

3. Following cessation of pollutant input, photolysis causes a much more rapid drop in pollutant concentration in the water column than in the bottom sediment. Consequently, the ratio of concentrations in the bottom sediment to the water column increases sharply with time until the flux rate out of the sediment equals the photolysis rate in the water column.[6] This result indicates that the sediment-water ratio for photoreactive compounds may sometimes exceed that predicted by the partition coefficient.

VOLATILIZATION AND ATMOSPHERIC PHOTOLYSIS

Finally, it should be noted that an important process for dissipation of certain aquatic pollutants involves volatilization followed by photochemical transformation in the atmosphere (Figure 1). This pathway is most likely to be important for relatively hydrophilic chemicals having high vapor pressures such as benzene or vinyl chloride. As noted above, the rate-limiting step for volatilization may be vertical mixing in certain lakes. This appears to be the case for tetrachloroethylene and other volatile organics in Lake Zurich.[52]

Mackay has shown that hydrophobic chemicals with low vapor pressures such as DDT also can volatilize from water at surprisingly large rates.[51,71] Assessment of the role of volatilization and subsequent atmospheric photolysis for this type of compound is complicated by several factors. First, such chemicals have a great tendency to sorb on suspended and bottom sediments. Sorption on sediments would, of course, decrease the net rate of volatilization. Moreover, once such a chemical volatilizes, it is probable that extensive sorption to atmospheric particulates would occur, although there is a great deal of uncertainty concerning this point.[72,73] The effect that such sorption would have on photolysis rates in the atmosphere is not clear at present. Conflicting reports have been presented concerning direct photolysis of sorbed chemicals in air. Moilanen and Crosby[74] have reported that the pesticide parathion photolyzes much more rapidly on dust than in the vapor state. On the other hand, Korfmacher and co-workers[75] have concluded that photolysis of polycyclic aromatics sorbed on fly ash is extremely slow. This conclusion is not in conflict with the half-life data shown in Table I. Water is a very polar reaction medium, and it is entirely possible that photochemical processes such as photoionizations or photonucleophilic reactions that occur in water would not occur on surfaces of fly ash. Thus, reaction quantum efficiencies (and photolysis rates) for some pollutants conceivably could be much larger in water than in air.

ASSESSING PHOTOLYSIS IN AQUATIC ENVIRONMENTS

Although the bulk of this chapter has been devoted to discussion of factors that influence photolysis rates, assessment of the role of photolysis in aquatic environments requires comparison of photolysis rates with kinetic data concerning other transport and transformation processes. Quantitative data concerning hydrolysis and other thermal processes are available,[76,77] but such data are almost nonexistent for biological transformation. In this section, kinetic data concerning photolysis, biolysis, and volatilization of the first four compounds in Table I are used to compute their likely fate in two types of aquatic systems. The computer model developed by Burns[70] was employed for the calculations.

Inputs for computation of the fate of the polycyclic aromatics are summarized in Table IV. The photolysis data are averaged over both night and day for a full year at latitude 40° N. Although near-surface values are listed light attenuation effects are taken into account in the computations. Second-

Table IV. Input Data for Computation of Fate of Selected Aromatic Hydrocarbons
in Aquatic Ecosystems

Parameter (units)	Naphthalene	Anthracene	Benz[a]-anthracene	Benzo[a]-pyrene
Direct photolysis rate constant (/hr)[a]	1.52×10^{-3}[e]	0.202[e]	0.249[c]	0.291[c]
Second order microbial degradation rate constant (ml/cell/hr)[b]	3.8×10^{-7}	1.3×10^{-7}	5.4×10^{-8}	0
Volatilization as fraction of reaeration rate (unitless)[c]	1.0[f]	0.14[g]	1.6×10^{-3}[c]	3.6×10^{-3}[c]
K_{oc}, sediment-water partition coefficient over fractional organic carbon (unitless)[d]	1.3×10^{3}[d]	2.6×10^{4}[d]	2.6×10^{5}[h]	6.9×10^{5}[h]
Bacteria-water partition coefficient (unitless)	4.6×10^{2}[i]	6.5×10^{3}[i]	3.6×10^{4}[c]	4.12×10^{5}[c]
Water solubility (mg/liter), 25°C	31.7[j]	4.5×10^{-2}[j]	9.4×10^{-3}[j]	1.2×10^{-3}[c]

[a]Near-surface rate constant averaged over dark and light periods for full year, latitude 40°N.
[b]Computed using Equation 14 and Herbes and Schwall's data.[79]
[c]Smith *et al.*[6b]
[d]Karickhoff *et al.*[55]
[e]Zepp and Schlotzhauer.[22]
[f]Computed to correspond with Mackay and Leinonen.[71]
[g]Computed to correspond with Southworth.[82]
[h]Computed from octanol-water partition coefficient.[55]
[i]Computed assuming equal to 0.2P.[55]
[j]May *et al.*[87]

order microbial rate constants k_{b2} [6,78] were computed from data published by Herbes and Schwall[79] employing Equation 14, which is based on studies of Steen.[80]

$$k_{b2} = \frac{k_b (1 + K_p\rho_b)}{[B]} \qquad (14)$$

where k_b is the first-order rate constant for degradation in the sediment.
[B] is the microbial population, which is about 5×10^{7} cells/ml for the petroleum-contaminated sediments studied by Herbes and Schwall[79]

K_p and ρ_b are the sediment-water partition coefficient and weight ratio of sediment to pore water, respectively;
ρ_b was assumed to be 2.7 and the K_p values assumes were: naphthalene, 49; anthracene, 990; and benz[a]anthracene, 9900.

Herbes and Schwall[79] and Smith *et al.*[6] were unable to isolate organisms from aquatic environments that were capable of degrading benzo[a]pyrene so a rate constant of zero was assumed. Steen[80] has found that Equation 14 applies to microbial degradation of a variety of nonionic pollutants in suspended bottom sediments. Sediment-water partition coefficients for the pond and river (Table V) were computed by multiplying fractional organic carbon in the sediments by the K_{oc} values listed in Table IV.[55] Bacteria-water partition coefficients are listed for information; accumulation by the biota has little effect on the results listed in Table V, however.

Results of the computed fate of the polycyclic aromatics are summarized in Table V. The model computes loss of chemicals from aquatic systems in which cessation of discharge was preceded by a long period of continuous contamination of the systems. Perusal of computed results for two types of systems indicates the following.

1. Biolysis and volatilization are the most important processes for naphthalene in the eutrophic pond, but in the riverine system, volatilization and export are strongly dominant. Photolysis is the most important process for the three other aromatics in both the pond and river. Calculations concerning lakes, a reservoir, and other types of rivers[70] tell the same story. In a study that is consistent with these computations, Thruston[81] has found photooxidation products of anthracene, benz[a]anthracene, and other aromatic compounds in a municipal water supply.

2. Water column and system half-lives increase sharply with molecular weight of the aromatics. This increase is mainly attributable to increases in the tendency of the compounds to sorb to the sediments. Also, as mentioned above, the assumption that the aromatics are not photoreactive when sorbed to suspended sediments greatly affects the computed half-lives of the higher molecular weight compounds. For example, if benzo[a]pyrene photolyzed at the same rate when sorbed as when dissolved in water, the computed half-lives in the pond would be 45 days in the water column (vs 353 days shown in Table V) and 100 days in the system (vs 435 days). Even so, it is apparent that long periods of time are required for benz[a]-anthracene and benzo[a]pyrene to be lost from water bodies like ponds and lakes, despite their short near-surface photolysis half-lives.

3. Rivers self-cleanse themselves via water-borne export and photolysis much more effectively than ponds or other impounded water bodies. Ex-

Table V. Analysis of Steady-State Fate of Several Aromatic Compounds Employing EXAMS Computer Model[70]; Results Expressed as Percent of Load Accounted for by Various Processes

	Biolysis[a]		Photolysis[b]	Volatili-zation	Waterborne Export	Half-Life, Days[c]	
	Water	Sediment				Water Column	System
Eutrophic Pond (2 m deep)							
Naphthalene	27	5.6	2.2	58	7.3	2.5	4.7
Anthracene	2.8	3.3	89	2.4	2.9	29	65
Benz[a]anthracene	0.97	1.2	90	0.02	7.1	153	224
Benzo[a]pyrene	0	0	86	0.04	13	353	435
River (50-km stretch, 3 m deep)							
N. phthalene	0.40	0.25	0.53	42	56	0.25	0.49
Anthracene	0.11	0.41	56	4.7	38	2.6	12
Benz[a]anthracene	0.04	0.17	61	0.05	38	6.4	23
Benzo[a]pyrene	0	0	60	0.09	40	14.7	31

[a]Microbial populations assumed: for pond, 3.7×10^7 cells/100 g (dry weight) of bottom sediment, 1×10^4 cells/ml in the water column; for rive , 1.85×10^7 cells/100 g bottom sediment, 1×10^3 cells/ml in water column.[70]

[b]Computed assuming chemicals do not photoreact when sorbed on suspended sediment and assuming the following reductions in average photolysis rate because of light attenuation: pond, 0.2; river 0.5.

[c]Computed as a first-order approximation of the kinetics via analysis of a dynamic computer simulation of the decrease in pollutant concentration after external loadings cease. The system half-life includes loss from the mixed layer of the bottom sediment, assumed to be 5 cm deep.

port is so effective that photolysis does little more than reduce the system half-lives of anthracene, benz[a]anthracene and benzo[a]pyrene by a factor of two compared to a "no photolysis" case. Calculations not shown in the table indicate that, in the river at steady state, photolysis reduces the concentration in a 50-km stretch by only a factor of two or so for these three compounds.

4. The calculations also reveal the seeming anomaly that volatilization of naphthalene accounts for a larger fraction of the loss in the pond than in the 50-km stretch of the river, although the latter is much more turbulent. This result is again attributable to the fact that volatilization must compete with rapid export in the river.

Results of these calculations differ from Southworth's recent assessment of the behavior of anthracene in natural waters.[82] His analysis indicated that microbial degradation is the most significant process affecting anthracene in a variety of systems. The main reason for our differences involves the microbial kinetic data. Southworth assumed that the rate constants observed for microbial degradation of anthracene in water from the same creek studied by Herbes and Schwall[79] were applicable to a wide variety of aquatic systems. The approach here was to use the second-order rate constants k_{b2} computed by Equation 14 to compute first-order rate constants for biolysis in a variety of environments having different microbial populations.[6,78] It is interesting to note that the k_{b2} values in Table IV are similar in magnitude to those computed by Smith et $al.$[6] for microbial degradation of other energy-related aromatic compounds by acclimated cultures of bacteria. Taking Southworth's first-order rate constant of 0.061 hr^{-1} for anthracene in the creek water and the k_{b2} value in Table IV, the microbial population in the creek water was calculated to be $10^5 - 10^6$ organisms per milliliter, about two orders of magnitude lower than in the creek sediment.[79] These populations and thus the biolysis rate constants employed by Southworth are one to two orders of magnitude larger than those assumed for the computations in Table V[70]

SUMMARY AND CONCLUSIONS

Evidence is accumulating that a variety of aromatic and unsaturated organic compounds and organometallic complexes react rapidly when exposed to sunlight in water. Although photolysis is rapid typically only in the surface region of water bodies, sufficient turbulence exists in many natural waters to insure rapid vertical mixing from the dark to the highly illuminated photic zone. Computations indicate that rapid photolysis of some chemicals occurs at depths in excess of 100 m in the open ocean and clear lakes, although the photic zone is probably much shallower in most inland water bodies.

Both direct and indirect processes can be involved in the photochemical transformation of aquatic contaminants. Equations have been defined that use reaction quantum yields, electronic absorption spectra, and simulated underwater scalar irradiance to compute photolysis rates in aquatic environments. Preliminary expressions that key on the ultraviolet absorbance of natural waters have been developed for the prediction of near surface photosensitized oxygenation and other indirect photochemical processes. Computations indicate that indirect photolysis involving energy transfer or electron transfer may be rapid for certain types of chemicals, but hydrogen transfer to triplet sensitizers is too slow to be of general importance. Light-induced free radical reactions are of undetermined importance at present.

Many gaps of information prevent adequate assessment of photochemical transformations in water. Assessment of photolysis requires research not only in the area of photochemistry but in a broad range of disciplines, including optical oceanography and limnology, chemical oceanography, hydrology, chemical engineering, and microbial kinetics. Some information that is needed is listed below.

1. Data concerning ultraviolet light in natural waters are lacking, and, more important, procedures for extrapolating UV light attenuation data from measured to unmeasured situations have not been developed.

2. Little is known about photoreactivity of hydrophobic chemicals sorbed to suspended sediments or microorganisms.

3. Studies have shown that the dissolved substances in natural waters alter photolysis rates, sometimes causing large increases compared to distilled water. An empirical relationship has been developed to predict photosensitized oxygenation, but it is based on a very limited data set that should be expanded. Research also should be conducted to learn more about other indirect photoprocesses such as those involving free radicals derived from photolysis of dissolved species in sea water.

4. Available evidence indicates that direct photolysis is the most generally important pathway for photochemical transformation in water, but little is known about rates and products of direct photoreactions of several industrially important classes of sunlight-absorbing compounds, such as dyes, trace metals, and organometallic complexes.

5. For highly photoreactive chemicals, vertical mixing in water columns and sediments probably is the rate limiting step for photochemical loss from certain aquatic systems such as lakes and coastal regions. Therefore,

results from research in this area are directly applicable to assessing photolysis in water.

6. Unlike the atmosphere where photolysis is dominant, photolysis in the aquatic environment competes with other important processes like hydrolysis and microbial degradation. In a general sense, therefore, photolysis cannot be evaluated without *quantitative* data and equations that describe the kinetics of these competing processes. In particular, procedures must be developed for predicting rates of microbial transformation.

ACKNOWLEDGMENTS

I thank Steven Broderius for the use of his data concerning photolysis of the iron cyanide complexes. Also, I thank Raymond C. Smith, Alex E.S. Green, George L. Baughman, Doris F. Paris and William C. Steen for prepublication copies of papers discussed here. I appreciate several helpful discussions with Lawrence A. Burns, who developed the EXAMS computer model that was employed for the calculations presented in Table V and Figure 13. Finally, I acknowledge that the impetus for several of the ideas presented here was derived from numerous discussions with other members of the Environmental Processes Branch, U. S. Environmental Protection Agency, Athens, Georgia.

REFERENCES

1. Leighton, P.A. *The Photochemistry of Air Pollution* (New York: Academic Press, 1961).
2. Demerjian, K.L., J.A. Kerr and J.G. Calvert. "The Mechanism of Photochemical Smog Formation," *Adv. Environ. Sci. Technol.* 4:1-262 (1974).
3. Glasgow, L.C., P.S. Gumerman, P. Meahin and J.P. Jesson. "The Fluorocarbon-Ozone Theory—IV. Fluorocarbon Mixing and Photolysis," *Atmos. Environ.* 12:2159 (1978).
4. Johnston, H.S. "Catalytic Destruction of Stratospheric Ozone by Nitrogen Oxides," *Adv. Environ. Sci. Technol.* 4:263-380 (1974).
5. Hill, J., H.P. Kollig, D.F. Paris, N.L. Wolfe and R.G. Zepp. "Dynamic Behavior of Vinyl Chloride in Aquatic Ecosystems," U.S. Environmental Protection Agency, Report No. EPA-600/3-76-001 (Athens, GA, 1976).
6. (a) Smith, J.H., W.R. Mabey, N. Bohonos, B.R. Holt, S.S. Lee, T.-W. Chou, D.C. Bomberger and T. Mill. "Environmental Pathways of Selected Chemicals in Freshwater Systems. Part I. Background and Experimental Procedures," U.S. Environmental Protection Agency, Report No. EPA-600/7-77-113 (Athens, GA 1977);
(b) Smith, J.H., W.R. Mabey, N. Bohonos, B.R. Holt, S.S. Lee, T.-W.

Chou, D.C. Bomberger and T. Mill. "Environmental Pathways of Selected Chemicals in Freshwater Systems. Part II. Laboratory Studies," U.S. Environmental Protection Agency, Report No. EPA-600/7-78-074 (Athens, GA, 1978).

7. Zepp, R.G., and D.M. Cline. "Rates of Direct Photolysis in Aquatic Environment," *Environ. Sci. Technol.* 11:359 (1977).

8. Zepp, R.G. "Quantum Yields for Reaction of Pollutants in Dilute Aqueous Solution," *Environ. Sci. Technol.* 12:327 (1978).

9. Zepp, R.G., and G.L. Baughman. In: *Aquatic Pollutants: Transformation and Biological Effects,* O. Hutzinger, I.H. van Lelyveld and B.C.J. Zoeteman, Eds. (New York: Pergamon Press, 1978), p. 237.

10. Turro, N.J., V. Ramamurthy, W. Cherry and W. Farneth. "The Effect of Wavelength on Organic Photoreactions in Solution. Reactions from Upper Excited States," *Chem. Rev.* 78:125 (1978).

11. Ross, R.D., and D.G. Crosby. "Photolysis of Ethylenethiourea," *J. Agr. Food Chem.* 21:335 (1973).

12. Zepp, R.G., N.L. Wolfe, J.A. Gordon, and G.L. Baughman. "Dynamics of 2,4-D Esters in Surface Waters: Hydrolysis, Photolysis, and Vaporization," *Environ. Sci. Technol.* 9:1144 (1975).

13. Ross, R.D., and D.G. Crosby. "The Photooxidation of Aldrin in Water," *Chemosphere* 4:227 (1975).

14. Draper, W.M., and D.G. Crosby. "Measurement of Photochemical Oxidants in Agricultural Field Water," presented in part at the 172nd National Meeting of the American Chemical Society, San Francisco, CA, September 1976.

15. Zepp, R.G., N.L. Wolfe, J.A. Gordon, and R.G. Fincher. "Light-induced Transformations of Methoxychlor in Aquatic Systems," *J. Agr. Food Chem.* 24:727 (1976).

16. Zepp, R.G., N.L. Wolfe, G.L. Baughman, and R.C. Hollis. "Singlet Oxygen in Natural Waters," *Nature* 267:421 (1977).

17. (a) Zafiriou, O.C., and M.B. True. "Nitrite Photolysis in Seawater by Sunlight," *Marine Chemistry* (in press, 1979);
(b) Zafiriou, O.C., and M.B. True. "Nitrite Photolysis as a Source of Free Radicals in Productive Surface Waters," *Geophys. Res. Lett.* (in press, 1979).

18. Zafiriou, O.C. "Marine Organic Photochemistry Previewed," *Marine Chemistry* 5:497 (1977).

19. Morowitz, H.J. "Absorption Effects in Volume Irradiation of Microorganisms," *Science* 111:229 (1950).

20. Morel, A., and R.C. Smith. "Relation Between Total Quanta and Total Energy for Aquatic Photosynthesis," *Limnol. Oceanog.* 19:591 (1974).

21. Foote, C.S. In: *Free Radicals in Biology, Vol. II,* W. Pryor, Ed. (New York: Academic Press, 1976), pp. 85-133.

22. Zepp, R.G., and P. Schlotzhauer. In: *Proceedings of Third International Symposium on Polynuclear Aromatic Hydrocarbons,* P.W. Jones and P. Lever, Eds. (Ann Arbor, MI: Ann Arbor Science Publishers, Inc., 1979), pp. 141-158.

23. (a) Turro, N.J., and R. Engel. "Quenching of Biacetyl Fluorescence and Phosphorescence," *J. Amer. Chem. Soc.* 91:7113 (1969);

(b) Turro, N.J. *Modern Molecular Photochemistry* (Reading, MA: Benjamin/Cummings Publishing Co., Inc., 1978), pp. 362-413.

24. Mill, T., H. Richardson, and D.G. Hendry. In: *Aquatic Pollutants: Transformation and Biological Effects,"* O. Hutzinger, I.H. van Lelyveld, and B.C.J. Zoeteman, Eds. (New York: Pergamon Press, 1978), p. 233.

25. Zika, R.G. "An Investigation in Marine Photochemistry," Ph.D. Thesis, Dalhousie University, Halifax, Nova Scotia.

26. Swallow, A.J. "Hydrated Electrons in Seawater," *Nature* 222:369 (1969).

27. Braslau, N., and J.V. Dave. "Effect of Aerosol on the Transfer of Solar Energy Through Realistic Model Atmospheres," *J. Appl. Meteor.* 12:601 (1973).

28. Mo, T., and A.E.S. Green. "A Climatology of Solar Erythema Dose," *Photochem. Photobiol.* 20:483 (1974).

29. Bener, P. "Approximate Values of Intensity of Natural Ultraviolet Radiation for Different Amounts of Atmospheric Ozone," U.S. Army, Report No. DAJA 37-68-C-1017 [U.S. Army Research and Development Group (Europe), Box 15, FPO NY 09510].

30. Calkins, J. "Measurements of the Penetration of Solar UV-B into Various Natural Waters." In: *Impacts of Climatic Change on the Biosphere, CIAP Monographs 5,* U. S. Department of Transportation, Report No. DOT-TST-75-55 (Springfield, VA: National Technical Information Service), pp. 2-267.

31. Smith, R.C., and K.S. Baker. "Penetration of UV-B and Biologically Effective Dose Rates in Natural Waters," *Photochem, Photobiol.* 27: 311 (1979).

32. Büttner, K. *Physik. Bioklimat,* Leipzig, (1938).

33. Schulze, R., and K. Grafe. In: *The Biologic Effects of Ultraviolet Radiation,* F. Urbach, Ed. (New York: Pergamon Press, 1069), pp. 359-373.

34. Spinhurne, J.D., and A.E.S. Green. "Calculation of the Relative Influence of Cloud Layers on Received Ultraviolet and Integrated Solar Radiation," *Atmos. Environ.* 13:in press (1979).

35. Nader, J.S. In: *The Biologic Effects of Ultraviolet Radiation,* F. Urbach, Ed. (New York: Pergamon Press, 1969), pp. 417-431.

36. Broderius, S.J., and L.L. Smith. "Direct Photolysis of Hexacyanoferrate Complexes in the Aquatic Environment," U.S. Environmental Protection Agency (Duluth, MN, 1979), in press.

37. Perry, F.M., Jr., E.W. Day, Jr., H.E. Magadanz, and O.G. Saunders. "Fate of Nitrosamines in the Environment: Photolysis in Natural Waters," presented in part at the 176th National Meeting of the American Chemical Society, Miami Beach, FL, September 1978.

38. Mancini, J.L. "Analysis Framework for Photodecomposition in Water," *Environ. Sci. Technol.* 12:1274 (1978).

39. Burkhard, N., D.O. Eberle, and J.A. Guth. "Model Systems for Studying the Environmental Behavior of Pesticides," *Environ. Qual. Safety* Suppl Vol. III:203 (1975).

40. Tyler, J.E., and R.W. Preisendorfer. In: *The Sea, Vol. I,* M.N. Hill, Ed. (New York: Interscience, 1966), pp. 397-451.

41. Smith, R.C., and J.E. Tyler. "Transmission of Solar Radiation into Natural Waters," *Photochem. Photobiol. Rev.* 1:117 (1976).

42. Miller, G.C., and R.G. Zepp. "Effects of Suspended Sediments on Photolysis Rates of Dissolved Pollutants," *Water Res.* 13:(in press, 1979).

43. Jerlov, N.G. *Marine Optics* (Amsterdam: Elsevier Scientific Publishing Co., 1976).

44. Zaneveld, J.R.V. "Penetration of Ultraviolet Radiation into Natural Waters." In: *Impacts of Climatic Change on the Biosphere, CIAP Monograph 5*, U.S. Department of Transportation, Report No. DOT-TST-75-55 (Springfield, VA:National Technical Information Service, 1975), pp. 2-108.

45. Ryther, J.H. "Photosynthesis and Fish Production in the Sea," *Science* 106:72 (1969).

46. Smith, R.C., and K.S. Baker. "Optical Classification of Natural Waters," *Limnol. Oceanog.* 23:260 (1978).

47. Smith, R.C., J.E. Tyler, and C.R. Goldman. "Optical Properties and Color of Lake Tahoe and Crater Lake," *Limnol. Oceanog.* 18:189 (1973).

48. Smith, R.C., and J. Calkins. "The Use of the Robertson Meter to Measure the Penetration of Solar Middle Ultraviolet Radiation (UV-B) into Natural Waters," *Limnol. Oceanog.* 21:746 (1976).

49. Holmes, R.W. "The Secchi Disk in Turbid Coastal Waters," *Limnol. Oceanog.* 15:688 (1970).

50. Poole, H.H., and W.R.G. Atkins. "Photo-electric Measurements of Submarine Illumination Throughout the Year," *J. Mar. Biol. Ass. U.K.* 16:297 (1929).

51. (a) Mackay, D.F. In: *Aquatic Pollutants: Transformation and Biological Effects*, O. Hutzinger, I.H. van Lelyveld and B.C.J. Zoeteman, Eds (New York: Pergamon Press, 1978), p. 175;
(b) Mackay, D., W.Y. Shiu and R.J. Sutherland. "Estimating Volatilization and Water Column Diffusion Rates of Hydrophobic Contaminants," *Proceedings of the Symposium on Dynamics, Exposure and Hazard Assessment of Toxic Chemicals in the Environment* (Ann Arbor, MI: Ann Arbor Science Publishers, Inc., 1979).

52. Giger, W., E. Molnar-Kubica, and S. Wakeham. In: *Aquatic Pollutants: Transformation and Biological Effects*, O. Hutzinger, I.H. van Lelyveld and B.C.J. Zoeteman, Eds. (New York: Pergamon Press, 1978), p. 101.

53. Csanady, G.T. *Turbulent Diffusion in the Environment.* (Boston, MA: D. Reidel Publishing Co., 1973), pp. 106-109.

54. Broecker, W.S., and T.-H. Peng. "Gas Exchange Rates Between Air and Sea," *Tellus* XXVI:21 (1974).

55. Karickhoff, S.W., D.S. Brown and T.A. Scott. "Sorption of Hydrophobic Pollutants on Natural Sediments," *Water Res.* 13:(in press, 1979).

56. Wolfe, N.L., R.G. Zepp, D.F. Paris, G.L. Baughman and R.C. Hollis. "Methoxychlor and DDT Degradation in Water," *Environ, Sci. Technol.* 11:1077 (1977).

57. Miller, G.C., and R.G. Zepp. "The Photoreactivity of Aquatic Pollutants Sorbed on Suspended Sediments," *Environ. Sci. Technol.*, 12: (in press, 1979).

58. Karickhoff, S.W., U.S. Environmental Protection Agency, Athens, GA, personal communication (1979).

59. (a) Paris, D.F., W.C. Steen and G.L. Baughman. "The Role of Physico-Chemical Properties of Aroclors 1016 and 1242 in Determining Their Fate and Transport in Aquatic Environments," *Chemosphere* 7:319 (1978); (b) Paris, D.F., D.L. Lewis, and J.T. Barnett. "Bioconcentration of Toxaphene by Microorganisms," *Bull. Environ. Contam. Toxicol.* 17:564 (1977); (c) Paris, D.F., and D.L. Lewis. "Accumulation of Methoxychlor by Microorganisms Isolated from Aqueous Systems," *Bull. Environ. Contam. Toxicol.* 15:24 (1976).

60. Baughman, G.L., and D.F. Paris. "Microbial Bioconcentration of Organic Pollutants from Aquatic Systems—A Critical Review," in preparation (1979).

61. O'Kelley, J.C., and T.R. Deason. "Degradation of Pesticides by Algae," U.S. Environmental Protection Agency, Report No. EPA-6009/3-76-022 (Athens, Georgia, 1976).

62. Anagnostopoulos, E., I. Scheunert, W. Klein and F. Korte. "Conversion of p-Chloroaniline-^{14}C in Green Algae and Water," *Chemosphere* 7:351 (1978).

63. Hill, J., D. Brockway, S. Durham and F. Stancil. "Dynamic Behavior of Methoxychlor in Aquatic Microcosms," U.S. Environmental Protection Agency, report in preparation (Athens, Georgia, 1979).

64. Ellgehausen, H., J.A. Guth and H.O. Esser. "Factors Determining the Bioaccumulation Potential of Pesticides in the Individual Compartments of Aquatic Food Chains," presented in part at the 4th International Congress on Pesticide Chemistry (IUPAC), Zurich, June 1978.

65. (a) Choi, W., and K.Y. Chen. "Associations of Chlorinated Hydrocarbons with Fine Particles and Humic Substances in Nearshore Surficial Sediments," *Environ. Sci. Technol.* 10:782 (1976); (b) Van Vleet, E.S., and J.G. Quinn. "Input and Fate of Petroleum Hydrocarbons Entering the Providence River and Upper Narragansett Bay from Wastewater Effluents," *Environ. Sci. Technol.* 11:1066 (1977).

66. Macek, K.J., S.R. Petrocelli, and B.H. Sleight. "Considerations in Assessing the Potential for, and Significance of, Biomagnification of Chemical Residues in Aquatic Food Chains." In: *ASTM Second Symposium on Aquatic Toxicology* (American Society for Testing Materials, 1977), in press.

67. Kearney, P.C., J.E. Oliver, C.S. Hilling, A.R. Isensee and A. Kontson. "Distribution, Movement, Persistence, and Metabolism of N-Nitrosoatrazine in Soils and a Model Aquatic Ecosystem," *J. Agr. Food Chem.* 25:1177 (1977).

68. Robbins, J.A., and D.N. Edgington. "Determination of Recent Sedimentation Rates in Lake Michigan using Pb-210 and Cs-137," *Geochim. Cosmochim. Acta* 39:285 (1975).

69. (a) Johnson, J.W., N.L. Guinasso and D.F. Schink. "Biological Mixing Rates in the Atlantic Abyssal Sediments Using Plutonium as a Tracer," presented at the 41st Annual Meeting of the American

Society of Limnology and Oceanography, Victoria, British Columbia, 1978;
(b) Robbins, J.A., J.B. Fisher, P.L. McCall and J. R. Krezoski. "Influence of Deposit Feeders on Migration of Radiotracers in Lake Sediments," presented at the 41st Annual Meeting of the American Society of Limnology and Oceanography, Victoria, British Columbia, 1978.

70. Burns, L.A. "EXAMS: An Exposure Analysis Modeling System for Toxic Organic Pollutants in Aquatic Ecosystems," U.S. Environmental Protection Agency, report in preparation (Athens, Georgia, 1979).

71. Mackay, D., and P.J. Leinonen. "Rate of Evaporation of Low-Solubility Contaminants from Water Bodies to Atmosphere," *Environ. Sci. Technol.* 9:1178 (1975).

72. Natusch, D.F.S., and B.A. Tomkins. In: *Carcinogenesis, Vol. 3: Polynuclear Aromatic Hydrocarbons,* P.W. Jones and R.I. Freudenthal, Eds. (New York: Raven Press, 1978), pp. 145-153.

73. Junge, C. In: *Fate of Pollutants in the Air and Water Environments,* I.H. Suffet, Ed. (New York: John Wiley and Sons, Inc., 1977), pp. 7-25.

74. Moilanen, K.W., D.G. Crosby, C.J. Soderquist, and A.S. Wong. In: *Environmental Dynamics of Pesticides* (New York: Plenum Press, 1975), pp. 45-60.

75. Korfmacher, W.A., D.F.S. Natusch, D.R. Taylor, E.L. Wehry and G. Mamantor. In: *Proceedings of Third International Symposium on Polynuclear Aromatic Hydrocarbons,* P.W. Jones and P. Lever, Eds. (Ann Arbor, MI: Ann Arbor Science Publishers, Inc., 1979), in press.

76. Mabey, W., and T. Mill. "Critical Review of Hydrolysis of Organic Compounds in Water Under Environmental Conditions," *J. Phys. Chem. Ref. Data* 7:383 (1978).

77. Wolfe, N.L., R.G. Zepp, G.L. Baughman, R.C. Fincher and J.A. Gordon. "Chemical and Photochemical Transformation of Selected Pesticides in Aquatic Systems," U.S. Environmental Protection Agency, Report No. EPA-600/3-76-022 (Athens, GA, 1976).

78. Paris, D.F., D.L. Lewis and N.L. Wolfe. "Rates of Degradation of Malathion by Bacteria Isolated from Aquatic Systems," *Environ. Sci. Technol.* 9:135 (1975).

79. Herbes, S.E., and L.R. Schwall. "Microbial Transformation of Polycyclic Aromatic Hydrocarbons in Pristine and Petroleum-Contaminated Sediments," *Appl. Environ. Microbiol.* 35:306 (1978).

80. Steen, W.C., U.S. Environmental Protection Agency, Athens, Georgia, Personal communication, 1979.

81. Thruston, A.D. "High Pressure Liquid Chromatography Techniques for the Isolation and Identification of Organics in Drinking Water Extracts," *J. Chrom. Sci.* 16:254 (1978).

82. Southworth, G.R. "Transport and Transformation of Anthracene in Natural Waters: Process Rate Studies." In: *ASTM Second Symposium on Aquatic Toxicology* (American Society for Testing Materials, 1977), in press.

83. Banerjee, S., H.C. Sikka, R. Gray, and C.M. Kelly. "Photodegradation of 3,3'-Dichlorobenzidine," *Environ. Sci. Technol.* 12:1425 (1978).

84. Wolfe, N.L., R.G. Zepp, J.A. Gordon, and R.C. Fincher. "N-Nitrosoatrazine Formation from Atrazine," *Bull. Environ. Contam. Toxicol.* 15:342 (1976).

85. Wagner, P.J., and R.A. Leavitt. "Photoreduction of α-Trifluoroacetophenone: Competitive Charge Transfer and Hydrogen Abstraction," *J. Amer. Chem. Soc.* 95:3669 (1973).

86. Parsons, G.H., and S.G. Cohen. "Effects of Polar Substituents on Photoreduction and Quenching of Aromatic Ketones by Amines. Fluorenone and Substituted Dimethylanilines," *J. Amer. Chem. Soc.* 96:2948 (1974).

87. May, W.E., S.P. Wasik and D.H. Freeman. "Determination of the Solubility Behavior of Some Polycyclic Aromatic Hydrocarbons in Water," *Anal. Chem.* 50:997 (1978).

10

ORGANIC PHOTOCHEMISTRY. XVI.
TROPOSPHERIC PHOTODECOMPOSITION OF
METHYLENE CHLORIDE[1,2] *

Wendell L. Dilling and Helmut K. Goersch

Environmental Sciences Research Laboratory
The Dow Chemical Company
Midland, Michigan 48640

INTRODUCTION

To assess the expected exposure and the hazard that a compound or its transformation products will present to living organisms, data on the persistence and fate of that compound in the environment are needed. The troposphere, or lower atmosphere, is the region of the environment where the major portion of a compound that has been released into the environment usually exists if that compound has a high vapor pressure and a low solubility in water.[3] Thus, the troposphere is the primary region of the environment that needs to be considered in determining the persistence and fate of such a compound.

METHODS FOR DETERMINING RATES OF DECOMPOSITION OF COMPOUNDS IN THE TROPOSPHERE

This section describes some methods used for determining the rates of photodecomposition of compounds in the troposphere and some problems associated with these measurements. CH_2Cl_2 will be used as a specific example in this discussion, but the principles apply to other compounds as well. CH_2Cl_2 is a widely used industrial solvent, much of which eventually is released to the troposphere.

---

*Paper No. B-600-489-78.

The following three methods are used for determining rates of decomposition of compounds in the troposphere.

1. at steady-state conditions, measurement of concentration of compound in troposphere and knowledge of anthropogenic release rate
2. measurements of rate constant of compound with tropospheric species (OH) and concentration of that tropospheric species
3. simulated tropospheric reactions

The first method can be applied to compounds that have been in use for several years but cannot be used for testing new compounds. The decomposition rate or half-life of a compound that is not a natural material can be calculated if its concentration in the troposphere and anthropogenic release rate are known. This calculation applies only under steady-state conditions where the input rate equals the loss rate.

A second method involves measuring the rate constant for the reaction of the compound with the specific tropospheric species primarily responsible for the decomposition of the compound in the troposphere. For CH_2Cl_2 and many other compounds, the hydroxyl radical appears to be that species, according to current theory. If the average concentration of that species in the troposphere is known, the rate of decomposition of the compound in the troposphere can be calculated.

The third method involves reactions under simulated tropospheric conditions where the rate of loss of the compound is measured directly. Each of these methods will be discussed as they apply to CH_2Cl_2.

Tropospheric Concentration (1975)[4-7]:

$$< 5 \text{ ppt} - \sim 35 \text{ ppt}$$

Release Rate (1974)[8]:

$$\sim 3.5 \times 10^{11} \text{ g yr}^{-1}$$

Decomposition Rate:

$$\tau = \frac{t_{1/2}}{0.693} = \frac{\text{mass of } CH_2Cl_2 \text{ in northern troposphere}}{\text{release rate}} \quad (1)$$

$$t_{1/2} = < \sim 20 \text{ days} - \sim 150 \text{ days}$$

The above shows the results of the calculation for CH_2Cl_2 using the first method. The tropospheric concentration of CH_2Cl_2 is not known precisely. There have been four reports in which values have been recorded[1-7] For example, Grimsrud and Rasmussen[4] reported in 1975 that they

could not detect CH_2Cl_2 at a detection limit of 5 ppt (molar or gas
volume basis). Other workers, Cox et al.[5] and Lovelock,[6] reported values
of approximately 30 to 35 ppt, also in 1975. Thus, the value for this
concentration is not known exactly but is probably in the range shown
above.

The anthropogenic release rate of CH_2Cl_2 was $\sim 3.5 \times 10^{11}$ g yr^{-1},[8]
a rate that was nearly constant for several years before 1974.

The decomposition rate was calculated by use of Equation 1 in which
τ, the lifetime, equals the mass of CH_2Cl_2 in the northern tropsophere
divided by the release rate. The northern troposphere is the only region
of the atmosphere considered in this calculation because the decomposition rate is fast enough so that only a small amount of CH_2Cl_2 can
escape from the northern troposphere. The lifetime is related to the
half-life as shown in Equation 1. By using the values of < 5 ppt or
$\sim$ 35 ppt as limits for the tropospheric concentration of CH_2Cl_2, a
half-life of < 20 days to $\sim$ 150 days is calculated. The accuracy of
this calculated half-life can obviously be no greater than that of the input data required for the calculation.

The second method of determining decomposition rates involves hydroxyl
radicals. The following shows a major route of formation of hydroxyl radicals in the atmosphere.[9]

$$NO_2 \xrightarrow{\quad h\nu \quad} NO + O(^3P)$$

$$O(^3P) + O_2 \xrightarrow{\quad M \quad} O_3$$

$$O_3 \xrightarrow{\quad h\nu\,(<310\ nm) \quad} O_2 + O(^1D)$$

$$O(^1D) + H_2O \xrightarrow{\quad\quad} 2 \cdot OH$$

Hydroxyl radicals are formed primarily by the reaction of excited oxygen
atoms, $O(^1D)$, with water. $O(^1D)$ is formed by the short wavelength photolysis of ozone, which is produced by the reaction of ground-state oxygen
atoms, $O(^3P)$, with O_2 in a termolecular reaction. One source of $O(^3P)$
is the photolysis of NO_2, a reaction that plays a major role in atmospheric
chemistry.

Rate Constant for

$$CH_2Cl_2 + \cdot OH \xrightarrow{\quad -8^\circ C \quad} \cdot CHCl_2 + H_2O$$

8.7×10^{-14} cm^3 molecule^{-1} sec^{-1} ($\pm \sim 20\%$)[10,12,13]

Average ·OH Concentration in Troposphere:[10,14]

$$10^6 \ cm^{-3} \ (\pm \ factor \ of \sim 2)$$

Decomposition Rate:

$$t_{1/2} = \frac{0.693}{k \ [\cdot OH]} = \sim 40 \ days - \sim 230 \ days \tag{2}$$

This shows the average of the rate constants measured in laboratory experiments by other workers for the reaction of CH_2Cl_2 with hydroxyl radicals. This reaction involves abstraction of a hydrogen atom from CH_2Cl_2 and the formation of a dichloromethyl radical and water. The average rate constant at -8°C, which Davis et al.[10,11] considered to be the average temperature of the troposphere, is $8.7 \times 10^{-14} \ cm^3$ molecule^{-1} sec^{-1} with a variation of $\pm$ 20% among the three different values.[10,12,13] Howard,[12] Perry[13] and their co-workers measured the value at room temperature, and Davis et al.[10] determined the temperature dependence for this rate constant. The rate constant shown above was derived by averaging all three values at room temperature and then converting this average rate constant to the corresponding value at -8°C according to the temperature dependence determined by Davis et al.[10]

The average hydroxyl radical concentration in the troposphere is about $10^6 \ cm^{-3}$ with an uncertainty factor of about 2 or 3. This value for the concentration was derived from experimental measurements[10] and from calculations.[14,15]

The half-life for the decomposition of CH_2Cl_2 can be calculated from the pseudo-first-order rate expression shown in Equation 2. If the values for the rate constant, k, and the hydroxyl radical concentration, which is assumed to be at steady-state, are substituted, half-lives ranging from $\sim$ 40 days to $\sim$ 230 days are obtained. The range arises from the variation of $\pm$ 20% for k and a $\pm$ factor of 2 for the hydroxyl radical concentration. These results are similar to those calculated on page 112.

The third method for determining rates of decomposition of compounds in the troposphere involves using simulated atmospheric conditions in which the compound, along with other tropospheric species, is subjected to artificial or real sunlight. The concentration of the compound is measured as a function of time.

Compounds may decompose by direct absorption of light. Figure 1 shows the vapor phase UV-absorption spectrum of CH_2Cl_2[15-18] which does not significantly absorb sunlight in the troposphere where the short wavelength cutoff for sunlight is $\sim$ 290 nm. Because the extinction coefficient of CH_2Cl_2 at $\geq$ 250 nm is $< 10^{-3}$, significant decomposition

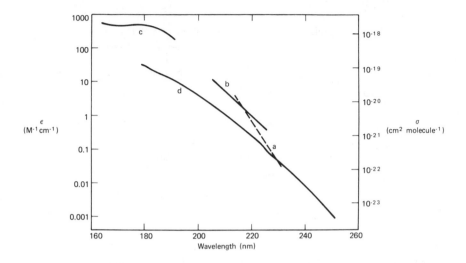

Figure 1. Vapor phase ultraviolet absorption spectrum of CH_2Cl_2; (a) Ref. 16; (b) Ref. 17, (c) Ref. 18, (d) Ref 15.

of CH_2Cl_2 by direct photolysis is not expected.

Table I shows a summary of work from several different laboratories on the simulated tropospheric reactions of CH_2Cl_2.[19-22] A goal of these experiments was the determination of the rate of decomposition of CH_2Cl_2 under simulated tropospheric conditions. A variety of conditions was used in these reactions. The initial concentration of CH_2Cl_2 ranged from 1 to > 1000 ppm. Usually some form of NO_x, NO or NO_2, was added, and various water vapor concentrations were used. Either purified air or, as in the last entry in Table I, ambient air was used. The light sources were either UV lamps or natural sunlight.

The half-lives determined in these experiments ranged from about one day to almost one year if it is assumed that in the experiments using artificial light sources the days were equivalent to 8 hr of bright sunlight and 16 hr of darkness. The 1.5 day half-life shown in the first entry[19] in Table I was extremely short compared to the other values. The second,[20] third,[21] and fourth[21] experiments gave only lower limits for the half-life. All four of these experiments were based on rather short reaction times, about 6-24 hr. These half-lives were calculated by assuming that the decomposition of CH_2Cl_2 was kinetically first-order in CH_2Cl_2. Because the reaction times were short compared to the calculated half-lives, the accuracy of these half-lives is low. The last entry in Table I is from an experiment in which CH_2Cl_2 was present in ambient air and natural sunlight was used as the light source.[22] The half-life was 230 days ± 50%, the

Table I. Simulated Tropospheric Reactions of CH_2Cl_2

$[CH_2Cl_2]$ (ppm)	$[NO_X]$ (ppm)	$[H_2O]$	Air	Light Source	$\sim t_{1/2}$	Reference
4	0.18 NO 0.02 NO_2	–	Purified	Fluorescent lamps	1.5 days	19
1	0.10 NO 0.15 NO_2	–	Purified	Fluorescent lamps	$\geqslant 70$	20
10	5 NO	35% RH	Purified	Mercury arc	> 90	21
10	16.8 NO_2	35%	Purified	Mercury arc	> 30	21
1300	?	?	Ambient	Sun	230 ($\pm$ 50%)	22

large error limits being caused by the variability of both sunlight intensity and the composition of ambient air. The last four values in Table I indicate that the half-life of CH_2Cl_2 is greater than a month, and similar to the half-lives calculated by the two methods that have already been discussed (see pages 112 and 114).

ANOMALOUS DECOMPOSITION REACTIONS UNDER SIMULATED TROPOSPHERIC CONDITIONS

The lower limits for the half-lives given in the third and fourth entries in Table I were derived from experiments performed several years ago.[21] No decomposition of the CH_2Cl_2 was observed within experimental error ($\pm$ 5%) in one day or less. Analyses were conducted by flame ionization/ gas chromatography. Recently similar experiments were performed over a longer period of time to determine more accurately the decomposition rate of CH_2Cl_2. The results of two preliminary experiments are shown in Figure 2.[21] In these experiments a higher concentration of CH_2Cl_2, 50 ppm, was used than in the earlier experiments (Table I). Dry cylinder air was used for these and subsequent simulated tropospheric reactions reported in this chapter. The concentration of NO_2 was 10 ppm and the relative humidity was 35%. The reactor consisted of a cylinder (11.5 liter) constructed of Pyrex® glass with two GE RS-275W sunlamps as light sources.

Figure 2 shows the results of duplicate experiments. The concentration of CH_2Cl_2 remained constant for approximately 4-5 days in the first experiment (a) and about 6-9 days in the second experiment (b). Then, in a few more days, the concentration of CH_2Cl_2 decreased to nearly zero. These unexpected results were similar to those observed for reactions

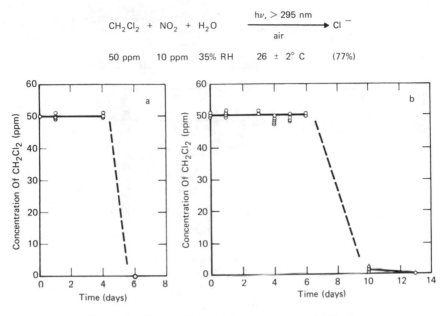

Figure 2. Simulated tropospheric reactions of CH_2Cl_2.

that exhibit induction periods in which, after the inhibitor is destroyed, the reaction proceeds. The CH_2Cl_2 was not lost by mechanical leakage, as shown by analyzing the contents of one reaction flask for chloride ion. By washing the flask with 0.01 N NaOH solution and titrating the wash solution we accounted for 77% of the chlorine in the original CH_2Cl_2 as chloride ion.

Figure 3 shows the results of a similar experiment performed later under slightly different conditions, but in which the same phenomenon was observed. In this reaction, 10 ppm of CH_2Cl_2 and 5 ppm of NO_2 in air were present; no water was added intentionally. CCl_3F, which is inert under these reaction conditions, was added as a tracer so that mechanical leaks could be detected readily. The same reactor that was used in the previous experiments was cleaned with chromic acid cleaning solution, rinsed well with distilled water, and evacuated.

The results of this photolysis were similar to those noted previously: the concentration of CH_2Cl_2 remained constant for about six days, and during the seventh day decreased rapidly. After the CH_2Cl_2 had been destroyed, at ~9.5 days an additional 10 ppm of CH_2Cl_2 was added. Continued irradiation rapidly decomposed the CH_2Cl_2 without any further induction or delay period. After the concentration of CH_2Cl_2 decreased to

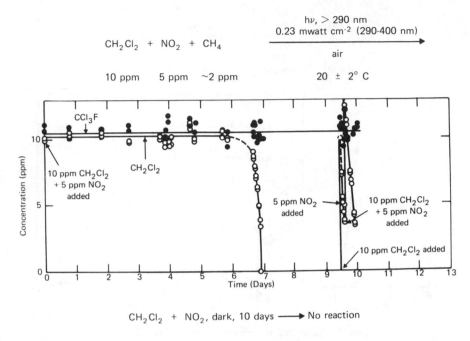

Figure 3. Simulated tropospheric reaction of CH_2Cl_2 (chromic acid-cleaned reactor).

~ 5 ppm, an additional 5 ppm of NO_2 was added, whereupon the decomposition of CH_2Cl_2 continued at about the same rate as before. Another 10 ppm of CH_2Cl_2 and 5 ppm of NO_2 were added, and again the decomposition of the CH_2Cl_2 proceeded at nearly the same rate.

Thus the induction period occurred only in a "clean" reaction vessel. After a reaction had taken place and more CH_2Cl_2 was added, no further induction period occurred. These results indicated that the induction period was not caused by an inhibitor present in one of the reactants. Otherwise the addition of fresh reactant should have caused an additional induction period.

Maintaining the mixture of reactants in the dark for 10 days produced no reaction (Figure 3), indicating that light was required for the decomposition of CH_2Cl_2.

Reactions or processes occurring on the reactor walls have been recognized as a problem by workers in the field of simulated atmospheric reactions. To investigate the possibility that wall or surface reactions occurred in our experiments, the same Pyrex glass reactor was used in which the walls were treated with concentrated NaOH solution after having been cleaned with chromic acid (Figure 4, upper diagram). The reactor was rinsed well with water and evacuated. An induction period of 3-4 days for the decomposition of CH_2Cl_2 was observed, a period significantly shorter than that obtained within the acid-washed reactor.

If the reactor, instead of being washed after one of the previous reactions, was evacuated at 0.05 mm Hg for 5 hr, the induction period decreased to slightly more than one day (Figure 4, lower diagram).

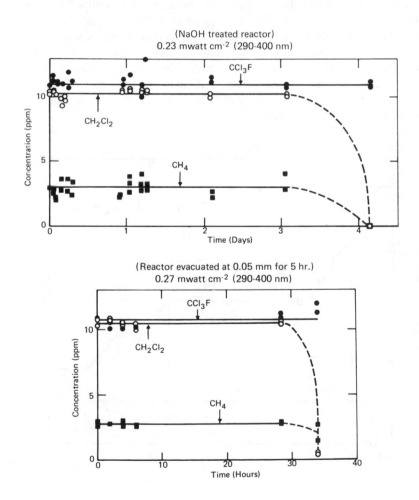

Figure 4   Photolysis of CH_2Cl_2 (10 ppm) + NO_2 (5 ppm) + air.

The different induction times resulting from various surface treatments indicated that some surface process may have been responsible for these induction periods. That the decomposition of CH_2Cl_2 was rapid ($t_{1/2} \ll 24$ hr in Figure 3) once the decomposition started suggests that the reaction may have occurred on the glass surface rather than in the gas phase. This decomposition was much faster than that expected for CH_2Cl_2, as judged from the data shown on pages 112, 114 and Table I.

Figure 5 shows that light is required for the reaction to proceed once the decomposition starts. In this figure are presented the results from the photolysis of 10 ppm of CH_2Cl_2 and 5 ppm of NO_2 in air in a reactor that contained the products from a previous reaction. Fresh charges of CH_2Cl_2 and NO_2 were added to the unilluminated reactor, whereby the concentration of CH_2Cl_2 remained constant. When the light was turned on, the CH_2Cl_2 started to decompose, and when the light was shut off, the decomposition of CH_2Cl_2 stopped. This light and dark cycle was repeated several times (Figure 5).

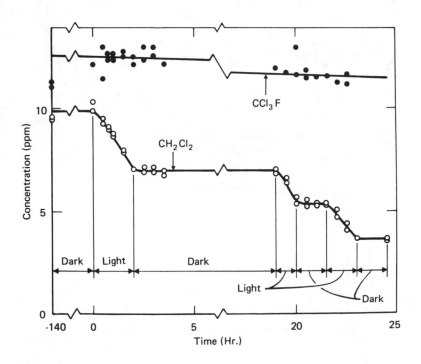

Figure 5. Photolysis of CH_2Cl_2 (10 ppm) + NO_2 (5 ppm) + air. (In presence of previous reaction product mixture; NaOH-treated reactor; 0.23 MW cm^{-2}, 290-400 nm).

One obvious way to avoid problems of surface reactions involves the use of a reaction vessel that is constructed of a material that is inert. Figure 6 shows the results from experiments performed with a reactor constructed of Teflon®* FEP film. This film is generally considered a more inert wall material than is glass for simulated atmospheric reactions.[23]

The two curves labeled CH_2Cl_2 (dark) represent runs in which the reactor was kept in the dark. A gradual decrease in the concentration of CH_2Cl_2 over several days was observed. This decrease indicated that CH_2Cl_2 diffused either into or through the Teflon film.

In the light [0.36 MW cm^{-2} (290-400 nm), 20 ± 2°C] the rate of loss of CH_2Cl_2 from the reactor was about the same as the rate in the dark (Figure 6), so that it was not possible to determine a reliable rate for the photochemical process. This result is consistent with the previous results (see page 112, 114 and Table I) that indicated CH_2Cl_2 had a half-life of at least a month. We would not expect to determine an accurate half-life from a 10-day run. The experiment could not be performed for longer times because of diffusion of the CH_2Cl_2 out of the container.

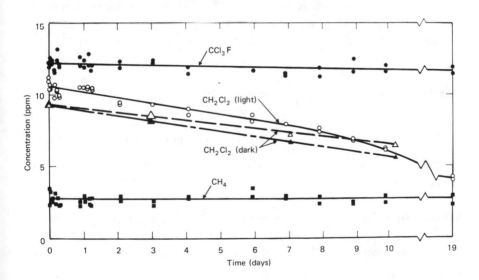

Figure 6. Photolysis of CH_2Cl_2 (10 ppm) + NO_2 (5 ppm) + air (Teflon FEP film reactor).

*Registered trademark of E.I. du Pont de Nemours and Company, Inc., Wilmington, DE.

Figure 6 shows that the rate of loss of CCl_3F was considerably slower than that of CH_2Cl_2 in the dark.

The results in Figures 2-6 show that care must be used in interpreting results of simulated atmospheric reactions. The absence of surface processes must be ascertained or compensation made for them. We are studying other halogenated compounds by using reactors constructed from Pyrex glass, quartz or Teflon film. Also, the surface/volume ratio is varied to determine whether reactions are accelerated or decelerated by the presence of a larger surface area.

DECOMPOSITION PRODUCTS FORMED UNDER SIMULATED TROPOSPHERIC CONDITIONS

Although only decomposition rates have been discussed here, identification of the decomposition products is also important. In principle, the products observed in simulated atmospheric reactions may have been formed by surface reactions. As far as we are aware, no systematic study has been reported on the importance of surface effects in these reactions.

The following shows results obtained in other laboratories on the products formed by the simulated tropospheric decomposition of CH_2Cl_2.[22,24,25]

$$CH_2Cl_2 + \text{ambient air} \xrightarrow[28°C]{h\nu,\ >290\ nm} CO_2 + HCl \tag{3}$$
$$(99\%)\ (95\%)$$

$$CH_2Cl_2 + \text{dry air} + Cl_2 \xrightarrow[5\ min]{h\nu} CO_2 + CO + HCl + COCl_2 \tag{4}$$

20 ppm   5 ppm   (12   (5   (38   (2

(19 ppm reacted)   ppm)   ppm)   ppm)   ppm)

$$CH_2Cl_2 + O_2 + Cl_2 \xrightarrow[32 \pm 2°C]{h\nu,\ 365.5\ nm} HCOCl + HCl + COCl_2 + CO \tag{5}$$

4-19   6-80   2-9
torr   torr   torr $\Phi = 49 \pm 3$ $\Phi = 4.4$

$\Phi_{-CH_2Cl_2} = 62$ Major products

Pearson and McConnell[22] reported (Equation 3) that CH_2Cl_2 photolyzed in ambient air to give mainly CO_2 and HCl.

Spence et al.[24] (Equation 4) irradiated CH_2Cl_2 in dry air in the presence of Cl_2. The Cl_2 served as a source of chlorine atoms which initiated the reaction of CH_2Cl_2 in much the same manner that hydroxyl radicals would presumably do in the real atmosphere. In addition to finding CO_2 and HCl, these workers observed small amounts of CO and $COCl_2$.

Sanhueza and Heicklen[25] (Equation 5) studied the chlorine-atom sensitized oxidation of CH_2Cl_2 in the presence of pure oxygen. The reaction proceeded by a chain reaction as indicated by the high quantum yield for the disappearance of CH_2Cl_2. The major product these workers observed was HCOCl, an unstable intermediate that decomposed to CO and HCl. Also, they found a small amount of $COCl_2$, which was formed with less than 0.10 the quantum yield of HCOCl.

Figure 7 shows the mechanism suggested by Sanhueza and Heicklen[25] for this oxidation of CH_2Cl_2. The reaction is initiated by photolysis of Cl_2 to chlorine atoms, which abstract hydrogen atoms from CH_2Cl_2, producing $\cdot CHCl_2$ and HCl. As mentioned previously, under real atmospheric conditions the hydroxyl radical presumably serves as the hydrogen-abstracting species and produces the same intermediate, $\cdot CHCl_2$. This latter radical can react with O_2 to give the peroxy radical, $CHCl_2O_2 \cdot$, which can decompose by two different routes according to the proposed mechanism. Both routes involve a bimolecular reaction of the peroxy radical and elimination of O_2 to form either $CHCl_2O \cdot$ or $(CHCl_2O)_2$. Neither of these reactions is expected to be a significant process under real atmospheric conditions because they are both bimolecular in a species that would be present at extremely low concentration.

Other reactions of the peroxy radical probably are more important; the reaction with NO to give the same radical mentioned previously, $CHCl_2O \cdot$, and NO_2 would probably occur. Thus, the same intermediate $CHCl_2O \cdot$ shown in Figure 7 would be formed, but by a different reaction.

The formation of $(CHCl_2O)_2$ is probably not significant in the troposphere because it requires a reaction that is bimolecular in two species, both of which are derived from CH_2Cl_2. Thus, the cycle on the right side of Figure 7 is not likely in the troposphere. Because $COCl_2$ is formed only in this cycle, according to the proposed mechanism, this product would not be expected from the tropospheric decomposition of CH_2Cl_2. The report by Pearson and McConnell[22] agrees with this reasoning, indicating that $COCl_2$ was not a significant product of the simulated tropospheric photolysis of CH_2Cl_2. Spence et al.[24] did not speculate on the source of $COCl_2$ in their experiments.

The alkoxy radical, $CHCl_2O \cdot$, loses a chlorine atom to form HCOCl. As well as decomposing into CO and HCl, HCOCl may be attacked by a chlorine atom to give HCl and $\cdot COCl$, which decomposes into CO and a chlorine atom, thus completing the cycle of the chain reaction (Figure 7).

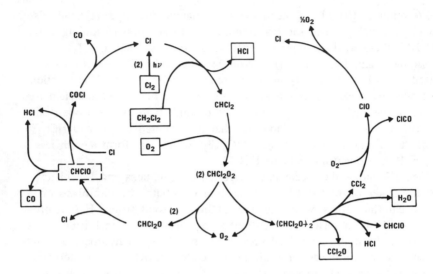

Figure 7. Cl-Sensitized photooxidation of CH_2Cl_2.[25]

SUMMARY AND CONCLUSIONS

The most important conclusions from this discussion are summarized below:

1. $t_{1/2}$ of CH_2Cl_2 in the troposphere is $\sim 40 - \sim 120$ days.
2. Very little CH_2Cl_2 escapes from the northern troposphere.
3. Major tropospheric photodecomposition products from CH_2Cl_2 are CO_2 and HCl.
4. Simulated tropospheric photodecomposition reactions of CH_2Cl_2 are subject to surface effects.

The half-life of CH_2Cl_2 in the troposphere probably is in the range of one to four months. This relatively short half-life, compared to those of other compounds such as the perhalogenated alkanes, indicates that only a very small fraction of the CH_2Cl_2 can excape from the northern troposphere.

The major tropospheric photodecomposition products from CH_2Cl_2 probably are CO_2 and HCl.

Simulated tropospheric reactions of CH_2Cl_2 in a Pyrex reactor, under the conditions that we used, are apparently subject to surface effects. These results suggest that one must be careful in interpreting the results of other similar studies. Surface effects must be avoided or compensated for in determining *gas-phase* tropospheric decompositoin rates.

Surface reactions may be significant under real tropospheric conditions in some instances. Solid and liquid surfaces on particulate matter and aerosols are present in the troposphere. Therefore, surface reactions could play a role in the decomposition of organic compounds in the troposphere. However, little is known about these processes.

REFERENCES

1. Dilling, W.L. "Organic Photochemistry − XV. Applications of Photo-cycloaddition Reactions in Natural Product Syntheses," *Photochem. Photobiol.* 25:605-621 (1977); 26:557 (1977).
2. Dilling, W.L., and H.K. Goersch. "Organic Photochemistry. XVI. Tropospheric Photodecomposition of Methylene Chloride," *Preprints of Papers Presented at the 176th National Meeting, ACS, Div. Environ. Chem.* 18(2):144-147 (1978).
3. Dilling, W.L. "Interphase Transfer Processes. II. Evaporation Rates of Chloro Methanes, Ethanes, Ethylenes, Propanes, and Propylenes from Dilute Aqueous Solutions. Comparisons with Theoretical Predictions," *Environ. Sci. Technol.* 11:405-409 (1977).
4. Grimsrud, E.P., and R.A. Rasmussen. "Survey and Analysis of Halo-carbons in the Atmosphere by Gas Chromatography-Mass Spectrometry," *Atmos. Environ.* 9:1014-1017 (1975).
5. Cox, R.A., R.G. Derwent, A.E.J. Eggleton and J.E. Lovelock. "Photochemical Oxidation of Halocarbons in the Troposphere," *Atmos. Environ.* 10:305-308 (1976).
6. Lovelock, J.E., and R.A. Rasmussen. Unpublished data cited in National Academy of Sciences. "Halocarbons: Effects on Stratospheric Ozone," p. 45 (Washington, D.C.: National Academy of Sciences, 1976).
7. Cronn, D.R., D.E. Harsch and E. Robinson. "Tropospheric and Lower Stratospheric Profiles of Halocarbons and Related Chemical Species," *Preprints of Papers Presented at the 176th National Meeting, ACS, Div. Environ. Chem.* 18(2):360-362 (1978).
8. National Academy of Sciences. "Halocarbons: Effects on Stratospheric Ozone," (Washington, D.C.: National Academy of Sciences, 1976), p. 37.
9. Demerjian, K.L., J.A. Kerr and J.G. Calvert. "The Mechanism of Photochemical Smog Formation." In: *Advances in Environmental Science and Technology, Vol. 4,* J.N. Pitts, Jr., R.L. Metcalf and A.C. Lloyd, Eds. (New York: John Wiley and Sons, 1974), pp. 1-262.
10. Davis, D.D., G. Machado, B. Conaway, Y. Oh and R. Watson. "A Temperature Dependent Kinetics Study of the Reaction of OH with CH_3Cl, CH_2Cl_2, $CHCl_3$, and CH_3D," *J. Chem. Phys.* 65(4):1268-1274 (1976).

11. Watson, R.T., G. Machado, B. Conaway, S. Wagner and D.D. Davis. "A Temperature Dependent Kinetics Study of the Reaction of OH with CH_2ClF, $CHCl_2F$, $CHClF_2$, CH_3CCl_3, CH_3CF_2Cl, and CF_2Cl $CFCl_2$," J. Phys. Chem. 81(3):256-262 (1977).

12. Howard, C.J., and K.M. Evenson. "Rate Constants for the Reactions of OH with CH_4 and Fluorine, Chlorine, and Bromine Substituted Methanes at 296K," J. Chem. Phys. 64(1):197-202 (1976).

13. Perry, R.A., R.Atkinson and J.N. Pitts, Jr. "Rate Constants for the Reaction of OH Radicals with $CHFCl_2$ and CH_3Cl over the Temperature Range 298-423°K, and with CH_2Cl_2 at 298°K," J. Chem. Phys. 64(4):1618-1620 (1976).

14. Crutzen, P.J., and I.S.A. Isaksen, National Oceanic and Atmospheric Administration, Boulder, CO. Private communication (1976).

15. Crutzen, P.J., I.S.A. Isaksen and J.R. McAfee. "The Impact of the Chlorocarbon Industry on the Ozone Layer," J. Geophys. Res. 83 (Cl):345-363 (1978).

16. Lacher, J.R., L.E. Hummel, E.F. Bohmfalk and J.D. Park. "The Near Ultraviolet Absorption Spectra of Some Fluorinated Derivatives of Methane and Ethylene," J. Am. Chem. Soc. 72:5486-5489 (1950).

17. Gordus, A.A., and R.B. Bernstein. "Isotope Effect in Continuous Ultraviolet Absorption Spectra: Methyl Bromide-d_3 and Chloroform-d," J. Chem. Phys. 22(5):790-795 (1954).

18. Tsubomura, H., K. Kimura, K. Kaya, J. Tanaka and S. Nagakura. "Vacuum Ultraviolet Absorption Spectra of Saturated Organic Compounds with Non-bonding Electrons," Bull. Chem. Soc. Japan 37(3): 417-423 (1964).

19. Joshi, S.B., and B. Dimitriades. "Reactivities of Organics Under Pollutant Transport Conditions," Preprints of Papers Presented at the 173rd National Meeting, ACS, Div. Environ. Chem. 17(1):20-22 (1977).

20. Wilson, K.W., and G.J. Doyle. "Investigations of Photochemical Reactivities of Organic Solvents," Stanford Research Institute, Project PSU-8029 (Springfield, VA: National Technical Information Service, U.S. Department of Commerce, September 1970).

21. Dilling, W.L., C.J. Bredeweg, and N.B. Tefertiller. "Organic Photochemistry. Simulated Atmospheric Photodecomposition Rates of Methylene Chloride, 1,1,1-Trichloroethane, Trichloroethylene, Tetrachloroethylene, and Other Compounds," Environ. Sci. Technol. 10 (4):351-356 (1976).

22. Pearson, C.R., and G. McConnell. "Chlorinated C_1 and C_2 Hydrocarbons in the Marine Environment," Proc. Roy. Soc. (London) B189:305-332 (1975).

23. Jaffe, R.J., F.C. Smith, Jr., and K.W. Last. "Study of Factors Affecting Reactions in Environmental Chambers—Final Report on Phase II," Lockheed Missiles and Space Company, Inc., Report No. EPA-650/3-74-004-a (April 1974).

24. Spence, J.W., P.L. Hanst, and B.W. Gay, Jr. "Atmospheric Oxidation of Methyl Chloride, Methylene Chloride, and Chloroform," J. Air Pollut. Control Assoc. 26(10):994-996 (1976).

25. Sanhueza, E., and J. Heicklen. "Chlorine-Atom Sensitized Oxidation of Dichloromethane and Chloromethane," J. Phys. Chem. 79(1):7-11 (1975).

11

ESTIMATING VOLATILIZATION AND WATER COLUMN DIFFUSION RATES OF HYDROPHOBIC CONTAMINANTS

Donald Mackay*, Wan Ying Shiu and
Russell J. Sutherland

Department of Chemical Engineering
and Applied Chemistry
University of Toronto,
Toronto, Ontario M5S 1A4.

INTRODUCTION

It is evident that successful management of environmental systems containing toxic compounds depends on having the capability of quantifying the environmental dynamics of the compound. Only when the relative importance of the various transformations, transport routes and storage compartments can be quantified is it possible to:

1. effect adequate and economic control measures,
2. elucidate the nature and extent of exposure of biota and humans to the compound and
3. predict probable future changes in exposure.

For some compounds, volatilization from water bodies may be a significant environmental pathway and may thus influence exposure. This chapter examines the mechanism of volatilization, as presently conceived; describes mathematical expressions which purport to describe these processes; identifies the major uncertainties in the predictive procedure· and describes

---

*To whom correspondence should be addressed.

and suggests laboratory and field techniques for obtaining the required data.

There have been several descriptions of the nature of air-water exchange processes, and equations have been suggested which permit estimation of the rate of diffusion from near-surface waters to the near-surface atmosphere. Notable has been the work of Liss,[1,2] Liss and Slater,[3] Mackay and Leinonen[4] and Mackay.[5] Examination is made here of the more general problem of quantifying volatilization rates when other processes such as diffusion from sediments or through stratified water columns may contribute significant resistances to the overall transport process. Indeed, in some situations, the overall rate of transfer to the atmosphere may not be significantly influenced by the air-water diffusion rate, since another process rate may control. A mathematical formulation for tackling this more general problem is suggested, and it is shown that the conventional two-resistance (Whitman) approach is a special case of this more general approach.

MECHANISM OF AIR-WATER EXCHANGE

Transport of a compound from the depths of a lake or from sediment involves several sequential stages, each of which has a rate, or diffusion velocity or resistance. Presumably, the overall rate is controlled by the stage with the slowest velocity (i.e., the stage with the greatest resistance). These stages may include (1) release from sediment, (2) diffusion in the hypolimnion, (3) diffusion through the thermocline, (4) diffusion through the epilimnion to the near-surface (ca. 1-mm-deep) region, (5) diffusion to the interface through the liquid surface "stagnant film," (6) transfer across the interface and (7) diffusion through the atmospheric film to the bulk of the atmosphere. Some of these stages can be expressed mathematically, but it is far from possible to predict all rates with acceptable accuracy.

For rivers, it is generally accepted that diffusion from bottom to surface is usually fast because of the interaction of the river current with the bottom. The principal resistance probably lies at the surface.

For shallow lakes, stages 5 and 7 are believed to be the controlling processes for diffusive transfer of organics, since they are slower than vertical diffusion through the epilimnion. They thus control how fast the surface waters are depleted of the contaminant.

Another mechanism, independent of diffusion, is transfer across the interface in association with bulk movement of matter. Bubbles rising through the water column from anaerobic decomposition processes may carry material directly to the surface. Liquid droplets propelled into the atmosphere from bursting bubbles or breaking waves may carry dissolved or particulate material. Precipitation (rain, snow or solid particulate matter) may be a significant method of depositing or redepositing material from the atmosphere. Contaminants may float or sink through the water column in association with live or dead biota, mineral or fecal matter. These processes merit quantification in order that their contribution can be assessed relative to diffusion.

Before any meaningful mathematical description can be advanced, it is essential that the mechanism of the dominant processes be fully understood. There is still doubt about some of these processes, especially the diffusion mechanism near the surface and the role played by surface layers in impeding transfer and accumulating hydrophobic solutes.

CONVENTIONAL TWO-RESISTANCE APPROACH

The conventional approach to quantifying air-water diffusive exchange is to apply the Whitman two-resistance model and express the overall flux in terms of gas and liquid mass transfer coefficients (K_G and K_L) and a Henry's law constant H. The resulting equation for flux N is

$$N = K_{OG} (HC - P)/RT = K_{OL} (C - P/H)$$

where $1/K_{OG} = 1/K_G + H/K_L RT$
or $1/K_{OL} = 1/K_L + RT/HK_G$
and $K_{OL} = H K_{OG}/RT$

The fundamental determinants of flux are thus K_G, K_L and H. R is the gas constant and T is temperature. The terms $1/K_G$ and $H/K_L RT$ are essentially resistances which are summed in series. It is instructive to define the conditions under which one or both resistances control.

The ratio of the liquid and gas resistances is

$$H K_G/RT K_L$$

For typical environmental conditions, K_G is 3000 cm/hr and K_L is 20 cm/hr; thus the ratio K_G/K_L is approximately 100 to 150. The group RT is 0.024 m^3 atm/gmol; thus the above ratio becomes approximately 6000 H where H has units of atm m^3/gmol and is essentially the ratio of vapor pressure (atm) to aqueous solubility (gmol/m^3). For solutes of high H (*i.e.,* > 10^{-3}), such as benzene or alkanes, the liquid phase resistance dominates. For solutes of low H (*i.e.,* < 10^{-4}) such as SO_2, the gas phase resistance dominates. For intermediate cases such as DDT both resistances are significant. It is thus essential to quantify H in order that the dominant resistance be identified.

Interestingly, when H is high and P is near zero the equations reduce to $N = K_L C$, in which the volatilization rate is independent of H. The reason is that the slow diffusion through the near-surface liquid layer controls the overall rate, the subsequent volatilization step whose rate is H-dependent being relatively fast and thus unimportant.

FUGACITY APPROACH

When considering more complex transport processes involving several resistances, it may be more convenient to replace concentration by fugacity. The total process is then considered as diffusion through a series of compartments, each of which has a resistance. Compartments could include sediment, a hypolimnion and a liquid or gas near-surface resistance.

Within a phase or compartment, mass diffuses because of concentration differences, the rate usually being expressed by Fick's first law. Between phases, equilibrium is usually established at different concentrations. For example, oxygen at equilibrium between air and water has a concentration of about 0.3 gmol/m^3 (10 mg/1) in the water and 8.0 gmol/m^3 in the air. Such equilibrium partitioning can be expressed by equating the chemical potential or fugacity of the substance in each phase. Unfortunately, chemical potential is a difficult concept to grasp and use, and it is preferable here to use the much simpler concept of fugacity.

Fugacity has two advantages in environmental science. First, it has units of pressure (say atm) and can thus be regarded as the pressure or "escaping tendency" of a substance from a phase. Second, under many environmental conditions fugacity is linearly proportional to concentration, whereas chemical potential is logarithmically related.

The concern here is concentration rather than fugacity; thus a simple linear relationship is postulated between fugacity (f) and concentration (C) by defining a proportionality constant F such that

$$f = F \, C \text{ (atm)}$$

F thus has units of atm/(gmol/m^3) and is dependent on temperature, pressure, the nature of the substance and the medium in which it is present. It may also be concentration-dependent, but this dependence proves to be slight at the high dilutions normally encountered environmentally.

The physical significance of F is that it quantifies the escaping tendency per unit of concentration. When F is large, a given concentration exerts a large fugacity, whereas when F is low, the same concentration exerts a lower fugacity. The result is that a diffusing pollutant tends to accumulate in regions of low F in which high concentrations can be tolerated. For

example, F for oxygen in water at ambient temperatures is 0.67 atm/
(gmol/m^3), whereas in air it is 0.024 atm/gmol/m^3). The result is that
oxygen adopts a higher concentration in air than in water by a factor of
about 27, the ratio of the two F values.

It is thus necessary to develop the capability of calculating F for a
volatilizing substance for all relevant phases. This proves to be relatively
simple, being merely a reorganization of existing phase equilibrium expressions for each phase. Justification for the equations used is given in
texts on phase equilibrium thermodynamics such as that by Prausnitz.[6]

Vapor Phase (*i.e.*, the atmosphere)

The fugacity is given by

$$f = y \, \phi \, P_T = P$$

where y is the mole fraction, P_T is the total pressure (atm) (here atmospheric pressure) and ϕ is the fugacity coefficient which is dimensionless
and is introduced to account for nonideal behaviour. Fortunately at atmospheric pressure ϕ is usually close to unity and can thus be ignored. The
exceptions are solutes such as carboxylic acids which associate in the
vapor phase. The fugacity is thus equivalent in most cases to the partial
pressure P (atm). It should be noted here that this equation assumes the
solute to be in a truly gaseous form, not associated with particulates.
Now concentration C (gmol/m^3) is related to partial pressure through the
gas law as

$$C = n/V = P/RT = f/RT = f/F$$

Thus F for vapors is simply RT which has a value of about 24 atm/(liter/
gmol) or 0.024 m^3 atm/gmol, corresponding for example to R of 82 x 10^{-6}
m^3 atm/gmol K and T of 293 K or 20°C. F is independent of the nature
of the solute or the composition of the vapor (for nonassociating, low
or atmospheric pressure conditions).

Water Phase

The fugacity is given by

$$f = x \gamma P^s$$

where x is the mole fraction, P^s is the vapor pressure of the pure liquid
solute at the system pressure and γ is the liquid phase activity coefficient
on a Raoult's law basis (not on a Henry's law basis). When x is unity,
γ is unity (by definition) and f becomes the pure component vapor pressure. Generally for non-ionizing substances γ increases as x decreases to

an "infinite dilution" value as x tends to zero. This relationship between x and γ is often of the form

$$\gamma = K (1 - x)^2$$

In most environmental situations x is quite small; thus γ can be equated to K without serious error. This near-constancy in γ leads to the very convenient near-constancy in F. The relationship between f and C to give F can be obtained by writing for infinite dilution conditions

$$F = f/c = P/C = H = fv/x = v\gamma P^s$$

where v is the molar volume (m^3/gmol) of the liquid (usually water). For liquids, F is simply the Henry's law constant H. This is in accord with the definition of H which is the constant by which liquid concentration is multiplied to give the partial pressure P which is here equal to fugacity.

Sorbed Phase

For sorbed material the simplest relation is of the form

$$X = KpC$$

where Kp is a sorption coefficient and X is the concentration in the solid (mols solute/mg dry solids). Kp is then numerically equal to the value in the equation.

X (ppm solute in dry sorbent) = KpC (ppm solute in solution). If the sorbent concentration is S g/m^3 (or ppm), then the concentration of the sorbed material C_s (mol/m^3) becomes

$$C_s = 10^{-6} SX = 10^{-6} SKpC \ mol/m^3$$

Its fugacity must equal that of the dissolved solute

$$f = H \ C = H \ C_s/(10^{-6} \ SKp)$$

or

$$C_s = f(10^{-6} SKp/H)$$

thus F becomes 10^6 H/SKp.

This analysis can be applied also to sorption on atmospheric particulate matter, but caution must be exercised in situations when the sorbed solute is physically trapped or enveloped in the particle. For example, this may occur to polynuclear aromatics formed during combustion and associated with (or inside) soot particles. Such materials are unable to exert their intrinsic fugacity outside the limits of the particle and thus are not in a truly equilibrium situation.

In summary, the values of F are

for vapor phase	RT
for liquid phase	H
for sorbed phase	$10^6 H/S \; Kp$

CALCULATION OF MASS FLUX

Figure 1 gives data for companion steady-state concentration and fugacity profiles for a solute which is diffusing from a sediment through a stratified water column to the atmosphere. The concentration profile has discontinuities at the water surface and the sediment-water interface, whereas the fugacity profile is continuous and monotonic (unless there is mass input or output). The diffusive process can be regarded as a series of resistances (in an electrical sense) with the overall flux (current) being driven by the terminal fugacity (voltage) difference.

For each compartment or region in which the fugacity slope is constant the flux equation can be written in terms of Fick's law as

$$N - D\Delta C/\Delta Z$$

where D is diffusivity, ΔC is concentration difference and ΔZ is diffusion distance. Eliminating ΔC in favour of fugacity gives

$$N = \Delta f D/F \Delta Z = \Delta f/r$$

where r is $\Delta ZF/D$ and is a resistance.

Near the air-water interface it is impossible to measure D or ΔZ accurately, and it is more convenient to combine them into a single parameter K defined as $D/\Delta Z$, the mass transfer coefficient. In such cases r becomes F/K.

There is thus a series of resistances r_1, r_2, r_3, etc. Rearranging the flux equation gives:

$$\Delta f_1 = r_1 N \quad , \quad \Delta f_2 = r_2 N \quad etc$$

But the sum of the Δf terms is the difference between the terminal fugacities Δf_T, for example between sediment and atmosphere, thus

$$\Delta f_T = \Sigma \Delta f_i = N(r_1 + r_2 + r_3 + etc) = N\Sigma r_i$$

or

$$N = \Delta f_T/(\Sigma r_i)$$

If each resistance term can be quantified, the dominant resistance can be identified and the steady-state flux calculated. The individual values of Δf can then be calculated as can the fugacities at the compartment boundaries

ATMOSPHERE

$f \times 10^{12}$ $c \times 10^{12}$
(atm)   (mol/m³)

- 0.100 - - - - - 4.2 - - -

AIR "FILM"   **WATER**
                           0.106   4.17
                SURFACE   35.3

WATER "FILM"
- 0.283 - - - - 94.3 - - -

EPILIMNION

- 0.?90   96.7

HYPOLIMNION

//// **SEDIMENT** //////////// 1.000 —— 333 ////
                                            10^{+6}

Figure 1. Schematic diagram of water column with fugacities and concentrations for case 2.

Interestingly, if this analysis is applied to the liquid and vapor "film" resistances r_L and r_G it can be shown that the total resistance r_T becomes

$$r_T = r_L + r_G = F_G/K_G + F_L/K_L$$

But F_G is RT and F_L is H; thus r_T becomes $(RT/K_G + H/K_L)$ and the flux is

$$N = (f_L - f_G)/r_T = (HC - P)/(RT/K_G + H/K_L) = K_{OG}(HC - P)/RT$$

where K_{OG} is the overall gas phase mass transfer coefficient defined earlier in the conventional Whitman two-resistance approach. If $r_L \gg r_G$ and f_G is small, the flux becomes as derived earlier

$$N = f_L/r_L = CH/(F_L/K_L) = K_L C$$

That approach is simply a particular case of the more general multiple resistance model derived here. The advantage of the general model is that it can identify and quantify situations in which the dominant resistances lie elsewhere than in the gas or liquid films.

THE ROLE OF SUSPENDED SOLIDS IN WATER COLUMN TRANSPORT

If the diffusing solute is present in both dissolved and sorbed state in the water column, the calculation of the resistance and hence mass flux is more complex. We treat first the situation in which the suspended matter has a negligible settling (or rising) velocity and thus diffuses at a velocity equal to that of the dissolved solute. It can be assumed that equilibrium (equal fugacities) exists between sorbed and dissolved state. Introducing separate values for F, C, r and N for sorbed (subscript S), dissolved (subscript D) and total (subscript T) states leads to the following equations.

$$C_S = f/F_S \qquad\qquad C_D = f/F_D$$

thus $C_T = C_S + C_D = f(1/F_S + 1/F_D) = f/F_T$

thus $1/F_T = 1/F_S + 1/F_D$

and since the values of ΔZ and D are common,

$$1/r_T = 1/r_S + 1/r_D$$

and

$$N_T = \Delta f/r_T = N_S + N_D = \Delta f/r_S + \Delta f/r_D$$

The total resistance and F value is thus less than that of the lower resistance or F value reflecting the parallel nature of the resistances in an electrical sense. The overall transfer rate N_T will thus tend to be controlled by the lower resistance (high F) state. The ratio of resistances is the ratio of the individual F values and is the inverse of the individual concentrations.

Sorbents will usually increase the overall transfer rate through a series of resistances by reducing the resistances of the sorbent-containing phases. If, however, all the phases contain sorbent there will be no net effect since although the resistances are reduced, the fugacities are also reduced.

Second is the situation in which the suspended matter diffuses and also has a falling (or rising) velocity with respect to the water column. The net velocity of the sorbed material is then the algebraic sum of the diffusion and particle velocities. It can be shown that the diffusion velocity V (m/s) is equivalent to (N/C). In the first case the velocities N_S/C_S, N_D/C_D and N_T/C_T are all equal to $\Delta fD/f\Delta Z$ or $\Delta fK/f$. In this case, if the mean particle velocity is V_P the sorbed state velocity becomes $(N_S/C_S + V_P)$ The total sorbed state mass flux is thus,

$$(\Delta fD/F\Delta Z + VC_s) = (\Delta f/r_s + VC_s)$$

and the total mass flux becomes

$$N_T = VC_s + \Delta f (1/r_s + 1/r_D)$$

This quantity is an algebraic sum and may be positive or negative depending on the direction of diffusion and of particle movement.

An interesting possibility is that the downward deposition velocity of the sorbed material equals or exceeds the upward diffusive velocity. It would then be possible for a fugacity gradient to exist in which there is no net transport or even in which transport is in the direction opposite to the gradient. This could occur in estuaries of rivers with high sediment loading in which the effect would be to "scavenge" the water column of the solute.

Comparison of the magnitude and sign of the three terms VC_s, $\Delta f/r_s$ and $\Delta f/r_D$ readily establishes the relative roles of sedimentation, diffusion of sorbed solute and diffusion of dissolved solute in contributing to the overall transport of the solute.

SORPTION-DESORPTION KINETICS

In this analysis it is assumed that equilibrium exists between the dissolved and sorbed forms, but this may not always apply. Sorption experiments generally show a fairly fast initial "outer surface" sorption (half-life of minutes) superimposed on a slower penetration of solute into intersticies (half-life of hours). Desorption is similar with fast "outer surface" desorption with slower release of the solute trapped more deeply in pores. These rates are of importance in volatilization since they may control whether or not the sorbed solute is volatilized from an element of water which eddies to the surface and remains there for a short time. If the "volatilization exposure half-life" at the surface is much shorter than the desorption time it can be assumed that only the dissolved solute will volatilize during exposure; thus, the driving force for volatilization is derived only from the dissolved state. If desorption is fast the sorbed material could also be volatilized after having been desorbed. Unsteady-state theory suggests that the mass transfer coefficient K_L will be related to diffusivity D and exposure time t by an equation of the form

$$K_L = \sqrt{4D/\pi t}$$

A typical diffusivity is 10^{-5} cm^2/sec. and typically K_L is 0.003 cm/sec; thus the exposure time t is typically 1.4 sec. It seems unlikely that appreciable desorption will occur during this short period. Under quiescent conditions when K_L is smaller, i.e., 0.001 cm/sec, t will be approximately 13 sec and some desorption may occur.

The general conclusions are that usually only the dissolved solute will volatilize at the surface with the sorbed material desorbing later in the bulk of the water column to establish a new equilibrium, and the sorbed material will probably not influence the transfer rate in the water film; i.e., it will not lower the resistance as may occur in the water column.

Kinetics may also influence the behavior of solids falling through the water column. If the falling velocity is low, it is likely that the sorbent will maintain close to its equilibrium amount of sorbate, whereas if the velocity is high (as may occur with mineral particles), there may be insufficient time for equilibrium to be reached.

The role of sedimenting sorbents is thus usually to act counter to upward diffusion (or to enhance downward diffusion) at a rate depending on sorbed concentration and settling velocity and to some extent on the sorption-desorption kinetics.

APPLICATION TO WATER COLUMN TRANSPORT

To illustrate this approach four hypothetical cases are examined involving partition and transport of a hydrophobic pollutant from sediment to water column to atmosphere. First is examination of an equilibrium (zero transport) system. Second is transport through a sorbent-free system. Third is transport through two systems with differing amounts of nonsettling sorbent. The fourth case examines the effect of adding a sorbent-settling velocity to the third. Although hypothetical, the cases are based on data for PCB transport from a pond reported by Paris, Steen & Baughman.[7] The pollutant properties are similar to an Aroclor 1242 ($C_{12}H_7Cl_3$) and are given below.

Molecular Wt	300 g/mol
Aqueous solubility	0.04 mg/l (1.3×10^{-4} mol/m^3)
Vapor Pressure	3×10^{-4} mm Hg (4×10^{-7} atm)
Henry's constant	3×10^{-3} atm m^3/mol
Kp on suspended sorbent	5000 (mol/g sorbent)(mol/g water)
Kp on sediment	1000 (mol/g sorbent)(mol/g water)

The system, illustrated in Figure 1, consists first of sediment which is assumed to be in equilibrium with the water immediately above it. This assumption is probably often invalid since it is likely that diffusion from and within the sediment will be rate controlling, but for simplicity this effect is ignored here. A 5-m hypolimnion with a vertical eddy diffusivity of 0.05 m^2/hr is overlain by a more mixed epilimnion 2 m deep with a diffusivity of 2 m^2/hr. Liquid and gas films contribute resistances corresponding to mass transfer coefficients of 0.04 and 10 m/hr respectively. The interface and the bulk of the atmosphere contribute negligible resistances. Two seston concentrations of 10 and 300 g/m^3 (or ppm) are considered, the concentrations being homogeneous throughout the entire water column. The sediment has a density of 3×10^6 g/m^3.

Case 1. Equilibrium

If a fugacity f of 10^{-12} atm is assumed to apply throughout the entire system, the concentrations can be readily calculated for each compartment as $C_i = f/F_i$. The values obtained are given in Table 1. The dissolved concentrations (mol/m^3) in the water column are about a factor of 8 higher than those in the atmosphere and a factor of 3000 less than those in the sediment. At a seston concentration of 10 ppm, most (95%) of the pollutant is in dissolved form, whereas at 300 ppm only 40% of sorbed concentration is in dissolved form. These values are consistent with those of Paris et al.[7] and illustrate the importance of sorbent concentration in controlling partition between dissolved and sorbed state. In both cases the concentration of pollutant is 5×10^{-4} $\mu g/g$ sorbent. This is higher by a factor of 5000 (i.e., Kp) than the water column concentration of 10^{-7} $\mu g/g$.

Case 2. Transport with No Sorption

If it is assumed that the atmospheric fugacity falls to 0.1×10^{-12} atm and that the sediment fugacity is unchanged, then a steady-state fugacity gradient develops and a mass flux is established dependent on the total resistance. Calculation of the compartment resistance is given in Table I and the resultant fugacity and concentration profiles are presented in Figure 1. The total flux is the total fugacity difference from sediment to atmosphere (0.9×10^{-12} atm) divided by the total resistance (0.3804), namely, 2.37×10^{-12} $mol/m^2 hr$. The dominance of the hypolimnion resistance is apparent. It is thus in the hypolimnion that most of the fugacity gradient is dissipated. Considering only the surface waters, most of the resistance lies in the liquid film with slight resistances occurring in the gas film and in the mixed surface waters. This latter condition is probably typical of river transport.

It is interesting to calculate the "half-life" or residence time of the pollutant from these waters. Considering only the surface waters (epilimnion), the average concentration is approximately 10^{-10} mol/m^3; thus, the average residence time calculated as (depth x concentration/flux) will be (2×10^{-10}/ 2.37×10^{-12}) or 84 hr. The corresponding hypolimnion residence time is 453 hr (19 days) as a result of the greater depth and higher concentrations. If the sediment is considered to be 10 cm deep, thus containing 10^{-7} mol/m^2, it would be able to support the above flux for 42,000 hr or 4.8 yr. In practice, of course, the flux would drop as the concentration drops with a sediment half-life of 3.3 yr. This illustrates the well known capacity of sediments to accumulate large quantities of toxic substances and then release them slowly over a period of years. The general picture which

Table I. F Values, Equilibrium Concentrations and Diffusion Resistances

	Equilibrium Concentration (f = 10^{-12} atm)			Individual Resistances				Total Resistances at Seston Concentrations		
	F	mol/m^3	ng/m^3	D (m^2/h)	Δz (m)	K (m/hr)	r	(0 ppm)	(10 ppm)	(300 ppm)
Atmosphere	0.024	4.2×10^{-11}	12.5	–	–	–	0	–	–	–
Air Film	0.024	4.2×10^{-11}	12.5	–	–	10	0.0024	0.0024	0.0024	0.0024
Water Film	0.003	3.3×10^{-10}	100	–	–	0.04	0.075	0.075	0.075	0.075
Epilimnion:										
Seston (0 ppm)	0.003	3.3×10^{-10}	100	2	2	–	0.003	0.003	–	–
Seston (10 ppm)	0.06	1.67×10^{-11}	5	2	2	–	0.06	–	0.0029	–
Seston (300 ppm)	0.002	5×10^{-10}	150	2	2	–	0.002	–	–	0.0012
Hypolimnion:										
Seston (0 ppm)	0.003	3.3×10^{-10}	100	5×10^{-2}	5	–	0.300	0.300	–	–
Seston (10 ppm)	0.06	1.67×10^{-11}	5	5×10^{-2}	5	–	6.0	–	0.286	–
Seston (300 ppm)	0.002	5×10^{-11}	150	5×10^{-2}	5	–	0.2	–	–	0.120
Sediment	10^{-6}	10^{-6}	300,000	–	–	–		–	–	–
Total								0.3804	0.3664	0.1986
Mass Flux (mol/m^2 h)								2.37×10^{-12}	2.46×10^{-12}	4.54×10^{-12}

emerges is that the pollutant is released slowly from the sediment, in which the residence time is several years, at a rate largely controlled by the hypolimnion, in which the residence time is 19 days. Once the pollutant reaches the surface waters, the residence time is only a few days before volatilization occurs. An implication of these findings is that sampling surface waters for pollutants may be quite misleading since concentrations will tend to be variable and considerably lower than those at greater depths.

Case 3. Transport with Sorption

Considering the two levels of sorbent, the various values of F and r can be calculated using the previously derived equations as shown in Table I. Because of kinetic considerations, sorption is assumed to have no effect on water film resistance. As expected, the other resistances fall, negligibly at 10 ppm but substantially at 300 ppm, the fall in resistance being dependent on the fraction of the substance sorbed. In the 10-ppm case the overall mass flux increases by 4%, whereas in the 300-ppm case the flux nearly doubles to 4.54×10^{-12} mol/m^{2}hr. The reason for this increase is that in the hypolimnion the higher total concentrations permit a greater mass flux for a given diffusive velocity. The residence times are affected relatively little since both flux and concentration increase. It can be concluded that sorption is likely to affect transport rates significantly only if an appreciable fraction of the total amount of pollutant present is in sorbed form and if the compartment in which this occurs is rate controlling.

Case 4. Transport with Sorption and Deposition

The simplest approach to determining if superimposing a settling velocity appreciably affects transport is to examine the relative magnitude of the term VC_s and the diffusive mass flux.

At the 10-ppm sorbent level the sorbed concentration is approximately 10^{-11} mol/m^3 and the mass flux is 2.46×10^{-12} mol/m^{2}hr. To cancel the upward diffusion exactly would require a falling velocity of 0.246 m/hr. At such a velocity it is likely that near-equilibrium would be established as the particles fall. At the 300-ppm level the sorbed concentration is approximately 30×10^{-11} mol/m^3 and the mass flux is higher at 4.54×10^{-12} mol/m^{2}hr. The canceling velocity is thus 0.015 m/hr, which is also so slow that equilibrium would be achieved during the fall.

It can be concluded that if the sorbent deposition rate is appreciable, corresponding to 0.25 m/hr at 10 g/m^3 (*i.e.*, 2.5 g/m^{2}hr) or 0.015 m/hr at 300 g/m^3 (i.e., 2.0 g/m^2 hr), this will be sufficient to cancel the upward diffusion in the water column completely. A rate of 2 g/m^2 hr corresponds to a sediment deposition of 1 mm of material of density 1.0 g/cm^3 in 20 days. In highly productive or eutrophic waters it is thus likely that organic matter could effectively scavenge the water column of the toxic substance.

CONCLUSIONS

An attempt has been made to establish a framework for undertaking calculations of the vertical dynamics of solutes in the water column and into the atmosphere. If the various compartments and regions of transport resistance can be assigned resistances in consistent units, the overall calculation of transport rate becomes considerably simpler and the dominant resistances can be readily identified. Obviously, the key parameters in any such calculation are the physical-chemical properties of the solute, especially its sorption and air-water partition properties, the kinetic terms such as transfer coefficients and diffusivities and the sedimentation velocity of the suspended solids. The role of sorption in modifying transport rates is quite complex in that by tending to increase concentrations it lowers resistances and increases diffusion rates. This effect may also be accompanied by a depositing effect, the magnitude of which may equal or exceed the diffusive rate.

It is suggested that progress toward understanding the environmental dynamics of toxic substances can only be made by taking measurements from the environment and interpreting them with some physically reasonable, quantitative model of environmental processes. It is hoped that the approach suggested here may assist in the formulation of such models.

REFERENCES

1. Liss, P.S. "Processes of Gas Exchange Across an Air-Water Interface," *Deep Sea Res.* 20:221-238 (1973).
2. Liss, P.S. "The Exchange of Gases Across Lake Surfaces," Proc. First Spec. Conf. on Atmos. Contribution to the Chemistry of Lake Waters Int. Assoc. Great Lakes Res. (1975), p. 88.
3. Liss, P.S., and P.G. Slater, "Flux of Gases Across the Air-Sea Interface," *Nature* 247:181-184 (1974).
4. Mackay, D., and P.J. Leinonen. "Rate of Evaporation of Low Solubility Contaminants from Water Bodies to the Atmosphere," *Envir. Sci. Technol.* 9:1178 (1975).
5. Mackay, D. "Volatilization of Pollutants from Water". *Proc. 2nd Intnl. Symp. on Aquatic Pollutants* (Amsterdam: Pergamon Press, 1977), p. 175.

6. Prausnitz, J.M. *Molecular Thermodynamics of Fluid Phase Equilibrium.* (New York: Prentice Hall, Inc., 1972).
7. Paris, D.R., Steen, W.C. and Baughman, G.L. "Role of Physical-Chemical Properties of Aroclors 1016 and 1242 in Determining Their Fate and Transport in Aquatic Environments," *Chemosphere* 4:319-325 (1978).

12

ASSESSMENT OF THE VAPOR BEHAVIOR OF
TOXIC ORGANIC CHEMICALS

William F. Spencer

USDA Science and Education Administration
Department of Soil and Environmental Sciences
University of California
Riverside, California 92521

Walter J. Farmer

Department of Soil and Environmental Sciences
University of California
Riverside, California 92521

INTRODUCTION

Our ability to predict the behavior of organic chemicals in the environment has fallen short of our needs. Pesticide behavior has received the most attention, and the ubiquitous distribution of pesticides in the environment has alerted the scientific community to the importance of developing predictive capabilities and methods for controlling the distribution of other potentially toxic chemicals.[1] Air transport is probably the principal method of their dispersion over wide areas and into bodies of water far removed from their manufacture, use or disposal.

Residues of organic chemicals can enter the environment during the manufacturing or use process or following disposal of wastes or consumable goods containing the chemical. Such wastes may be applied to land as sewage sludge, secondary effluents or other forms of organic wastes and subsequently mixed into the soil. Also, many potentially hazardous wastes are disposed of in landfills, which may be either covered with uncontaminated soil or left uncovered and in so-called "acid pits" or lagoons in

143

which the organic chemical is dissolved in, or stored under, water and the water is allowed to evaporate before final disposal of the chemical.[2,3] Organic pollutants enter the atmosphere principally by volatilization, by direct injection, such as in pesticide sprays or from stacks of industrial plants, and by movement of wind-blown dust particles. The soil obviously becomes a residence for a large portion of the chemicals that find their way into the environment, either by inadvertent spillage or intentional application to the land, or by release into the air. One management requirement for the use of land as a waste disposal medium is the ability to control volatilization of the chemicals in the waste, whether in landfill or mixed into the soil after application.

This paper discusses transfer of toxic organic chemicals into the atmosphere—the mechanisms involved, factors influencing rates of vapor transfer and procedures for determining vapor pressure and vapor loss rates. It emphasizes the importance of reliable values for physicochemical properties in developing predictive models for estimating fate of the chemical in the environment and presents information on vapor loss rates from surface deposits, from soil-incorporated chemicals and from simulated landfills in laboratory and field studies to demonstrate the importance of volatilization in the dissipation of chemicals from solid and liquid surfaces. Vaporization and atmospheric transport of pesticides is emphasized, but the same principles apply to other organic pollutants.

Postapplication volatilization of pesticides under field conditions was thoroughly reviewed by Taylor.[4] The reader is also referred to other recent reviews on pesticide volatilization by Spencer *et al.*,[5] Hamaker,[6] Wheatley,[7] Guenzi and Beard[8] and Plimmer.[9]

PHYSICOCHEMICAL PROPERTIES AND VAPOR BEHAVIOR

Potential volatility of a chemical is related to its inherent vapor pressure, but actual vaporization rates will depend on environmental conditions and all factors that attenuate the effective vapor pressure or behavior of the chemical at a solid-air or liquid-air interface. Vaporization from surface deposits depends only on the vapor pressure of the chemical and its rate of movement away from the evaporating surface. Vaporization from aqueous systems depends not only on the vapor pressure of the chemical, but also on its water solubility; vaporization from soil is controlled by solubility and adsorption as well as vapor pressure. Consequently, no single physicochemical property can describe and predict the probable vapor behavior and fate of a chemical in the environment or its likely method of transport in the atmosphere. However, relative vaporization rates useful for environmental indices can be calculated from basic physical properties of vapor pressure, water solubility, adsorption and persistence, if reliable values are known for each of these properties at various temperatures.

The vapor pressures of many organic chemicals of environmental interest increase three- to fourfold for each 10°C increase in temperature. Consequently, reliable values for vapor pressures at various temperatures are necessary to estimate vapor losses of the chemical from surface deposits, to predict their partitioning between soil, water and air in order to predict volatility from water solutions and from wet soils, and to calculate atmospheric residence times of chemicals in droplets and aerosols.

A search of available literature on the vapor pressure of 49 commonly used pesticides indicated a lack of vapor pressure data on many of the pesticides and a wide variability in vapor pressure of the same compound reported by different authors.[10] No vapor pressure data could be found for 18 of the 49 pesticides. Vapor pressures were found only in compilations of vapor pressure data or as quoted vapor pressure values, without reference to the source of information or methods used to determine them, for 19 of the pesticides. Vapor pressure data with the method of measurement reported were found for only 12 of the 49 pesticides, and, of these 12, the vapor pressures of only 4, i.e., trifluralin, lindane, dieldrin and parathion, were determined by more than one author. There was little agreement among the reported values for vapor pressures for three of these four pesticides. For example, vapor pressures of lindane at 20°C were reported by five authors, with the highest value 3,000 times that of the lowest one; dieldrin vapor pressures reported by six authors varied 16-fold (Table I); parathion vapor pressures reported by four authors varied 9-fold; trifluralin vapor pressures reported by three authors were in fairly good agreement. For each of the compounds, the most recently determined values using a specific method for analytically determining the pesticide agreed fairly well as illustrated by the last three values for dieldrin in Table I.

Volatility from surface deposits is directly proportional to vapor pressure; consequently, predicted vaporization rate of dieldrin could differ 16-fold, depending upon whether the lowest or highest vapor pressure value reported for dieldrin is used. Furthermore, when more than one physicochemical property is combined to estimate relative volatility, prediction errors are greatly magnified. For example, volatility from water surfaces is proportional to vapor pressure and solubility, whereas volatility from wet soils is proportional to vapor pressure, solubility and the adsorption coefficient. Assuming a reported solubility difference of 0.1 to 0.25 ppm for dieldrin and a difference in adsorption coefficient of between 25 and 100 depending upon the soil and method of measurement, the

Table I. Variability in the Reported Values for Vapor Pressure of Dieldrin at $20°C$

Vapor Pressure (mm Hg)	Reference
1.78×10^{-7}	11
2×10^{-7}	12
7.78×10^{-7}	13
1.9×10^{-6}	14
2.6×10^{-6}	15
2.9×10^{-6}	16

predicted volatility could vary 40-fold from water and 160-fold from wet soil, depending upon which values of the parameters were used in the calculation. Such calculations illustrate the importance of reliable values for all physicochemical properties to be used in predicting the fate of chemicals in the environment.

The variability in reported vapor pressure values attests to the need for accurately determining vapor pressures of pesticides and other toxic organic chemicals by a standard procedure, preferably one that relies upon a specific analytical method to eliminate impurity effects. The gas saturation method has proved a reliable method of measuring vapor pressures of pesticides[15,17,18] and should be equally reliable for other organic chemicals with vapor pressures in the range of those of most pesticides. When the method recently was used to measure the vapor pressure of technical grade and purified ethyl and methyl parathion, no significant differences in vapor pressure were observed between technical grade or purified materials of either chemical.[17] The observed vapor pressures that indicated methyl parathion is approximately twice as volatile as ethyl parathion are consistent with the relative persistence of these pesticides on foliage.[19,20] In contrast, the most frequently quoted value for vapor pressure of parathion[21] is about four times higher than that of methyl parathion.[22]

Measurements of vapor loss rates from various surfaces under the same conditions to establish the relationship between vapor loss rate and vapor pressure are also helpful in comparing relative volatility of potentially toxic chemicals. This can be done by a procedure such as that described by Spencer et al.[17] or by Guckel et al.[22]

RELATIVE VOLATILITY FROM INERT SURFACES

Theoretically, vaporization of a chemical from a complete surface of constant area is independent of the depth of the chemical layer, and under constant conditions the loss rate will be constant until so little substance remains that it no longer can cover the surface. According to Hartley,[23] the evaporation rate of a substance is determined solely by its vapor pressure, or vapor density, and its rate of diffusion through the air closely surrounding the substance. Volatilization from deposits on inert surfaces or from any accumulation of a chemical, such as in disposal sites, storage areas, or manufacturing sites is controlled by the saturation vapor density, or vapor pressure, of the chemical and its rate of movement away from the surface. The rate of movement away from the evaporating surface is diffusion controlled. Air movement is relatively nil close to the evaporating surface and the vaporized substance is transported from the surface through this stagnant air layer only by molecular diffusion. Since molecular diffusion coefficients of organic compounds in air are inversely proportional to the square root of molecular weight, the actual rate of mass transfer by molecular diffusion will be proportional to $p(M)^{1/2}$. Thus, the vapor flux of a chemical from an inert surface can be described by Equation 1.

$$J = kp(M)^{1/2} \tag{1}$$

where:

- J = vapor flux,
- p = vapor pressure,
- M = molecular weight, and
- k = a proportionality constant.

The magnitude of k depends mainly on external conditions that control air exchange rates near the surface, such as geometry of volatilization chambers, surface roughness, wind speed, etc. The rate of movement over the evaporating surface is important in controlling evaporation since the depth of the stagnant air layer depends on the air flow rate. Farmer and Letey[24] confirmed the applicability of Equation 1 in predicting relative volatilization rates of several pesticides under controlled laboratory conditions when the pesticides were applied at high rates to inert sand. They indicated that a k value determined for one compound could be used in predicting volatilization rates of other chemicals under the same conditions.

To eliminate adsorption effects when comparing relative volatility of chemicals, Spencer et al.[17] used a procedure whereby a sufficient depth of the chemical was maintained on a glass surface so that the chemical was vaporizing from its own surface throughout the measurement period. They used an air flow rate of 1 liter/min through a rectangular volatilization chamber, which provided an average wind speed across the vaporizing surface of 1 km/hr and a change of atmosphere in the head space above the surface of 167 times/min. Table II shows vaporization rates of several pesticide chemicals measured with their technique and k values calculated

Table II. Relationship Between Vapor Pressures of Various Pesticides and Their Vaporization Rates from Inert Surfaces as Affected by External Conditions of Measurements

Chemical	Temp (°C)	Vapor Pressure (mm Hg)	M	Air Flow Rate (liter/min)	Head Space Over Surface (ml)	Flux ($\mu g/cm^2/hr$)	k^a	Reference
Parathion	25	0.942×10^{-5}	291	1.0	6	0.210	1304	17
Methyl Parathion	25	1.80×10^{-5}	263	1.0	6	0.390	1363	17
Trialate	25	1.03×10^{-4}	305	1.0	6	5.71	1696	25
Dieldrin	25	5.07×10^{-6}	381	1.0	6^b	0.133	1338	Spencer et al (Unpublished)
Mean				1.0			1425	
p,p'-DDT	30	7.26×10^{-7}	354	0.48	24	0.004	283	24
p,p-DDE	30	6.49×10^{-6}	320	0.48	24	0.034	201	24
Lindane	30	1.28×10^{-4}	291	0.48	24	0.596	273	24
Trifluralin	30	2.42×10^{-4}	335	0.48	24	1.292	292	24
Dieldrin	30	1.0×10^{-5}	381	0.48	24	0.051	263	24
Mean				0.48	24^c		280	
Dieldrin	30	1.0×10^{-5}	381	0.12	214^c	0.020	110	24
Dieldrin	30	1.0×10^{-5}	381	0.48	214^c	0.041	211	24
Dieldrin	30	1.0×10^{-5}	381	0.96	214^c	0.063	320	24
		Field Conditions^d				1.70	17,180	26
HCB	25	1.91×10^{-5}	285	0.77	6^b	0.363	1,124	27
Trifluralin	30	2.42×10^{-4}	335	0.80	36^e	1.67	376	28

a Vaporization rate constant calculated from Equation $1 - J = kp(M)^{1/2}$.
b Rectangular volatilization chamber with chemical on glass surface or uncovered.
c Chemical on treated sand at 1000 μg/g.
d Dieldrin applied to grass pasture at 5.6 kg/ha. Calculations based on flux during first 2-hr sampling period after application.
e Cylindrical glass chamber with chemical on surface of wet soil.

using Equation 1, along with vaporization rates and calculated k values
for measurements made under other laboratory conditions and in the field.
The data indicate some of the factors affecting vaporization rates under
controlled conditions and illustrate the higher loss rate usually observed
in the field where the air exchange rate is much greater than in the laboratory.

The volatilization rate constant, k, in Table II varied from approximately 100 to 1700 under laboratory conditions, but exceeded 17,000 for field
measurement of dieldrin volatilization during the first 2 hours after application of 5.6 kg/ha dieldrin to grass pasture.[26] Since Equation 1 is valid
only for comparing relative volatility of chemicals vaporizing from complete surfaces of the chemical, it probably holds only for a very short
period after application of materials, such as pesticides to foliage. Pesticides
are usually applied at rates of less than 10 kg/ha, which results in either
incomplete surface coverage or a complete surface that can be maintained
only for minutes after application. Consequently, Equation 1 would not be
expected to predict relative volatilization rates of pesticides or other chemicals applied at these low dosages for extended periods after application.
Vaporization rates are extremely high compared to the amounts applied,
even for chemicals such as dieldrin with a vapor pressure of approximately
5×10^{-6} at 25°C. For example, Taylor et al.[26] could not account for
82% of the heptachlor and 66% of the dieldrin applied at 5.6 kg/ha only
3 hr after application.

We think it is safe to conclude that any accumulation of organic chemicals, such as in disposal dumps, storage areas, or manufacturing or use
sites that are exposed to the atmosphere, sunlight and other environmental
extremes, will vaporize at a rate which can be roughly approximated by
using Equation 1 and a k value calculated from field measurements, such
as those of Taylor et al.,[26] during the first 2 hr after application. To re-
iterate, the k value will depend almost entirely on air exchange rate and
other environmental factors controlling movement away from the surface,
and greater or lesser values of k should be used for such calculations depending upon the expected conditions relative to those reported by Taylor
et al.[26]

For surface deposits, the decrease in volatilization rate when the chemical no longer covers the entire surface is influenced by many factors. The
rate begins decreasing as soon as the layer becomes thin and discontinuous.
As the exposed area of chemical progressively becomes less, the loss rate
would be expected to follow some type of exponential curve. This was
shown to be the case by Phillips,[29] who reported that decreasing vaporization rate can be expressed by the relatively simple exponential equation

$$y = Ae^{-kt} \qquad (2)$$

which indicates the loss rate of the chemical at any time, t, is proportional to the amount, y, remaining on the surface.

VAPOR LOSS FROM WATER

Relative volatility of various organic chemicals from water can be estimated from the basic physical properties of vapor pressure and water solubility.[6] Mackay and Leinonen[30] extended this concept by using liquid and vapor phase mass transfer coefficients to estimate the absolute vaporization rate of low-solubility organic contaminants from water bodies. They based their calculations on a physical model of well-mixed air and water phases separated by an interface with nearby stagnant films of air and water on either side in which most of the resistance to mass transfer would occur. Whether the liquid or vapor phase resistance controls the vapor loss rate depends on the water-air partition ratio of each compound calculated from its vapor pressure and solubility. Calculations indicated that most organic compounds of low solubility will vaporize rapidly from water in the absence of clay or organic colloids. Many of the well-known organic contaminants had half lives of less than 12 hr from a 1-m depth or water.[30] Such estimates require accurate values of vapor pressure and solubility to prevent large errors in predicted vaporization rates.

VAPOR LOSS FROM SOIL

Vaporization of chemicals following applications to soil can be predicted from considerations of the physical and chemical factors controlling concentrations at the soil surface.[5,24] Chemicals present on the soil surface at high concentrations, such as might result from inadvertent spills, intentional dumping or even pesticides sprayed on the soil surface, will result in vaporization losses approaching those expected from inert surfaces until the concentration is reduced below that which would result in a saturated vapor density of the chemical at the soil surface. However, volatilization rate of chemicals deposited on the soil surface usually will be at a reduced rate dependent upon the degree of vapor pressure reduction from adsorption on the surface. Adsorption reduces the chemical activity, or fugacity, below that of the pure compound; this is then reflected in decreases in vapor pressure of the chemical.

Spencer and Cliath[15] developed a method for measuring vapor pressure of pesticides in soil that enabled them to evaluate directly the effects of soil on potential volatility of individual organic chemicals. They found the magnitude of the adsorption effect, or reduction in vapor pressure, depended mainly on the nature and concentration of the chemical, the soil water content, and soil properties such as organic matter and clay content.[5] The vapor pressure of weakly polar compounds in soil increased greatly with increases in concentration and temperature, but decreased markedly when the soil water content decreased below about 1 molecular layer of water.[28,31-33] They found soil-water content effects were especially important, and their data indicated that greater vaporization of relatively nonpolar organic chemicals from wet than from dry soils was due mainly to an increase in vapor pressure resulting from displacement of the chemical from the soil surface by water. Volatility would undoubtedly increase for all weakly polar compounds with which water can compete for adsorption sites on any type of a sorbing surface.

Spencer and Cliath[32] compared vapor phase desorption of lindane from Gila silt loam with desorption in 1:5 soil:water suspensions. Desorption isotherms, relating adsorbed lindane to relative vapor density and to relative solution concentration, were described by the same line. This means that Henry's law can be used to predict pesticide behavior in the soil system, and that the same amount of adsorbed pesticide is needed for any degree of saturation of solution as for saturation of vapor. Hence, soil-water desorption isotherms can be used to calculate relative vapor densities in the soil atmosphere or, conversely, relative vapor densities can be used to calculate soil-water desorption isotherms. Assuming that vapor pressure and solubility of a pesticide varies the same with temperature in soil as without soil, the minimum data needed to evaluate relative volatility in soil are vapor pressure and solubility of the pesticide itself at various temperatures and desorption isotherms relating soil pesticide concentrations to either vapor densities (d) or soil solution concentrations (c). From the known relationship between vapor pressure, solubility, and temperature, a desorption isotherm at one temperature would be adequate for predicting (d) and (c) at other temperatures. In this way, the comparison of pesticide behavior based on vapor pressure measurements was extended to yield estimates of relative volatility from water or volatility from moist soil by Hamaker[6] and Goring.[34]

Incorporation of an organic chemical into the soil decreases its concentration at the evaporating surface, thereby greatly reducing volatilization rate. When a chemical is mixed into the soil, it must desorb from the soil and move upward to the soil surface before it can vaporize into the atmosphere. Initially, the volatilization rate is a function of the vapor pressure of the chemical as modified by adsorption on the soil surface. Volatilization rate decreases rapidly as the concentration at the surface is depleted and soon becomes dependent upon rate of movement of the chemical to the soil surface.[35-37] Chemicals move to the evaporating surface by diffusion and by mass flow in evaporating water. In the field, water and the chemical usually vaporize at the same time, unless the chemical is essentially water insoluble or has a sufficiently high vapor pressure to result mostly in vapor phase movement through the soil pores rather than movement within the soil solution.

In the absence of evaporating water, volatilization rate depends on rate of movement of the chemical to the soil surface by diffusion. If diffusion coefficients for the chemical in the soil are known, diffusion equations can be used to predict changes in concentration of the chemical within the soil and its loss rate at the soil surface. Mayer et al.[38] and Farmer and Letey[24] obtained good agreement between volatilization rates predicted from mathematical models using diffusion coefficients and those observed in the absence of evaporating water.

When water evaporates from the soil surface, the resulting suction gradient causes appreciable water to move upward to replace that evaporated, and any chemical in the soil solution will move upward by mass flow in the water. Spencer and Cliath[35] demonstrated that this phenomenon, called the wick effect by Hartley,[23] accelerated volatilization of lindane and dieldrin. Figure 1 shows lindane vaporization rate with and without water loss from the soil surface. During periods of 100% humidity, the movement of lindane to the soil surface was controlled by diffusion; at relative humidities of less than 100%, water was evaporating and the chemical was moving to the evaporating surface by mass flow in the water. Data obtained for both dieldrin and lindane indicated that vaporization rates due to the mass flow effect can be estimated from the relationship:

$$J_p = J_w(c) \qquad (3)$$

where:

J_p = the pesticide flux,
J_w = the water flux, and
c = the concentration of the chemical in the soil water.

To use such a relationship to estimate loss of an organic chemical due to the mass flow effect, data for the rate of water loss are needed as well as adsorption coefficients relating chemical concentrations in the soil to soil solution concentrations. The magnitude of the wick effect will depend on the adsorption characteristics and water solubility of the chemical and other factors affecting partitioning between the water, air, and solid phases in the soil. For example, with lindane the flux due to mass flow ranged from 18 to 71% of the total lindane flux, whereas with dieldrin it ranged from 3 to 33% of the total dieldrin vapor flux.

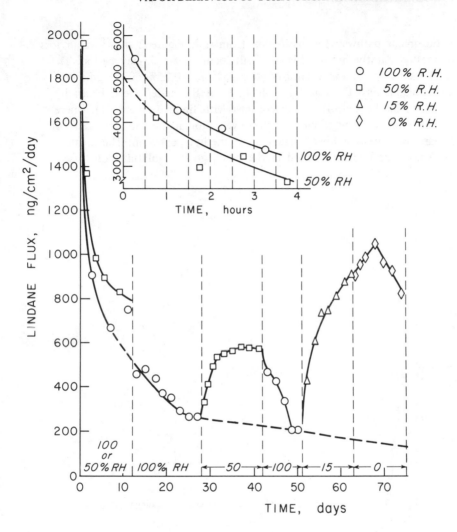

Figure 1. Lindane volatilization rate as related to relative humidity of the N_2 gas passing over the soil surface at $30°C$ with 10 ppm lindane mixed in Gila silt loam at a water suction of 50 millibars (from Reference 35).

In addition to factors directly affecting vapor behavior, how much of the total amount of the chemical in the soil is lost to the atmosphere will depend on the resistance of the chemical to degradation and leaching. Any models for predicting vapor behavior must take into account the degradation rate of the chemical as compared with its movement into the atmosphere. With methyl parathion[17] and parathion[39] biological degradation is the major pathway for dissipation from soil. Figure 2 shows vapor loss of methyl parathion from Flanagan silt loam over a 33-day period. The total methyl parathion volatilized averaged only 0.25% of the amounts incorporated into the soil, but only 0.5% of the added methyl parathion remained in the soil. This is in contrast with lindane (Figure 1), where volatilization rates were much higher even though the vapor pressure of lindane is somewhat lower than that of methyl parathion. The greater persistance of lindane resulted in the much greater volatilization losses.

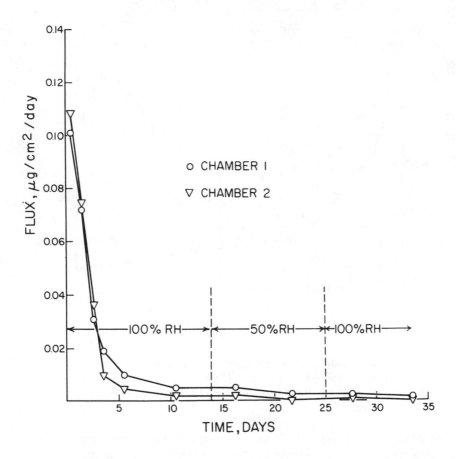

Figure 2. Vapor loss rates of methyl parathion incorporated at 10 **ppm** into Flanagan silt loam as related to time and relative humidity at 25°C (From Reference 17).

VAPOR LOSS FROM WASTES IN LANDFILLS

Many industrial wastes are disposed of in landfills, which subsequently may, or may not, be covered with a layer of soil to prevent or reduce vapor loss of any potentially toxic organic chemicals in the waste. Volatilization of the chemicals from uncovered wastes will follow the principles outlined in the section on Vapor Losses from Inert Surfaces and the actual rate of loss will depend on weather variables affecting vapor pressure and air turbulence at the waste site. Vapor loss of chemicals from covered landfill will be much slower, in accordance with principles discussed in the section on Vapor Loss from Soil. Volatilization losses will be related to the rate of movement of the chemical through the soil cover into the atmosphere.

One of the management tools required for the utilization of landfill as a waste disposal site is the ability to control the volatilization of any hazardous chemicals from the wastes. Farmer, et al.[27,40] recently completed a study designed to gather data useful to the landfill planner in limiting the vapor loss of hexachlorobenzene (HCB) through a soil cover to the surrounding atmosphere to an acceptable level. Their research findings are applicable to developing predictive equations for use in designing landfill covers for other potentially volatile chemicals, and are presented below.

Hexachlorobenzene is a persistent, water-insoluble, fat-soluble organic compound. Farmer et al.[27,40] used a simulated landfill to determine the parameters necessary to predict HCB volatilization from industrial waste samples containing HCB (hex wastes) placed under a soil cover. Covering hex wastes with only 1.8 cm of soil reduced HCB vapor flux from 363 to 5 ng/cm²/hr, indicating that soil is a very effective cover. They found that soil depth and soil-air-filled porosity were the two prime factors controlling HCB vapor flux.

The relationship between vapor flux, soil depth and air-filled porosity is predictable based on known factors controlling vapor behavior in soil. Since HCB is essentially insoluble in water, mass flow will not be important in moving HCB to the soil surface; consequently, diffusion in soil pores will be the only mechanism available. Because of the water-air partition ratio of HCB and the higher rate of diffusion in air than in water, HCB diffusion will be primarily in the vapor phase through the soil pores. Thus, HCB vapor loss can be considered as a diffusion-controlled process involving only vapor phase diffusion through air-filled soil pores.

The rate at which compounds will volatilize from the soil surface will be controlled by the rate at which they diffuse through the soil cover over the waste. Assuming no degradation of the compound and no transport in moving water, the volatilization can be predicted using Fick's First Law for steady-state diffusion:

$$J = -D_s(C_2 - C_s)/L \qquad (4)$$

where:

    J = the vapor flux from the soil surface ($ng/cm^2/day$),
    D_s = the apparent stead-state diffusion coefficient (cm^2/day),
    C_2 = the concentration of the volatilizing material in air or vapor density at the bottom of the soil layer (ug/l),
    C_s = the concentration of the volatilizing material in air or vapor density at the bottom of the soil layer (ug/l), and
    L = the soil depth.

To use Equation 4 for predicting volatilization, one must determine the apparent diffusion coefficient, D_s. Millington and Quirk[41] suggested an apparent diffusion coefficient that includes a porosity term to account for the geometric effects of soil on diffusion.

$$D_s = D_0(P_a^{10/3}/P_T^2) \qquad (5)$$

where:

    D_0 = the vapor diffusion coefficient in air (cm^2/day),
    P_a = the soil air-filled porosity (cm^3/cm^3), and
    P_T = the total soil porosity.

Combining Equations 4 and 5 yields the following expression:

$$J = -D_0(P_a^{10/3}/P_T^2) (C_2 - C_s)/L \qquad (6)$$

The validity of Equation 6 was experimentally verified for hexachlorobenzene using a simulated landfill and HCB-containing industrial waste as the volatilizing compound.[40]

Equation 6 can be used to determine the most efficient combination of soil porosity and soil depth to produce an acceptable value of vapor flux through a soil landfill cover. Since soil porosity is controlled mainly by bulk density and water content, vapor flux can be decreased by increasing compaction and soil water content as well as by increasing soil depth. Figure 3 from Farmer et al.[40] illustrates the use of their procedures in predicting vapor flux through various depths of soil cover at varying water contents. HCB vapor flux through a soil cover of 30 cm or greater would be less than 1 $ng/cm^2/hr$ (1 kg/ha/year), compared with a vapor flux of 363 $ng/cm^2/hr$ from the uncovered waste under the same conditions. Increasing the soil water content greatly decreased HCB volatilization rate by decreasing the air-filled porosity and, consequently, the vapor phase diffusion through the soil. For compounds that are more water soluble, increasing soil water content may increase or decrease vapor loss, depending upon the potential for downward or upward movement of the chemical in the water phase.

Equation 6 was used by Farmer et al.[40] as the basis for a suggested stepwise procedure intended to assist a planner in designing a landfill cover that minimizes the escape of HCB or other vapors into the atmosphere. After an acceptable flux value has been established, either directly by a regulatory agency or indirectly by use of an air dispersion model, Equation 6 is used to determine what soil conditions would limit flux to this value. Alternatively, they suggested Equation 6 may be used to assess the effectiveness of an existing landfill cover for limiting HCB vapor flux.

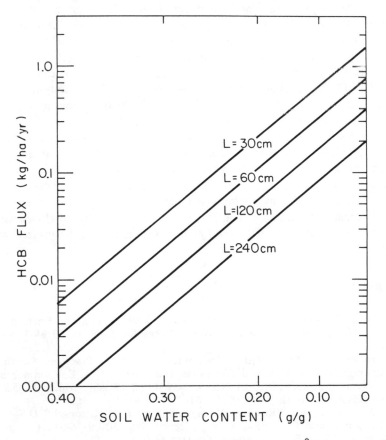

Figure 3. Predicted hexachlorobenzene volatilization fluxes at 25°C through a soil cover of various water contents and thicknesses (L). Soil bulk density is 1.2 g/cm³ (From Reference 10).

SUMMARY

Some of the physicochemical principles governing vapor behavior of organic chemicals have been discussed. The vapor behavior of new chemicals can be estimated from a knowledge of their physicochemical properties relative to those of compounds of known behavior. At least, the likely contribution of vapor loss and atmospheric transfer as a determinant of their persistence and fate can be estimated with sufficient precision to alert us to potential hazards from vapor transfer. Vapor losses of organic compounds under field conditions have been measured using micro-climate techniques in which concentration profiles of the chemical vapor over soil or crop surfaces were measured and fluxes calculated from meteorological data over the treated area. The limited data from field measurements indicate that the same factors affecting vapor pressure and volatilization rates in the laboratory also control volatilization rates under field conditions, *i.e.,* temperature, windspeed, soil water content, chemical concentration and soil properties. Consequently, comparative behavior can be estimated, but the absolute vapor loss rates under varying conditions of natural environments can only be predicted approximately from a comparison of the known behavior of well-studied compounds.

Success in predicting the relative importance of vapor transfer depends upon the availability of reliable values for vapor pressure, water solubility, adsorption characteristics and persistence or degradability of the organic chemical in environmental systems, such as soils, waters and sediments. Standard procedures for measuring vapor pressure and vapor loss rates for toxic organic chemicals of continuing interest should be developed and adopted so that data gathered by different investigators can be utilized with confidence.

REFERENCES

1. Haque, R. and V.H. Freed. "Behavior of Pesticides in the Environment:Environmental Chemodynamics," *Residue Reviews* 52:89-116 (1974).
2. Gruber, G.I. "Assessment of Ind. Hazardous Waste Practices: Organic Chemicals, Pesticides, and Explosives Industries," U.S. Environmental Protection Agency, Report No. 530-SW-118C (Washington, DC: U.S. Government Printing Office, 1975).
3. Fuller, W.H., Ed. "Residual Management by Land Disposal," U.S. Environmental Protection Agency, Report No. 600/9-76-015 (Washington, DC: U.S. Government Printing Office, 1976).
4. Taylor, A.W. "Post-Application Volatilization of Pesticides Under Field Conditions," *J. Air Poll. Control Assoc.* 28(9)922-927 (1978).

5. Spencer, W.F., W.J. Farmer, and M.M. Cliath. "Pesticide Volatilization," *Residue Reviews* 49:1-47 (1973).
6. Hamaker, J.W. In: *Organic Chemicals in the Soil Environment*, C.A.I. Goring and J.W. Hamaker, Eds. (New York: Marcel Dekker, 1972), pp. 341-397.
7. Wheatley, G.A. In: *Environmental Pollution by Pesticides*, C.A. Edwards, Ed. (London, England: Plenum Press, 1973), pp. 365-408.
8. Guenzi, W.D., and W.E. Beard. In: *Pesticides in Soil and Water*, W.D. Guenzi, Ed. (Madison, WI: Soil Science Society of America, Inc., 1974), pp. 108-122.
9. Plimmer, J.R. In: *Herbicides: Chemistry, Degradation and Mode of Action, Vol II*, P.C. Kearney and D. D. Kaufmann, Eds. (New York: Marcel Dekker, 1976), pp. 891-934.
10. Spencer, W.F. In: *A Literature Survey of Benchmark Pesticides*. (Washington, DC: Washington University Medical Center, 1976), pp. 72-165.
11. Martin, H., Ed. *Pesticide Manual: Basic Information on the Chemicals Used as Active Components of Pesticides* (Oxford, England: British Crop Protection Council, 1971).
12. Harris, C.R., and J.H. Mazurek. "Comparison of the Toxicity to Insects of Certain Insecticides Applied by Contact and in Soil," *J. Econ. Entomol.* 57:698-702 (1974).
13. Porter, P.E. In: *Analytical Methods for Pesticides, Plant Growth Regulators, and Food Additives II*, G. Zweig, Ed. (New York: Academic Press, 1964), p. 143-163.
14. Zimmerli, B. and B. Marek. "Modellversuche zur Kontamination von Lebensmitteln mit Pestiziden via Gasphase," *Mitt Gebiete Lebensm Hyg* 65:55-64 (1974).
15. Spencer, W.F. and M.M. Cliath. "Vapor Density of Dieldrin," *Environ. Sci. Technol.* 3:670-674 (1969).
16. Atkins, D.H.F., and A.E.J. Eggleton. "Studies of Atmospheric Wash-Out and Deposition of γ-BHC, Dieldrin, and p,p'-DDT Using Radio-Labelled Pesticides," In: *Proc. Symp. on Nucl. Tech. Environ. Poll.* (Vienna: International Atomic Energy Agency, 1971), p. 521-533.
17. Spencer, W.F., T.D. Shoup, M.M. Cliath, W.J. Farmer and R. Haque. "Vapor Pressures and Relative Volatility of Ethyl and Methyl Parathion," *J.Agric. Food Chem.* 27(2):273-278 (1979).
18. U.S. Environmental Protection Agency. "Pesticide Programs—Guidelines for Registering Pesticides in the United States," *Federal Register.* 40(123):26829 (1975).
19. Ware, G.W., B.J. Estesen and W.P. Cahill. "Organophosphate Residues on Cotton in Arizona," *Bull. Environ. Contam. Toxicol.* 8:361-362 (1972).
20. Ware, G.W., D.P. Morgan and B.J. Estesen. "Establishment of Reentry Intervals for Organophosphate-Treated Cotton Fields Based on Human Data," *Arch. Environ. Contam. Toxicol.* 2:117-129 (1974).
21. Bright, N.F.H., J.C. Cuthill and N.H. Woodbury. "Vapor Pressure of Parathion and Related Compounds," *J. Sci. Food Agric.* 1:344-348 (1950).

22. Guckel, W., G. Synnatschke and R. Rittig. "A Method for Determining the Volatility of Active Ingredients Used in Plant Protection," *Pestic. Sci.* 4:137-147 (1973).
23. Hartley, G.S. "Evaporation of Pesticides," *Adv. Chem. Series* 86:115-134 (1969).
24. Farmer, W.J. and J. Letey. "Volatilization Losses of Pesticides from Soils," U.S. Environmental Protection Agency, Report No. 660/2-74-054 (Washington, DC: U.S. Government Printing Office, 1974).
25. Grover, R., W.F. Spencer, W.J. Farmer and T.D. Shoup. "Triallate Vapor Pressure and Volatilization from Glass Surfaces," *Weed Sci.* 26(5):505-508 (1978).
26. Taylor, A.W., D.W. Glotfelty, B.C. Turner, R.E. Silver, H.P. Freeman and A. Weiss. "Volatilization of Dieldrin and Heptachlor Residues from Field Vegetation," *J. Agric. Food Chem.* 25:542-548 (1977).
27. Farmer, W.J., M. Yang, J. Letey and W.F. Spencer. In: "Residual Management by Land Disposal," W.H. Fuller, Ed., U.S. Environmental Protection Agency Report No. 600/9-76-015 (Washington, DC: U.S. Government Printing Office, 1976), pp. 177-185.
28. Spencer, W.F., and M.M. Cliath. "Factors Affecting Vapor Loss of Trifluralin from Soil," *J. Agric. Food Chem.* 22:987-991 (1974).
29. Phillips, F.T. "Persistence of Organochlorine Insecticides on Different Substrates Under Different Environmental Conditions. I. The Rate of Loss of Dieldrin and Aldrin by Volatilization from Glass Surfaces," *Pestic. Sci.* 2:255-266 (1971).
30. Mackay, D., and P.J. Leinonen. "Rate of Evaporation of Low-Solubility Contaminants from Water Bodies to Atmosphere," *Environ. Sci. Technol.* 9:1178-1180 (1975).
31. Spencer, W.F., M.M. Cliath and W.J. Farmer. "Vapor Density of Soil-Applied Dieldrin as Related to Soil-Water Content, Temperature, and Dieldrin Concentration," *Soil Sci. Soc. Am. Proc.* **33:509-511 (1969)**.
32. Spencer, W.F., and M.M. Cliath. "Desorption of Lindane from Soil as Related to Vapor Density," *Soil Sci. Soc. Am. Proc.* 34:574-578 (1970).
33. Spencer, W.F., and M.M. Cliath. "Volatility of DDT and Related Compounds," *J. Agric. Food Chem.* 20:645-649 (1972).
34. Goring, C.A.I. In: *Organic Chemicals in the Soil Environment,* C.A.I. Goring and J.W. Hamaker, Eds. (New York: Marcel Dekker, 1972) pp. 793-863.
35. Spencer, W.F., and M.M. Cliath. "Pesticide Volatilization as Related to Water Loss from Soil," *J. Environ. Qual.* 2:284-289 (1973).
36. Farmer, W.J., K. Igue, W.F. Spencer and J.P. Martin. "Volatility of Organochlorine Insecticides from Soil. I. Effect of Concentration, Temperature, Air Flow Rate, and Vapor Pressure," *Soil Sci. Soc. Am. Proc.* 36:443-447 (1972).
37. Farmer, W.J., K. Igue and W.F. Spencer. "Effect of Bulk Density on the Diffusion and Volatilization of Dieldrin from Soil," *J. Environ. Qual.* 2:107-109 (1973).
38. Mayer, R., W.J. Farmer and J. Letey. "Models for Predicting Pesticide Volatilization of Soil-Applied Pesticides," *Soil Sci. Soc. Am. Proc.* 38:563-568 (1974).

39. Yang, M.S. 'Processes of Adsorption, Desorption, Degradation, Volatilization, and Movement of O,O-diethyl O-p-nitrophenol Phosphorothioate (Parathion) in Soils," Ph.D. Dissertation, University of California, Davis, CA (1974).
40. Farmer, W.J., M. Yang, J. Letey, W.F. Spencer and M.H. Roulier. In: *Land Disposal of Hazardous Waste. Proceedings of the Fourth Annual Research Symposium,* U.S. Environmental Protection Agency, Report No. 600/9-78-016 (Washington, DC: U.S. Government Printing Office, 1978), pp. 182-190.
41. Millington, R.J., and J.M. Quirk. "Permeability of Porous Solids," *Trans. Faraday Soc.* 57:1200-1207 (1961).

13

DETERMINING THE ROLE OF HYDROLYSIS IN THE FATE OF ORGANICS IN NATURAL WATERS

N. Lee Wolfe

Environmental Research Laboratory
Office of Research and Development
U.S. Environmental Protection Agency
Athens, Georgia 30605

INTRODUCTION

As our knowledge of the fate and transport processes of pollutants increases, it is logical that the emphasis on these processes should evolve from a observation to prediction. Studies in which a given amount of pollutant is added to a system and the concentration is monitored irrespective of the individual transport and transformation processes are inadequate, however, as it becomes increasingly desirable to predict exposure concentration.[1] The monitoring experiments are generally system specific and as such the results cannot be extrapolated to other aquatic ecosystems. On the other hand, the predictive approach requires a mathematical description of the individual fate processes, including the rate constants and functions describing the properties of the ecosystem that act on the pollutant.

This requirement is particularly true for hydrolysis because several hydrolytic pathways may compete over the common pH ranges of aquatic ecosystems.[2] Thus, it is necessary to determine the rate constants and concentration effects of reactants for each mechanism and combine them into a general expression that can be used to predict hydrolysis under reaction conditions that are characteristic of aquatic ecosystems.

These measurements are often considered to be straight-forward but a close examination of the environmental literature does not support this supposition.

The overriding factor affecting hydrolysis at a given temperature is generally hydrogen or hydroxide ion concentration.[2] Detailed studies of the pH dependency, although desirable, are not always necessary. The minimum requirement for environmental application is a value of the observed rate constant at a specific pH that usually lies in the pH range of 5 to 9.[3] A pH-rate profile will often provide this information.[4]

Given that the correct rate expression can be determined in distilled water, it must be decided whether the expression adequately predicts hydrolysis in natural waters. Several different types of experiments have been performed in attempts to confirm the validity of this approach. These include field tests, model ecosystems studies, tests using natural water samples, and addition of naturally occurring substances to distilled water.

In many cases, an estimate of a rate constant is all that is needed, particularly if the rate constant is very large or very small. Such a value may sometimes be obtained from structure reactivity relationships.[5,6] These linear free energy relationships (LFER) have been widely used in fields such as organic chemistry[7,8] or pharmacokinetics[9] and consequently are well established. More recently, LFER's are being developed by environmentalists as tools to predict bioconcentration factors,[10] sediment-sorption partition coefficients,[11] octanol/water partition coefficients[12] and even toxicity to aquatic organisms.[13] For hydrolysis reactions LFER's offer a convenient way to estimate rate constants, to predict hydrolytic stability of classes of compounds, and to serve as a check on reported experimental values.

DISCUSSION

Hydrolysis Rate Data

When searching the literature for hydrolysis rate constants for use in evaluating hydrolytic transformation in natural waters, two types of studies are generally encountered. The first is the detailed mechanistic study in which the goal is to elucidate the intricacies of bond making and breaking. These studies, usually found in the chemical literature, are often concerned with solvent effects, ionic strength, salt effects and catalytic effects of various species. For many of these studies, only a limited pH range is investigated. The second type of study consists of simply adding a quantity of the compound to water and observing the change in concentration with time irrespective of factors such as water solubility, temperature or pH, all of which influence the hydrolysis rate constant. These studies, in which some specific phenomena such as persistence was of interest, are often found in the environmental literature.

From the mechanistic studies, it is sometimes possible to extrapolate the data to obtain a rate constant that is applicable to reaction conditions of interest because the rate constants are usually expressed as a function of the independent variables of the system. The rate data from these studies are generally system specific and consequently the rate constants cannot be extrapolated to any other reaction conditions, let alone those of natural waters.

In general, for simple nonequilibrium reactions the required data to describe hydrolysis for most freshwaters consists of the first and/or second-order rate constants for acid (k_H), base (k_{OH}) and neutral (k_{H_2O}) hydrolysis pathways.[2] In differential equation form, this is given as:

$$-\frac{d[P]}{dt} = (k_H[H^+] + k_{H_2O} + k_{OH}[OH]) \, [P] \tag{1}$$

where [P] is the concentration of the hydrolysate. From Equation 1, it is readily seen that the observed rate constant (k_{obs}) at any pH is given as:

$$k_{obs} = k_H[H^+] + k_{H_2O} + k_{OH}[OH^-] \tag{2}$$

For environmental considerations, the values of k_{obs} over the pH 5 to 9 range are usually required.[14]

Values for k_{obs} over this pH range can be obtained in two ways. The first is adjust the pH so that each of the three potential processes can be studied independently to determine each rate constant.[15] Then using Equation 2, k_{obs} can be calculated at any pH. The advantage of this procedure is primarily one of convenience because the rate constants can be measured in predetermined time intervals by a suitable choice of pH. Also, k_{obs} can be calculated over a wider pH range than 5 to 9.

The second method consists of determining the disappearance rate constant (k_{obs}) at several pH over the pH range of 5 to 9 and plotting log k_{obs} versus pH.[2] (Figure 1) This method provides a pH-rate profile that can be used in graphical extrapolation of other values of k_{obs} at other pH over this range. The advantage of this method is that it requires no knowledge of the mechanism and is applicable to more complex hydrolysis reactions. The disadvantage is lack of control of the time intervals for reaction other than by temperature changes.

Hydrolysis in Natural Waters

Once the quantitative expression describing hydrolysis has been obtained from laboratory studies, it is necessary to demonstrate that the expression

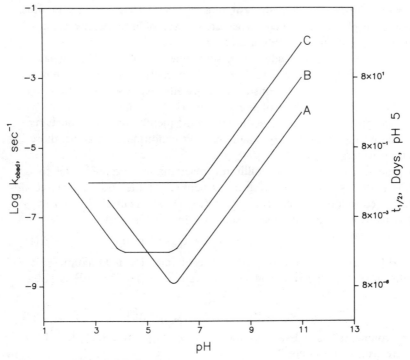

Figure 1. pH-Rate profile for three carboxylic acid esters of varying reactivity.

is applicable to natural waters. Several types of studies are often carried out that, ideally, should allow for a comparison, either directly or indirectly, of predicted and measured values of k_{obs}. In field studies, the compound is generally added to an ecosystem and the concentration is monitored with respect to time. Unfortunately, one cannot very often draw any concrete conclusions from this comparison because the k_{obs} is usually a first-order decay rate constant that includes contributions from other processes besides hydrolysis. Put another way, some functions of these other processes operate on the concentration term [P] in Equation 1 so that the rate constant (k_{obs}) cannot be compared with the predicted value.

To circumvent the inherent complexity of the field study, other approaches are sometimes pursued. One method is to bring natural water samples into the laboratory, carry out hydrolysis studies, and compare the results with those obtained in distilled water. Although this approach is always subject to the criticism that the system is not "realistic," even more important is that it may still be "real enough" that other processes are still competing with hydrolysis.

Another approach is to select some isolatable component of natural waters that might influence hydrolysis and study its effect usually as a function of elevated concentrations in laboratory water. In this case, it is necessary to provide a kinetic expression so that extrapolation can be made to environmental concentrations of reactants.

Many examples are available to provide some qualitative insight of hydrolytic behavior, but they generally preclude any quantitative comparison between predicted values of k_{obs} and those observed in natural waters. In some instances, results of these studies are interpreted to support either rate enhancement or retardation. No attempt is made here at a comprehensive hydrolysis review, but a few reports do deserve some comments.

Field Studies

In many of these aquatic field studies, the objective has been to evaluate persistence of compounds. Usually the field data are compared with laboratory hydrolysis data. In one such study, the persistence of the propylene glycol butyl ester of 2-(2,4,5-trichlorophenoxy) propionic acid was addressed in three outdoor ponds having varying hydrogen ion concentrations to determine the disappearance rate constants.[16,17] Transformation of the ester was substantiated by monitoring the formation of 2-(2,4,5-trichlorophenoxy) propionic acid. Using the first order rate constants and the pH of the pond water, second-order alkaline hydrolysis rate constants (k, m^{-1}-sec^{-1}) are calculated to be: pH 6.07, k, 2×10^3; pH 6.09, k, 3×10^3; and pH 6.25, k, 2×10^3. A rate constant for the ester was not obtained in distilled water but an estimate of 2×10^1 m^{-1}-sec^{-1} can be made by use of results from later studies.[18,19] Although two orders of magnitude difference might suggest enhanced hydrolysis, biolysis, which would likely result in acid formation, cannot be excluded.[20] Also, assuming it is chemical hydrolysis, it is not possible to distinguish between sediment and aqueous phase catalysis.

In a similar field study, 21 pesticides were applied as emulsifiable concentrates to small ponds, and the concentration of each pesticide was determined at given time intervals.[21] For comparison, the rates of disappearance were followed in distilled water at pH 8 at 35°C in the laboratory. Comparison of the data showed that decay in the natural water was much faster than in distilled water. No conclusion can be drawn, however, concerning enhanced chemical hydrolysis because sorption, photolysis, and biolysis processes were not accounted for.

Ecosystems

Synthesized ecosystems allow one to couple or isolate selected features of environmental compartments and study fate and transport processes. In one such study, the hydrolysis of carbaryl was carried out in a marine water-sediment system and the results compared to a marine water system.[22] In sea water alone (pH 8.0, 8°), the half-life was 38 days, which gives a second-order alkaline hydrolysis rate constant of 2.1×10^{-3} m^{-1}-sec^{-1}. The extrapolated value at this temperature based on other studies in distilled water, is about 0.4 m^{-1}-sec^{-1}.[23-25] No data are provided in this study to allow one to resolve this discrepancy, however. In the presence of marine sediments, the authors state that carbaryl persisted longer because of sorption. Careful examination of the data does not permit one to dichotomize the degradation into the sediment and water phases.

The results of a study of the hydrolysis of a carboxcylic acid ester in tanks with and without fish indicated that with fish present ester hydrolysis was accelerated.[26] The hydrolysis of the *n*-butoxy ethyl ester of 2,4-D was carried out in a tank of well water, pH 7.0 to 7.2 at 2°C. Based on data from this experiment, the second-order alkaline hydrolysis rate constant calculated from a reported half-life of 90 hours and pH 9.09 is 21 m^{-1}-sec^{-1}. This value is higher than the value of 13 m^{-1}-sec^{-1} reported by other workers.[19] When fish were present (1.5 g/l), the half-life was reduced to 24 hr (k, 80 m^{-1}-sec^{-1}). Although transformation was enhanced, any chemical hydrolysis effect cannot be elucidated because biologically mediated hydrolysis was not accounted for.[20]

Natural Water Samples

Degradation studies of organic compounds in natural water samples are probably the most common of the four types of studies being addressed. Many of these studies were carried out using natural water samples, but for all practical purposes, might as well have been done in distilled water.[27,28] The data obtained from these experiments do not allow calculation of the required rate constants. Sometimes, the data do allow for the relative comparison of rates of disappearance when a group of compounds are hydrolyzed in the same natural water sample, however.

Even though some of the fate processes are eliminated in this type of system, biolysis, volatilization, and photochemical degradation can still compete with hydrolysis. For example, in a study of the persistence of some organophosphate and carbamate pesticides in distilled and raw water samples, it was found that transformations occurred faster in the raw water samples.[29] It is not possible to make a quantitative comparison in the two systems, however, because the pH of the raw water varied from 7.3 to 8.3 while the pH for the distilled water was not reported.

More detailed studies such as those carried out with three phosphorus pesticides, dichlorovos, malathion and parathion, showed mixed results.[30] Degradation experiments in distilled water and four different river water samples indicated that only dichlorovos was degraded at an enhanced rate in one of the river water samples. Further work on buffer catalysis by phosphate, carbonate and borate, as well as suspended organics, did not provide an explanation.[31] The authors considered biolysis as a possibility but did not carry out the definitive experiments to confirm or disprove it.

Another detailed study that offers some insight addresses alkaline hydrolysis[32] and biodegradation[33] of a group of carbamate pesticides. Hydrolysis rate constants were determined in distilled water studies and the results compared to those obtained in river water samples (pH 7.0 ± 0.1). No change in concentration was detected in the river or sewage waters after 4 months. Although a quantitative comparison of rate constants cannot be made, at least the results were consistent with predictions based on the laboratory studies.

The few detailed studies that do allow for a quantitative comparison of data suggest that if catalyzed hydrolysis processes are operative they are slow in natural waters.[19,23,34,35] Table I includes observed and predicted hydrolysis half-lives for several organic compounds. In all cases, the rate constants were obtained under experimental conditions that excluded volatilization, photolysis, and biolysis. It should be noted that the water samples were filter sterilized by use of 0.22-μm millipore filters prior to use,[20] and sterility was confirmed by plate counts at the end of the experiments.

Agreement between predicted and measured values is quite satisfactory. For the compounds with short half-lives, agreement was good to about ± 10%. For the compounds for which no degradation was observed, the results are at least consistent with the predictions.

These results suggest that catalytic effects on the hydrolysis of organics in natural waters are not large. It should be pointed out, however, that in cases where the half-lives were short, the contribution of alkaline hydrolysis may have swamped out the catalyzed hydrolysis process. Additional studies at different pH's in which acid and/or base mediated reactions are slowed should be pursued for similar types of compounds.

Table I Comparison of Predicted and Experimental Half-Life of Selected Organic
Compounds in Natural Water Samples

Compound	Water Sample[a]	pH	Half-Life	
			Predicted	Found
2,4-D Octyl Ester[b,f]	Withlacoochee River	8.1	4.9 hr	4.9 hr
Malathion[b,g]	Withlacoochee River	8.2	22 hr	20 hr
Atrazine[b,h]	Okafenokee Swamp	3.8	>1 yr	NR[d]
	Gulf of Mexico	8.1	>1 yr	NR[d]
	Aucilla River	7.4	>1 yr	NR[d]
Carbaryl[b,i]	Hickory Hills Pond	6.7	32 hr	30 hr
	USDA Pond	7.2	12 hr	10 hr
Chlorpropham[c,j]	Hickory Hills Pond	6.7	>1 yr	NR[e]
	USDA Pond	7.2	>1 yr	NR[e]

[a]Filter sterilized water.
[b]Temperature, 30 ± 2°.
[c]Temperature, 67 ± 0.02°.
[d]No detectable reaction after 30 days.
[e]No detectable reaction after 1 week.
[f]Reference 19.
[g]Reference 34.
[h]Unpublished data.
[i]Reference 23.
[j]Reference 23.

Metal Catalysis

Laboratory studies indicate that for certain types of compounds solution phase metal ion catalysis may be important. Because the rate constants for these processes can be quite large, catalysis may be likely even at very low concentrations of metal ions. Two such reports suggest the importance of this process. One is for the catalyzed hydrolysis of organophosphate esters by cupric ion[36] at concentrations down to 10^{-6} M of the ion. The other is sulfide oxidation by trace metal ions.[37] Additional kinetic studies are required in this area to evaluate any contribution to hydrolysis in natural waters. If experiments with natural waters indicate no catalysis, these studies might be excluded.

Naturally Occurring Organics

Several studies suggest that naturally occurring organic constituents of fresh waters can catalyze the hydrolysis of certain organic compounds. These studies include both suspended and dissolved organic fractions. Although it is difficult to extrapolate these results to the concentration of organics that exists in the water column, the process may be important in bottom sediments for hydrophobic compounds.

Sodium humate enhanced the hydrolysis of aliphatic esters of 2,4-D and 2,4,5-T in laboratory studies.[18] With a 5 mg/l suspension of sodium humate, a rate enhancement of about 3 was calculated. This calculation is based on the value of the observed disappearance rate constant which is the sum of the sorbed and solution rate constants. No attempt was made to differentiate between surface and solution catalysis.

Acid catalyzed hydrolysis by humic acid has also been reported. The hydrolysis of atrazine in distilled water was catalyzed by three different samples of humic acid.[38] The concentration of humic material varied from 5 g/l to 40 g/l with the pH adjusted to 4 using either strong acid or base. A plot of half-life versus concentration of humic acid was non-linear,[39] but a similar plot using data for acetic acid in place of humic acid was linear. The reason for this difference is not obvious but may be caused by the high concentrations of humic acid that were used. From the acetic acid studies, the general acid catalyzed rate constant is calculated to be about 3×10^{-4} m^{-1}-sec^{-1} indicating a slow reaction.

Fulvic acid also catalyzed atrazine hydrolysis.[40] In an extensive study, pseudo-first-order hydrolysis rate constants were determined as a function of pH (2.9 to 7.0) and fulvic acid concentration (0.5 to 5 mg/l). A plot of the hydrolysis rate constant versus concentration of fulvic acid at a given pH is also non-linear. The reason for this non-linearity might be the complexity of the atrazine-acid-catalyzed hydrolysis mechanism. The reaction is not first-order in hydrogen ion concentration, which suggests an equilibrium step prior to the rate determining step. These results suggest the need for additional studies for classes of compounds that are known to be susceptable to general acid catalysis. Also the role of fulvic acid as a general base should not be overlooked.

Use of Linear Free Energy Relationships

Linear free energy relationships (LFER) are valuable tools for predicting rate constants for use in assessing the fate of pollutants in aquatic ecosystems. They can be used to provide estimates of these constants when experimental values are not available,[41,42] to evaluate hydrolytic reactivity of classes of compounds[43] and to serve as an independent check on reported experimental values.[7,8] In many cases a guess can be made as to whether a compound will undergo acid or base mediated hydrolysis based on the functional moieties, but little can be said about the magnitude of the rate constant. This is the role of LFER s. Two examples of the utility of LFER s are given using data for alkene hydration and amide hydrolysis.

The alkenes comprise a large class of organic compounds that are widely used in the polymer, pesticide, and dye industries.[44] Large quantities of 2,2-substituted alkenes are used as starting materials and as a result significant quantities of these compounds are anticipated to enter the aquatic environment.

In assessing the fate and transport of these compounds in aquatic ecosystems, hydrolysis rate constants are needed because alkenes undergo an acid catalyzed reaction as shown below.[4,5]

$$R_1 R_2 C = CH_2 \xrightarrow[H_2O]{H^+} R_1 R_2 \underset{\underset{OH}{|}}{C}\text{-}CH_3 \tag{3}$$

Other investigators have reported a correlation of the second-order acid-catalyzed hydrolysis rate constant (k_{H+}) and the σ^+ substituent constant.[46,47] This correlation includes compounds spanning 16 orders of magnitude reactivity. Figure 2 is a plot of these data along with calculated half-lives at pH 5 assuming pseudo-first-order kinetics. The equations that describe the relationship are

$$\log k_{H+} = \rho\sigma^+ + \log k_o \tag{4}$$

$$= -12.1 \ \sigma^+ - 10.1 \tag{5}$$

where ρ is the slope of the line and k_o is the intercept.

The usefulness of this relationship is shown in Table II in which rate constants and calculated half-lives are presented for some selected monomers that are important to the polymer industry. When available, experimental values have been included for comparison with the predicted values. For example, in the case of styrene, the predicted and experimental values agree within about an order of magnitude. In view of the long hydrolysis half-life (7×10^7 days) at pH 5, an order of magnitude does not affect the obvious conclusion. At the other extreme for compounds like 2,2-diethoxyethylene that hydrolyze with half-lives of a few seconds at pH 5, an estimate is also satisfactory and will generally suffice at higher pH's where the half-life is increased but still short. On the other hand for moderately reactive compounds such as isobutylene that has a calculated half-life of 300 days at pH 5, an order of magnitude in the rate constant might well be significant and acceptance of the estimated value must be based on the contribution of competing processes.

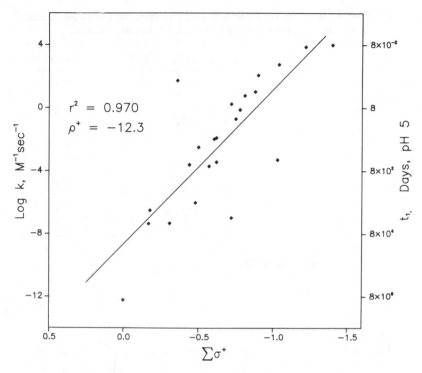

Figure 2.   Linear free energy relationship for the acid catalyzed hydrolysis of
2,2-substituted alkenes.[46]

Two useful LFER's are established from literature data for the hydrolysis
of amides.[48] These two correlations demonstrate the difference in sensitivity with changes in substituents (R) (equation 6) for different types of
reactions.

$$RCONH_2 \xrightarrow[H_2O]{H^+ \text{ or } OH^-} RCO_2H + NH_3 \qquad (6)$$

The acid catalyzed reaction has a slope of near zero.[49,50] This indicates
that acid hydrolysis rate constants of most primary amides will not vary
significantly with structure and are all quite small.

Alkaline hydrolysis rate constants (k_{OH}) for primary amides are more
sensitive to structural features and are estimated using equation 7

$$\log k_{OH} = 1.6 \ \sigma^* - 3.17 \qquad (7)$$

where σ^* is the Taft substituent constant (Figure 3). Table III contains
some rate constants and estimated half-lives for some industrially important amides

Table II Estimated Acid Catalyzed Hydrolysis Rate Constants and Calculated Half-Lives
at pH 5 in Water for Some Widely Used 2,2-Substituted Alkenes

Compound	k $(m^{-1}\text{-sec}^{-1})$	$t_{1/2}$ (days pH 5^a)
Vinyl Chloride	3.3×10^{-12}	2×10^{11}
Vinylidene Chloride	1.4×10^{-13}	6×10^{12}
Propylene	4.6×10^{-7} $(5.0 \times 10^{-8})^b$	2×10^6
Vinylidene fluoride	6×10^{-10}	1×10^9
Isobutylene	2.7×10^{-3} $(3.7 \times 10^{-4})^b$	3×10^2
Styrene	1.2×10^{-8} $(3.3 \times 10^{-7})^b$	7×10^7
Vinyl Acetate	7.7×10^{-9}	1×10^8
Methyl methacylate	6.3×10^{-13}	1×10^{12}

[a]Half-life calculated assuming pseudo-first-order kinetics.
[b]Experimental values from reference 46.

SUMMARY

The quantitative approach to hydrolytic transformation processes requires either a general kinetic expression and values for the rate constants
or a pH-rate profile over the pH range of interest (usually pH 5 to 9).
The observed rate constant at the pH or pH's of interest can then be calculated. Unfortunately, few qualitative studies support the extrapolation
of laboratory hydrolysis data to natural waters. A few careful studies in
natural water samples, however, are in agreement with this supposition
and suggest that if there are catalytic effects they result in slow processes.
The fact that they are slow is supported by studies with humic and fulvic
acids in distilled water. The contribution of trace metal catalysis to the
fate process for organics is difficult to evaluate. It is very likely that these
catalyzed hydrolyses will be very compound specific. For compounds that
do undergo simple hydrolysis reactions, estimates of these rate constants
are sometimes available from LFER's. The relationships do not presuppose
experimental values and are often limited by an inadequate data base.

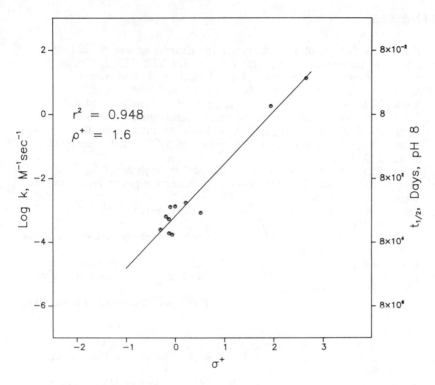

Figure 3. Linear free energy relationship for the alkaline hydrolysis of primary amides.

Table III Estimated Alkaline Hydrolysis Rate Constants and Calculated Half-Lives
at pH 8 in Water for Some Commonly Used Primary Amides

Compound	k (m^{-1}-sec^{-1})	$t_{1/2}'$(days pH 8)[a]
Acetamide	6.8×10^{-4} (1.3×10^{-3})[b]	1×10^3
Acrylamide	2.7×10^{-3}	3×10^2
Methylacrylamide	2×10^{-3}	3×10^2
Propionamide	2.6×10^{-3} (1.3×10^{-3})[b]	3×10^2

[a]Half-lives calculated assuming pseudo-first-order kinetics.
[b]Experimental values from reference 49.

REFERENCES

1. Baughman, G.L., and R.R. Lassiter. In: *Estimating the Hazard of Chemical Substances to Aquatic Life, ASTM STP 657,* J. Cairns, Jr., K.L. Dickson, and A.W. Maki, Eds. (Philadelphia, PA: American Society for Testing and Materials, 1978), pp. 35-54.
2. Wolfe, N.L., R.G. Zepp, G.L. Baughman, R.C. Fincher and J.A. Gordon. "Chemical and Photochemical Transformation of Selected Pesticides in Aquatic Systems," U.S. Environmental Protection Agency, Report No. EPA-600/3-76-067 (Washington, DC: U.S. Government Printing Office, 1977).
3. Faust, S.D., and H.M. Gomma. "Chemical Hydrolysis of Some Organic Phosphorous and Carbamate Pesticides in Aquatic Environments," *Environmental Letters* 3:171 (1972).
4. Benson, S.W. *The Foundation of Chemical Kinetics* (New York: McGraw-Hill Book Co., 1960).
5. Shorter, J. *Correlation Analysis in Organic Chemistry* (Oxford, England: Clarendon Press, 1973.
6. Wolfe, N.L., R.G. Zepp and D.F. Paris. "Use of Structure-Reactivity Relationships to Estimate Hydrolytic Persistence of Carbamate Pesticides," *Water Res.* 12:561 (1978).
7. Jaffê, H.H. "A Reexamination of the Hammett Equation," *Chem. Rev.* 53:191 (1953).
8. Taft, R.W. In: *Steric Effects in Organic Chemistry,* M.S. Newman, Ed. (New York: John Wiley and Sons, Inc., 1956), Chapter 13.
9. Redl, G., R.D. Cramer and C.E. Berkoff. "Qualitative Drug Design," *Chem. Soc. Revs.* 3:273 (1974).
10. Neely, W.B., D.R. Branson and G.E. Blau. "Partition Coefficients to Measure Bioconcentration Potential of Organic Chemicals in Fish," *Environ. Sci. Technol.* 8:1113 (1974).
11. Karickhoff, S.W., D.S. Brown and T.A. Scott. "Sorption of Hydrophobic Pollutants on Natural Sediments," *Water Res.* 13(2):241-248 (1979).
12. Leo, A., C. Hansch and D. Elkins. "Partition Coefficients and Their Uses," *Chem. Revs.* 71:525 (1971).
13. Veith, G.D., and D.E. Konasewich, Eds. In: *Proceedings of the Symposium "Structure Activities Relationships in Studies of Toxicity and Bioconcentration with Aquatic Organisms,"* (Burlington, Ontario, Canada: International Joint Commission, 1975), pages 1-72.
14. Stumm, W., and J.J. Morgan. *Aquatic Chemistry* (New York: Wiley-Interscience, 1970).
15. Kirby, A.J. In: *Ester Formation and Hydrolysis and Related Reactions,* C.H. Bamford and C.F.H. Tipper, Eds. (New York: Elsevier Publishing Co., 1972), Chapter 2.
16. Bailey, G.W., A.D. Thurston, Jr., J.D. Pope, Jr. and D.R. Cochrane. "The Degradation Kinetics of an Ester of Silvex and the Persistence of Silvex in Water and Sediments," *Weed Sci.* 18:413 (1970).
17. Cochrane, D.R., J.D. Pope, Jr., H.P. Nicholson and G.W. Bailey. "The Persistence of Silvex in Water and Hydro-soil," *Water Resources Res.* 3:517 (1967).

18. Struif, B., L. Weil and K.E. Quentin. "Verhalten von herbiziden Phenoxyessigsäuren und ihrer Ester in Gewässer," *Vom Wasser* 45:53 (1975).
19. Zepp, R.G., N.L. Wolfe, J.A. Gordon and G.L. Baughman. "Dynamics of 2,4-D Esters in Surface Waters: Hydrolysis, Photolysis, and Vaporization," *Environ. Sci. Technol.* 9:1144 (1975).
20. Paris, D.F., D.L. Lewis, J.T. Barnett, Jr. and G.L. Baughman. "Microbial Degradation and Accumulation of Pesticidies in Aquatic Systems," U.S. Environmental Protection Agency Report No. EPA-660/3-75-007 (Washington, DC: U.S. Government Printing Office, 1975).
21. Mulla, M.S. "Persistence of Mosquito Larvicides in Water," *Mosquito News.* 23:234 (1963).
22. Karinen, J.F., J.G. Lamberton, N.E. Stewart and L.C. Terriere. "Persistence of Carbaryl in the Marine Estuarine Environment. Chemical and Biological Stability in Aquarium Systems," *J. Agr. Food Chem.* 15:148 (1967).
23. Wolfe, N.L., R.G. Zepp and D.F. Paris. "Carbaryl, Propham, and Chlorpropham: A Comparison of the Rates of Hydrolysis and Photolysis with the Rate of Biolysis," *Water Res.* 12:565 (1978).
24. Wauchope, R.D., and R. Haque. "Effects of pH, Light, and Temperature on Carbaryl in Aqueous Media," *Bull. Environ. Contam. Toxicol.* 9:257 (1973).
25. Aly, O.M., and M.A. El-Dib. "Studies on the Persistence of Some Carbamate Insecticides in the Aquatic Environment - I: Hydrolysis of Sevin, Baygon, Pyrolan, and Dimetilan in Waters," *Water Res.* 9:257 (1973).
26. Rodgers, C.A., and D.L. Stallings. "Dynamics of an Ester of 2,4-D in Organs of Three Fish Species," *Weed Sci.* 20:101 (1972).
27. Weiss, C.M., and J.H. Gakstatter. "The Decay of Anti-cholenesterase Activity of Organic Phosphorus Insecticide on Storage in Waters of Different pH," *Advances in Water Poll. Res.* 1:83 (1965).
28. Ruzicka, J.H., J. Thomson and B.B. Wheals. "The Gas Chromatographic Determination of Organophosphorus Pesticides: Part II. A Comparative Study of Hydrolysis Rates," *J. Chromatog.* 31:37 (1967).
29. Eichelberger, J.W., and J.J. Lichtenberg. "Persistence of Pesticides in River Water," *Environ, Sci. Technol.* 5:541 (1971).
30. Drevenkar, V., K. Fink, M. Stipčević and B. Štengl. "The Fate of Pesticides in Aquatic Environment. I. The Persistence of Some Organophosphorus Pesticides in River Water," *Arh. Hig. Rada.* 26:257 (1975).
31. Drevenkar, V., K. Fink, M. Stipčević and B. Tkalčević. "The Fate of Pesticides in Aquatic Environment. II. Dichlorovos in a Model System and in River Water," *Arh. Hig. Rada.* 27:297 (1976).
32. El-Dib, M.A., and O.A. Aly. "Persistence of Some Phenylamide Pesticides in the Aquatic Environment - I. Hydrolysis," *Water Res.* 10:1047 (1976).
33. El-Dib, M.A., and O.A. Aly. "Persistence of Some Phenylamide Pesticides in the Aquatic Environment - III. Biological Degradation," *Water Res.* 10:1055 (1976).

34. Wolfe, N.L., R.G. Zepp, J.A. Gordon, G.L. Baughman and D.M. Cline. "Kinetics of Chemical Degradation of Malathion in Water," *Environ. Sci. Technol.* 11:88 (1977).

35. Wolfe, N.L., R.J. Lai, S.A. Whitlock and G.M. Jett. "Physical and Chemical Parameters That Influence the Rates and Products of Triazine Pesticide Degradation in Water," 174th Meeting of the American Chemical Society, Chicago, IL, August 1977.

36. Ketelaar, J.A.A., H.R. Gersmann and M.M. Beck. "Metal-catalysed Hydrolysis of Thiophosphoric Esters," *Nature* 177:392 (1956).

37. Hoffman, M.R., and B.C.H. Lin. "Kinetics and Mechanisms of the Oxidation of Sulfide by Oxygen: Catalysis by Homogeneous Metal-Phthalocyanine Complexes," *Environ. Sci. Technol.* In press, 1979.

38. Li, G.C., and J.T. Felbeck, Jr. "Atrazine Hydrolysis as Catalyzed by Humic Acids," *Soil Sci.* 114:201 (1972).

39. Jencks, W.P. *Catalysis in Chemistry and Enzymology* (New York: McGraw-Hill Book Co., 1969).

40. Khan, S.U. "Kinetics of Hydrolysis of Fulvic Acid Solution," *Pestic. Sci.* 9:39 (1978).

41. Wolfe, N.L., W.C. Steen and L.A. Burns. "Use of Linear Free Energy Relationships and An Evaluative Model for Phthalate Transport and Fate Estimates," submitted for publication (1979).

42. Hill, J. IV, H.P. Kollig, D.F. Paris, N.L. Wolfe, and R.G. Zepp. "Dynamic Behavior of Vinyl Chloride in Aquatic Ecosystems," U.S. Environmental Protection Agency Report No. EPA-600/3/76-001 (Washington, DC: U.S. Government Printing Office, 1976).

43. Wolfe, N.L. "Organophosphate and Organophosphorothionate Esters: Application of Linear Free Energy Relationships to Estimate Hydrolysis Rate Constants for Use in Environmental Assessment," *Chemosphere*, in press.

44. Verschueren, K. *Handbook of Environmental Data on Organic Chemicals*, New York: Van Nostrand Reinhold Co., 1977).

45. Bolton, R. In: *Comprehensive Chemical Kinetics*, C.H. Bamford and C.F.H. Tipper, Eds. (New York: Elsevier Scientific Publishing Co., 1973).

46. Oyama, K., and T.T. Tidwell. "Cyclopropyl Substituent Effects on Acid Catalyzed Hydration of Alkenes. Correlation by σ^+ Parameters," *J. Amer. Chem. Soc.* 98:947 (1976).

47. Brown, H.C., and Y. Okamoto. "Electrophilic Substituent Constants," *J. Amer. Chem. Soc.* 80:4979 (1958).

48. Charton, M., and B.T. Charton. "Application of the Hammett Equation to Nonaromatic Unsaturated Systems - IX. Electrophilic Addition to Olefins. X. Nucleophilic Addition to Olefins," *J. Org. Chem.* 38:1631 (1973).

49. Mabey, W., and T. Mill. "Critical Review of Hydrolysis of Organic Compounds in Water Under Environmental Conditions," *J. Physical and Chemical Reference Data* 7:383 (1978).

50. Charton, M. "Steric Effects. 6. Hydrolysis of Amides and Related Compounds," *J. Org. Chem.* 41:2906 (1976).

14

BIODEGRADATION OF TOXIC CHEMICALS
IN WATER AND SOIL

Martin Alexander

Laboratory of Soil Microbiology
Department of Agronomy
Cornell University
Ithaca, New York 14853

Organic chemicals introduced into water or soil, either inadvertently or deliberately, are subject to nonbiological and biological changes. Although significant alterations in structure and properties of organic molecules may result from nonbiological processes, the major and often the only mechanism by which such compounds are converted to inorganic products is biological. Moreover, the chief and possibly the sole biological agents for the total conversion of organic compounds to inorganic products in water and soil appear to be microorganisms. Incomplete degradation is frequently of environmental concern because the products of these partial reactions may be (1) more toxic than the original substance, (2) toxic, whereas the parent molecule was nontoxic at environmental concentrations, (3) more persistent and able to endure longer than the original substance, or (4) subject to biomagnification or other biological changes different from those undergone by the precursor molecule. Moreover, toxicology, like other scientific disciplines, is imperfect, and its data base is ever increasing; a putatively innocuous product that is generated in water or soil from a synthetic chemical may, with the progress in toxicological knowledge, show itself to be toxic instead. Because of these legitimate concerns with possible toxicity, persistence and susceptibility to biomagnification, as well as the perennial state of incomplete knowledge of environmental hazards of products of partial breakdown, microbial mineralization must be deemed a real virtue, at least from the environmental viewpoint.

Only recently has attention been given to detailed investigations of the microbial transformations of toxic and other organic molecules in natural waters and soil, and most of the available information in natural ecosystems is derived from investigations of pesticides. These studies are difficult because the prevailing concentrations of the chemicals are extremely low, the biochemical pathways for their catabolism are frequently unknown, the responsible microorganisms are affected by neighboring populations, and substances in the natural ecosystem interfere in the analytical procedures or identifications. For these reasons, studies usually have been conducted using pure cultures of microorganisms and/or concentrations of the chemical far higher than are observed in natural ecosystems. These investigations have resulted in generalizations on how some of these molecules are transformed and on the relative rates of their biodegradation.

The data in this report are derived largely from investigations in our laboratory. Such a bias toward these studies does not reflect a view that these are the best or even the only results that illustrate specific problems or approaches to characterizing the biodegradation of toxic chemicals, but rather the obvious fact that the author is more familiar with details of work done in his than in other laboratories. The research presented was performed by postdoctoral associates and former graduate students without whose efforts this brief summary would not have been possible.

The term *biodegradation* is unfortunate because it includes, and sometimes obscures, a series of distinctly different kinds of processes of toxicological importance in natural ecosystems. In their action on synthetic or natural compounds, the microorganisms of waters and soils may bring about one or more of the following: mineralization, detoxication, cometabolism, activation, change in spectrum of toxicity or defusing. *Mineralization* is the conversion of an organic compound to inorganic products. When an organic molecule is mineralized in soil and waters, the transformation is usually the result of microbial action. *Detoxication* refers to the conversion of a toxicant to innocuous metabolites. Although mineralization is characteristically a detoxication, many other kinds of detoxication are known. *Cometabolism* is the metabolism by microorganisms of a compound that the responsible organisms cannot use as a nutrient. Cometabolism does not result in mineralization; hence, organic products remain (Figure 1). In *activation*, a nontoxic molecule is converted to one that is toxic, or a molecule with low potency is made into a product of greater activity against some species. *Defusing*, the conversion of a potentially hazardous substance into an innocuous metabolite before the potential for harm is expressed, is so far known only in laboratory cultures of individual microorganisms.

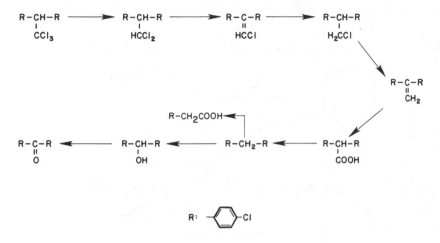

Figure 1. Cometabolism of DDT by a bacterium in culture.

Some of these types of reactions take place among the phenoxy herbicides. Thus, activation is operative when 4-(2,4-dichlorophenoxy)butyrate is acted upon by microorganisms in soil to yield 2,4-dichlorophenoxyacetate[1] (2,4-D) because the latter but not the former is phytotoxic, and 2,4-D is detoxified as it is converted to 2,4-dichlorophenol[2] but is mineralized when it is converted to CO_2. On the other hand, the removal of the butyrate moiety of 4-(2,4-dichlorophenoxy)butyrate to yield 2,4-dichlorophenol[3] is defusing since the herbicidal influence that would have appeared were 2,4-D formed never becomes evident (Figure 2).

Assessments of biodegradation are typically performed using models of natural ecosystems. The models represent convenient and inexpensive test systems to evaluate the kinetics of chemical loss in soils and waters, either aerobic or anaerobic. Simplicity in design and ease of performing analytical determinations are also necessary for a procedure allowing for the screening of groups of compounds. To reduce the cost and the number of operations, often only a single concentration is tested; however, the chemical in the model study frequently is used at toxic levels, and misleading conclusions have sometimes been reached because the rate of decomposition in natural environments is greater than is observed in the laboratory. On the other hand, we have found that chemicals are decomposed at far slower rates at environmental concentrations than at the higher levels usually used in laboratory studies. With some compounds, the reduction in rate is directly proportional to the lower concentration of the chemical, as would be expected from Michaelis-Menten kinetics (Figure 3). With 2,4-D

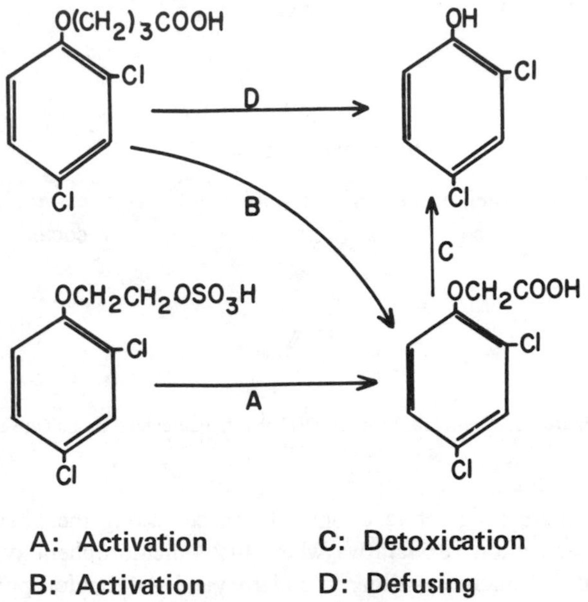

A: Activation   C: Detoxication
B: Activation   D: Defusing

Figure 2. Activation, detoxication and defusing reactions among phenoxy herbicides.

in natural waters, by contrast, the decomposition does not proceed at very low concentrations, possibly because the active populations do not derive energy from 2,4-D metabolism at a sufficiently rapid rate to satisfy their requirements for maintenance energy; hence, the population does not replicate. Such findings that chemicals in waters are decomposed appreciably more slowly than the same compounds at the higher concentrations commonly used in laboratory models have considerable practical significance in sewage, wastewaters and fresh and marine waters, and also in developing and evaluating laboratory tests of biodegradation.

Activation has long been known as a critical feature of insecticide toxicity, but only recently has evidence been obtained that it may occur, directly or indirectly, by microbial action in waters and soils. The formation of 2,4-D from the corresponding butyrate cited above yields a phytotoxin, a reaction known for some time. The methylation of arsenicals to yield

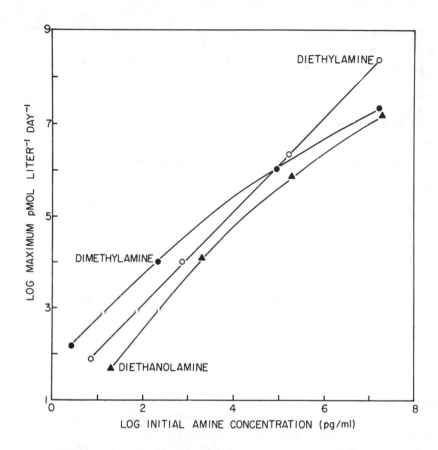

Figure 3. Effect of initial concentration of three amines on the rate of their mineralization in stream water (R.S. Boethling and M. Alexander, manuscript submitted for publication).

trimethylarsine is a second example.[4] More recently, we have become aware of the microbial role in the formation of nitrosamines, $R(R')N-N = 0$, in these environments. Nitrosamines are of significance because many are carcinogenic, mutagenic and teratogenic. Moreover, the few dialkylnitrosamines that have been tested are persistent to some degree in soil and even more so in fresh water (Figure 4), and at least one, dimethylnitrosamine, can be assimilated from soil by plants or be leached through the soil so it may enter ground waters,[6] which are widely used for human consumption. The potential for their formation increases because of the ubiquity of the precursors,

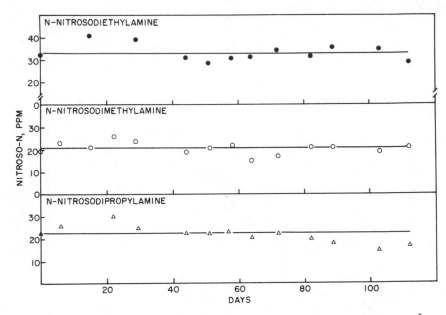

Figure 4. Persistence of three dialkylnitrosamines in samples of fresh water.[5]

The immediate precursors of nitrosamines are secondary amines and nitrite. Secondary amines are found among the synthetic chemicals used for many purposes, and they are readily formed microbiologically from tertiary amines. Thus, a simple dealkylation of trimethylamine is accomplished microbiologically in laboratory medium[7] or in models of natural environments (Figure 5). Quaternary ammonium compounds must give rise to such amines during their degradation, and evidence also exists for microbial N-alkylation so that the formation of secondary amines from a primary amine introduced into soil or water is likely. Although nitrite rarely is deliberately introduced into waters or soils, it is continuously being formed during ammonium oxidation or nitrate reduction by microorganisms; the concentration at any one time is usually quite low, but occasionally high nitrite levels appear as ammonium is oxidized in environments at pH values just slightly above neutrality or as nitrate is being reduced.

The final step in the activation to give rise to nitrosamines has been demonstrated to occur in samples of soil, sewage and fresh water.[8,9] The process can be catalyzed by microbial enzymes or effected by microbial cell components under highly artificial conditions,[10] but the nitrosation in models of natural conditions is more probably the result of a nonenzymatic process. Nevertheless, it appears that microbial products or cell

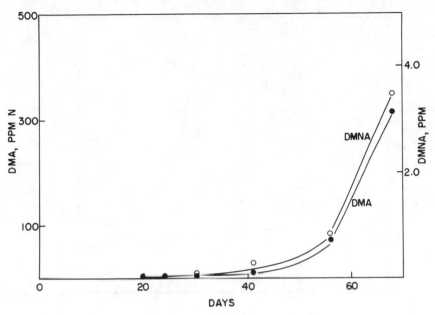

Figure 5. Formation of dimethylamine (DMA) and dimethylnitrosamine (DMNA) in lake water receiving trimethylamine and nitrite.[8]

constituents, if not their enzymes, are implicated. This nitrosation, more-over, takes place at neutral pH values (Figure 6), which was not expected on the basis of studies of the effect of pH on nitrosation in simple aque-ous solutions.

Microorganisms are also known to bring about the activation of insecti-cides. Such conversions may be effected by reactions involving a conversion of P = S to P = O or by an oxidation of sulfur-containing molecules to the corresponding sulfoxide.

One of the most dramatic cases of the microbial conversion of a chem-ical toxic to one group of organisms to a product inhibitory to entirely dissimilar species comes from investigations initiated because of practical problems from the use of pentachlorobenzyl alcohol. This chemical was introduced into Japan for the control of a fungus causing a major disease of rice; i.e., the chemical is antimicrobial. However, once the plant re-mains entered the soil and were subject to microbial decay, the antimicro-bial agent was dehalogenated and the $-CH_2OH$ converted to $-COOH$; the resulting halogenated benzoic acids were not antimicrobial but they were phytotoxic.[12] A case known so far only from studies of soils in the labo-ratory is the conversion of the fungicide thiram to dimethylnitrosamine,

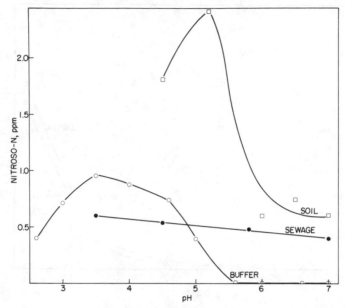

Figure 6. Accumulation of dimethylnitrosamine in soils, sewage, and buffer solutions at different pH values. Each sample received 250 μg dimethylamine and 100 μg of nitrite-N/ml.[11]

$(CH_3)_2N-N = 0.$[9] The role of microorganisms in the conversion of this fungicide to the carcinogen is unknown.

$$[(CH_3)_2N-C(S)S-]_2$$

Microbial populations also may act on a nonpersistent chemical and convert it to a persistent product. For example, dimethylamine is readily metabolized and disappears quickly from soil and waters, but the nitrosamine formed from it is much more long-lived than the amine.[5] With pentachlorobenzyl alcohol, DDT, dieldrin, heptachlor and other chlorinated pesticides, by contrast, the original molecule is persistent but is acted on nevertheless by microorganisms; the metabolites they generate themselves endure so that a persistent parent molecule goes to a long-lived product. Such longevities point to the folly of assuming that a microbiologically effected loss of a chemical in soil or water necessarily removes the potential for environmental hazard. If biodegradation is taken to mean merely the microbiologically induced loss of a substance (that is, the parent molecule), then the occurrence of biodegradation is not *a priori* evidence that the possible danger is eliminated. If, to the contrary, biodegradation is taken to mean mineralization, then one can assume the absence of a problem once the reaction goes to completion. It is the confusion in the use of these terms that has led to a favoring of the words mineralization and cometabolism in place of biodegradation.

By comparing the rates of disappearance of organic compounds in samples of natural environments that have been sterilized with comparable samples that have not received the sterilization treatment, evidence has been obtained for the microbial destruction of a large number of organic molecules. In many instances, however, no microorganism can be isolated that is able to grow on the organic compound and use it as a source of energy, carbon or some other nutrient element. Nevertheless, microorganisms can be obtained which transform the molecule, although not being able to use it for growth. This is a reflection of cometabolism, the metabolism by a microorganism of a compound which it cannot use as a nutrient source. Work in this laboratory on cometabolism in microbial cultures has focused on these reactions, among others,

$$m\text{-chlorobenzoate} \rightarrow 4\text{-chlorocatechol}^{[13]}$$
$$m\text{-nitrophenol} \rightarrow \text{nitrohydroquinone}^{[14]}$$
$$4,4'\text{-dichlorodiphenylmethane} \rightarrow 4,4'\text{-dichlorobenzophenone}^{[15]}$$

Examples from studies in other laboratories could also be cited, the compounds representing insecticides, herbicides, industrially important chemicals, model compounds, etc. In Figure 7 is presented a typical instance suggesting cometabolism in fresh water and sewage.

In cometabolism, the population density of the responsible species does not increase because it has no selective advantage in the presence of the compound and cannot use it as a nutrient. Because an increase in population density of species growing on chemicals is generally paralleled by an increase in degradation rate, cometabolism is characterized by the lack of an increase in disappearance rate with time after introduction of the chemical into a particular environment. These compounds thus are frequently persistent. The likeliest explanation of cometabolism is that the parent molecule is transformed by enzymes of low substrate specificity to yield a product that is not a substrate for other enzymes of the organism, and the product of the first enzymes then accumulates. Products of cometabolism are structurally similar to the original substances, and thus they might be as toxic as the original class of molecules. Molecules with chlorine, nitro and other substituents are frequently subject to cometabolism.

The product of the cometabolism by one species may sometimes be utilized for the growth of another. For example, DDT is cometabolized to yield p-chlorophenylacetic acid (Figure 1), but the latter compound can be used for growth and thus will be destroyed quickly.[16] This phenomenon has not yet been shown to take place in natural waters and soils if it

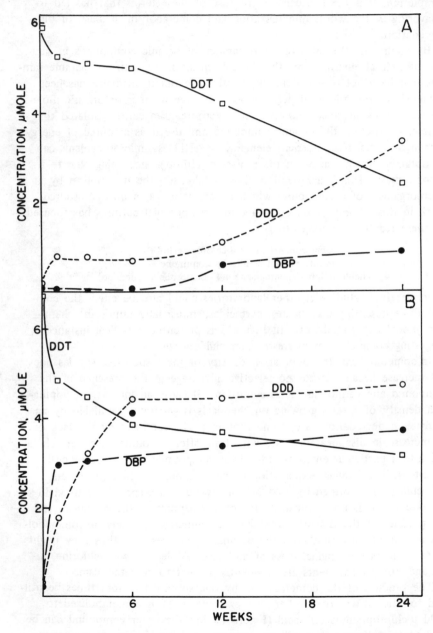

Figure 7. Conversion of DDT to 1,1-dichloro-2,2-*bis*(*p*-chlorophenyl)ethane (DDD) and 4,4'-dichlorobenzophenone (DBP) in fresh water (A) and sewage (B).[16]

does operate there, the reaction of the first organism would proceed slowly, but the final product of the cometabolizing organisms would not accumulate because of its rapid utilization for growth.

Microorganisms are remarkably versatile in their catabolic activity, and a wide array of substances is subject to microbial mineralization or cometabolism in water and soil. However, microorganisms are not omnivorous, and their enzymes are unable to catalyze reactions involving every one of the synthetic organic molecules of economic importance. Should microorganisms have no enzymes active in a needed catabolic sequence or should the substrate be protected from microbial attack because of special properties of the environment, the compound tends to persist. Such persistence may be for several months, but it sometimes can be for many years. A number of hypotheses have been advanced to account for the resistance of these recalcitrant molecules to microbial attack.[17] In some instances, persistence may be attributable to the absence of the requisite enzymes or to the lack of permeability of microorganisms to the substrate. Other mechanisms of recalcitrance have been postulated to explain the longevity of those chemicals for which the resistance is inherent in the molecule itself. Furthermore, frequently biodegradable compounds are destroyed quickly in one environment but are partially or wholly refractory in another environment; in these instances, a physical or chemical characteristic of the environment underlies the resistance to mineralization or cometabolism. Examples have been collected to illustrate some of the environmental characteristics associated with slow microbial attack.[17]

The foregoing has only highlighted some of the major issues concerned with biologically effected changes in organic chemicals, and no attempt has been made to go into experimental details or to review the literature. The purpose has been to cite some of the problems and concepts that have come to the fore as environmental scientists, microbiologists and toxicologists view the vast array of chemicals that have been, are being, or are likely to be introduced into fresh waters, marine environments and soils. Such a general and brief outline cannot define the field, but it is hoped that it will serve as an introduction to a subject of ever-growing importance as society seeks to rid itself of unwanted chemicals and endeavors to prevent the deterioriation of both aquatic and terrestrial ecosystems.

REFERENCES

1. Gutenmann, W.H., M.A. Loos, M. Alexander and D.J. Lisk. "Beta-Oxidation of Phenoxyalkanoic Acids in Soil," *Soil Sci. Soc. Am. Proc.* 28:205-207 (1964).

2. Loos, M.A., R.N. Roberts and M. Alexander. "Formation of 2,4-dichlorophenol and 2,4-dichloroanisole from 2,4-dichlorophenoxyacetate by *Arthrobacter* sp.," *Can. J. Microbiol.* 13:691-699 (1967).

3. MacRae, I.C., M. Alexander and A.D. Rovira. "The Decomposition of 4-(2,4-dichlorophenoxy)butyric acid by *Flavobacterium* sp.," *J. Gen. Microbiol.* 32:69-76 (1963).

4. Cox, D.P., and M. Alexander. "Production of Trimethylarsine Gas from Various Arsenic Compounds by Three Sewage Fungi," *Bull. Environ. Contam. Toxicol.* 9:84-8 (1973).

5. Tate, R.L., and M. Alexander. "Stability of Nitrosamines in Samples of Lake Water, Soil and Sewage," *J. Nat. Cancer Inst.* 54:327-330 (1975).

6. Dean-Raymond, D., and M. Alexander. "Plant Uptake and Leaching of Dimethylnitrosamine," *Nature (London)* 262:394-396 (1976).

7. Tate, R.L., and M. Alexander. "Microbial Formation and Degradation of Dimethylamine," *Appl. Environ. Microbiol.* 31:399-403 (1976).

8. Ayanaba, A., and M. Alexander. "Transformation of Methylamines and Formation of a Hazardous Product, Dimethylnitrosamine, in Samples of Treated Sewage and Lake Water," *J. Environ. Qual.* 3:83-89 (1974).

9. Ayanaba, A., W. Verstraete and M. Alexander. "Formation of Dimethylnitrosamine, a Carcinogen and Mutagen, in Soils Treated with Nitrogen Compounds," *Soil Sci. Soc. Am. Proc.* 37:565-568 (1973).

10. Mills, A.L., and M. Alexander. "N-Nitrosamine Formation by Cultures of Several Microorganisms," *Appl. Environ. Microbiol.* 31:892-895 (1976).

11. Mills, A.L., and M. Alexander. "Factors Affecting Dimethylnitrosamine Formation in Samples of Soil and Water," *J. Environ. Qual.* 5:437-440 (1976).

12. Ishida, M. In: *Environmental Toxicology of Pesticides*, F.M. Matsumura, G.M. Boush and T. Misato, Eds. (New York: Academic Press, Inc., 1972), pp. 281-306.

13. Horvath, R.S., and M. Alexander. "Cometabolism: A Technique for the Accumulation of Biochemical Products," *Can. J. Microbiol.* 16:1131-1132 (1970).

14. Raymond, D.G.M., and M. Alexander. "Microbial Metabolism and Cometabolism of Nitrophenols," *Pestic. Biochem. Physiol.* 1:123-130 (1971).

15. Subba-Rao, R.V., and M. Alexander. "Cometabolism of Products of 1,1,1-trichloro-2,2-*bis*(*p*-chlorophenyl)ethane (DDT) by *Pseudomonas putida*," *J. Agric. Food Chem.* 25:855-858 (1977).

16. Pfaender, F.K., and M. Alexander. "Extensive Microbial Degradation of DDT *in Vitro* and DDT Metabolism by Natural Communities," *J. Agric. Food Chem.* 20:842-846 (1972).

17. Alexander, M. "Nonbiodegradable and Other Recalcitrant Molecules," *Biotechnol. Bioeng.* 15:611-647 (1973).

15

TRANSPORT AND DIFFERENTIAL ACCUMULATION OF TOXIC SUBSTANCES IN RIVER-HARBOR-LAKE SYSTEMS

Walter J. Weber, Jr., James D. Sherrill,
Massoud Pirbazari, Christopher G. Uchrin
and Tin Y. Lo

Environmental and Water Resources Engineering
Department of Civil Engineering
The University of Michigan
Ann Arbor, Michigan   48109

INTRODUCTION

Increased awareness of the environmental impacts of halogenated hydro-carbons such as DDT, Aldrin, Dieldrin, PCBs and PBBs has led to severe restrictions on the use of these materials. Nonetheless, for a variety of reasons, the distribution and accumulation of such substances in the aquatic environment continues. Compounds of this type have extremely long half-lives because of their resistance to biological oxidation. Past applications provide sources which continue to leach into groundwater and surface runoff. PCBs and certain other selected chlorinated organics have a long history of prior use in diverse applications and are incorporated in many finished goods; thus, their distribution to the environment continues as the finished goods deteriorate and are discarded. Further increases result from time lags in environmental transport and dissipation.

Chlorinated organic compounds such as PCBs can thus enter a water body from a variety of point and nonpoint sources, including surface run-off, direct industrial discharge, municipal discharge, aerial fallout, use of marine antifouling paints, etc. Once in the system, regardless of source, transport becomes critical with respect to distribution and environmental impact.

In general, such materials are hydrophobic and of limited solubility and therefore tend to sorb onto detritis, clays, seston, biomass and other suspended solids driving to establish a dynamic equilibrium with the fraction dissolved in the water column.

Suspended solids, although commonly heterogeneous in nature, can be divided into three major categories: (1) inorganic clay and silts which are crystalline in form and which, depending on their nature, exhibit a range of sorption characteristics[1]; (2) organic particulates, which often provide nutrients for bottom fauna[2]; and (3) floc-forming microbes which adsorb pesticide-like materials from a water column and thereby effect a natural purification process.[3]

Chlorinated hydrocarbons have been reported to have a somewhat greater affinity for hydrophobic sites on organic particulates than for hydrophilic solids.[4] It has been noted further that fine particle fractions in sediments are closely associated with organic matter in the form of organomineral complexes. Kaplan and Nissenbaum[5] suggested that humic substances are the major components of the organic matter contained in bottom sediments. Other studies have shown that humic substances have a significant effect on the adsorption of chlorinated hydrocarbons.[6-9]

Virtually all water quality models presently used to describe the behavior and fate of pollutants account only for the transport and distribution dynamics of the homogenous phase, or dissolved pollutants. Pollutants sorbed on or contained within suspended solids constitute a separate phase in a heterogeneous system and can be expected to behave quite differently —chemically, biochemically, hydrodynamically and toxicologically—from dissolved pollutants. Accurate description of the environmental distribution and accumulation of solids-associated pollutants, of the impact of these pollutants on the food webs of the aquatics environment, and of their toxicologic implications to man must take account of the dynamics of transport of the suspended solids.

Of particular concern in this regard are the dredged and enlarged embayments, turning basins and harbors of large rivers tributary to estuaries or large freshwater lakes, such as the Great Lakes. Such rivers transport high burdens of suspended solids during periods of high flow. As river discharges enter the enlarged areas of harbors and embayments, flow velocities decrease and differential sedimentation of suspended solids begins.

Work by Palmer and Munson[10] on the transport of chlorinated hydrocarbons in the Upper Chesapeake Bay showed that comparison of bottom sediment data to (dry) suspended solids data for concentrations of PCB and chlorodane yielded 4 to 10 times higher concentrations in the suspended material. These studies support the hypothesis that suspended materials are likely to be the major transport medium for chlorinated hydrocarbons in river-estuary and river-lake systems. Any toxicological considerations or modeling efforts must take this into account to reasonably represent the real system.

Definition of the transport and distribution of solids-associated pollutants in such harbors, dredged rivers, turning basins and nearshore regions must therefore deal with more than apparent or superficial convective flow phenomena. Because of the small net flow velocities involved, intrusion flow, thermal stratification and resultant complex circulation patterns in harbors, embayments and nearshore regions, suspended matter with its sorbed pollution burden can be differentially accumulated by sedimentation in local flow shear planes.

SCOPE OF STUDY

The Environmental and Water Resources Engineering Laboratory at the University of Michigan is presently conducting a multifaceted investigation of the association of chlorinated hydrocarbons such as PCBs with suspended solids in aquatic environments. This program includes development of a comprehensive mathematical model for simulation and prediction of the distribution and differential accumulation of suspended solids and associated pollutants in river-lake and river-estuary systems.

Quantitative modeling and description of the transport of suspended solids and associated materials such as PCBs requires first the development and calibration-verification of appropriate submodels for: (1) the sorption and exchange dynamics between such materials of interest and the suspended solids and (2) the gravitational or vertical transport of the solids. The first submodel relates to sorption phenomena, and the second to sedimentation dynamics. These submodels must then be incorporated in a comprehensive hydrodynamic model describing hydraulic (advective and dispersive) transport of the solids. The last step is the collection of field data to calibrate and verify the overall modeling project.

The present paper describes results to date from investigations designed to delineate and quantify sorption-desorption equilibria and kinetics for a selected model toxic compound, Aroclor 1254, on different types of suspended solids and bottom sediments. It further presents data from a parallel investigation of the physical characteristics and settling dynamics of these solids. Lastly, a brief theoretical description of the overall system model currently being developed is also presented.

ADSORPTION AND EXCHANGE DYNAMICS SUBMODEL DEVELOPMENT

Selected PCB and Clay Materials

The PCB, Aroclor 1254, was chosen because of its predominance in natural systems, more specifically, in the particular natural system under study, the Saginaw River. Solids included a clay from the Rouge River drainage basin, bottom sediment from the Saginaw River, suspended solids from the Saginaw River, kaolinite clay and montmorillonite clay. The kaolinite and montmorillonite clays are crystalline reference materials, each representing a distinct class of clay structure and surface characteristics.

Solubility Study

The approximate aqueous solubility of Aroclor 1254 at 20°C was determined by continuously agitating 500 mg of the chemical in a sealed 2-liter volumetric flask filled with deionized-distilled water over an extended period of time. An aliquot of aqueous sample was withdrawn periodically, filtered through a 0.45-μ glass fiber filter and analyzed for the PCB. A saturation point of $\sim$ 50 ppb was reached after two months. This approximate solubility limit accords with the value of $\sim$ 56 ppb reported earlier by Hague et al.[11]

Aqueous solutions of Aroclor 1254 were prepared by spiking water samples with the PCB dissolved in pesticide-grade acetone. To illustrate qualitatively the relative solubilities of various isomers, typical chromatograms for an aqueous solution extract of Aroclor 1254 dissolved in hexane are presented in Figure 1. It is evident from comparison of relative peak heights that there are some differences in the aqueous solubilities of the isomers.

Volatilization Studies

Volatilization of Aroclor 1254 was examined at two temperatures. Open 3-liter glass containers containing aqueous solutions of Aroclor 1254 at initial concentrations of 40 ppb were agitated at 4°C and 20°C and sampled periodically over a 20-day period. Figure 2 illustrates concentration decay due to volatilization at each of these two temperatures. It is evident that volatilization losses are substantially greater at higher temperature. It was further determined by examination of various peak areas that the lower chlorine-bearing isomers volatilize most rapidly.

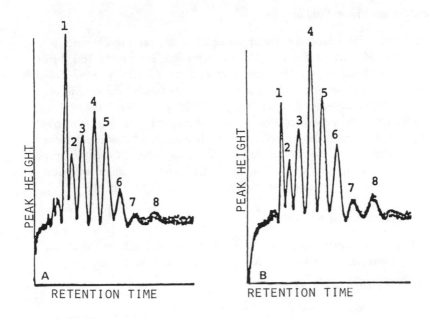

Figure 1. Typical chromatograms for Aroclor 1254: (A) aqueous solution extract; (B) standard solution.

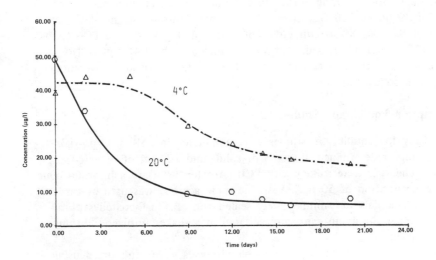

Figure 2. PCB (Aroclor 1254) volatilization rates at 4°C and 20°C.

Head-Space Vapor Studies

To determine head-space vapor concentrations, varying amounts of Aroclor 1254 were spiked into 125-ml hypo-vials containing 100 ml of deionized-distilled water. The vials were then sealed tightly using Teflon® septums and crimped-on aluminum caps, and agitated until equilibrium between the aqueous and vapor phases was achieved. Analyses of 1-ml samples from the headspace were then performed by gas chromatography (GC). At this point, only qualitative conclusions can be inferred from the results. The various isomers of Aroclor 1254 vaporize differently; specifically, the vapor appears to contain larger fractions of isomers with low chlorine content than does the standard Aroclor 1254 mix.

Size Distribution of Solids

Particle size distributions for all solids were obtained by employing sieve, pipette and Coulter Counter techniques. The results are presented in Figure 3.

PCBs Associated with Solids

A modification of the extraction procedure suggested by Goerlitz and Law[12] was used to determine background PCB contents of the Saginaw River sediment and Rouge River drainage basin solids. Hexane:acetone extracts were analyzed by a Varian 2700 Sc[3]H electron capture gas chromatograph (ECGC) using a 150-cm x 2-mm i.d. glass column packed with 3% SE 30 on 80/100 Gas Chrom Q. A typical chromatogram of the extract of Saginaw River sediment is presented in Figure 4. The PCB (Aroclor 1254) level in the sediment was estimated to be 1.45 ppm. No Aroclor 1254 residue was found in extracts from the Rouge River drainage basin solids.

Adsorption Equilibrium Studies

Adsorption equilibrium studies were conducted in 150-ml hypo-vials containing predetermined amounts of solids and 100 ml of organic-free water. The vials were spiked with PCB (Aroclor 1254) to achieve an aqueous concentration of 50 μg/l. Pesticide-grade acetone was used as carrier solvent. The vials were immediately sealed with Teflon®-coated septums and crimped aluminum caps, and agitated in a shaker until equilibrium was reached.

To determine residual concentration, the vials were centrifuged using an International Model IJ Centrifuge to separate solids. A 10- to 20-ml

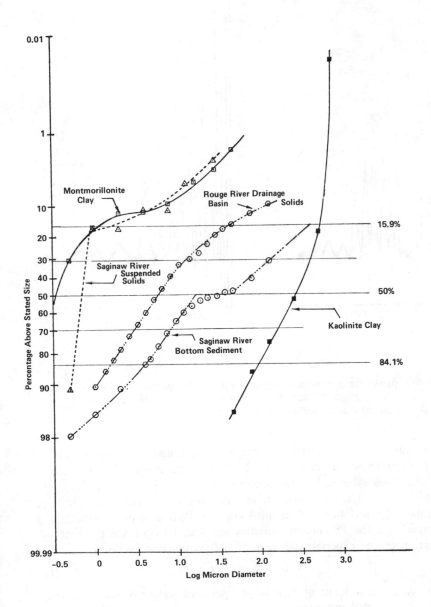

Figure 3. Particle size distributions for the solids studied.

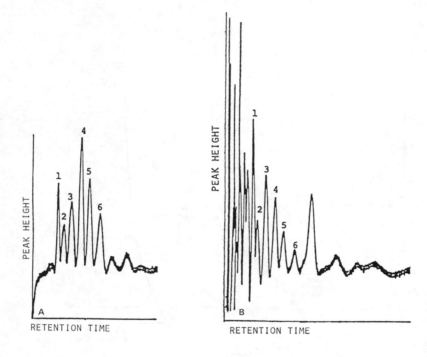

Figure 4. Typical chromatograms of Saginaw River sediment extract compared with
that of a standard solution: (A) standard Aroclor 1254 solution;
(B) Saginaw River sediment extract.

sample was then withdrawn from each vial and transferred to a 6-dram
vial. The sample was extracted with 2 ml of pesticide-grade hexane, and
the extract was subjected to GC analysis.

Several theoretical and empirical relationships were investigated for
mathematical description and quantification of PCB adsorption equilibria
and capacities. The Freundlich equation was found to provide the best
description of the experimental data. This equation has the form

$$q_e = K_F C_e^{1/n} \qquad (1)$$

where: q_e is the amount of PCB sorbed per unit weight of solid, C_e is
the amount of PCB remaining in solution at equilibrium, and K_F and
$1/n$ are characteristic constants relating to sorption capacity and intensity,
respectively. To quantify these sorption isotherm parameters, experimental

data are normalized by plotting the logarithm of q_e vs the logarithm of C_e. The data are then statistically fitted with a straight line of slope $1/n$ and intercept log K_F.

Figure 5 shows equilibrium data and fitted Freundlich isotherms for sorption of PCB (Aroclor 1254) on Saginaw River sediment, Rouge River drainage basin solids, and the reference clays, montmorillonite and kaolinite. Figure 6 presents data and fitted isotherms for natural and organic-stripped Saginaw River sediments and Rouge River drainage basin solids, the former representing a PCB-contaminated sample and the latter a PCB-free surface runoff solid. Stripping of the organic layer was accomplished by utilizing a hydrogen peroxide procedure described by Royse.[13] Comparison of the isotherms given in Figure 6 illustrates the important role played by the organic layer of solids with respect to PCB sorption. Freundlich isotherm parameters are summarized in Table I.

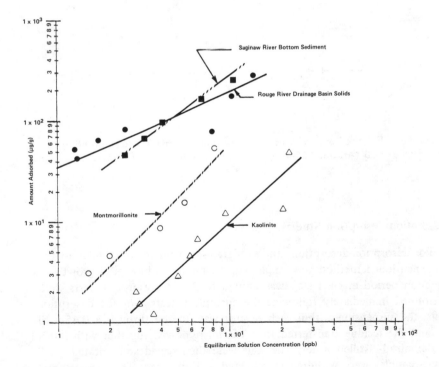

Figure 5. Freundlich isotherms for adsorption of Aroclor 1254 on different solids.

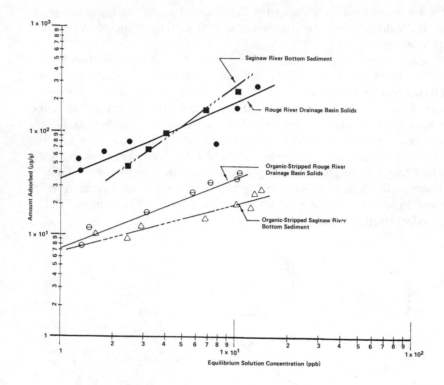

Figure 6. Effect of organic content on Freundlich isotherms for adsorption of PCB (Aroclor 1254) on two types of solids.

Adsorption-Desorption Studies

For adsorption-desorption studies a step-spike procedure similar to the aforementioned method was employed. To minimize biological growth, the agitation period at each step was limited to 24 hr. Desorption tests were performed immediately following the adsorption tests. For the desorption tests, the suspension within each reactor was filtered through a 0.45-μ filter and the residue transferred to a 150-ml hypo-vial together with 100 ml of deionized-distilled water. The vials were then sealed and agitated for 3 days. Samples were withdrawn, extracted and analyzed. The experimental adsorption-desorption data, presented in Figure 7, suggest a reasonably quantitative reversible equilibrium.

Table I. Freundlich Isotherm Parameters
(for q_e in $\mu g/g$ and C_e in $\mu g/l$

Type of Solids	K_F	$1/n$
Kaolinite	0.324	1.470
Montmorillonite	1.342	1.43
Rouge River drainage basin (natural composite sample)	40.047	0.577
Organic-stripped Rouge River drainage basin (natural composite sample)	7.890	0.70
Saginaw River sediment (freeze-dried sample)	14.060	1.158
Organic-stripped Saginaw River sediment (freeze-dried sample)	8.185	0.392
Saginaw River sediment (wet sample, biological floc apparent)	41.687	1.416

Rate Studies

Completely mixed batch (CMB) rate studies were conducted in 2.6-liter airtight Pyrex reactors having no head space. For each study a reactor was filled with organic-free distilled-deionized water, and PCB (Aroclor 1254) injected into the solution to achieve an aqueous concentration of 50 $\mu g/l$, using pesticide-grade hexane as carrier solvent. The reactor was then agitated for 5 min using a glass rod equipped with a Teflon® paddle to accomplish complete mixing. Preliminary studies indicated that 15-20% of the PCB was adsorbed on the reactor surface within a few minutes. After an equilibrium was reached between the solute and the reactor surface, a predetermined amount of solids was introduced to the reactor. At predetermined time intervals 15-ml samples were withdrawn and transferred to centrifuge tubes for solids separation. A 10-ml aliquot of the sample was then placed in a 6-dram vial, together with 2 ml of pesticide-grade hexane. The vial was immediately sealed with a Teflon®-coated screw-on cap, the sample extracted, and the extract subjected to GC analysis. The reactors were equipped with volume-displacement Teflon® rods which were adjusted each time a sample was withdrawn to ensure that no head space developed.

The typical concentration decay profile shown in Figure 8 indicates that a significant fraction of the Aroclor 1254 was adsorbed onto the solids

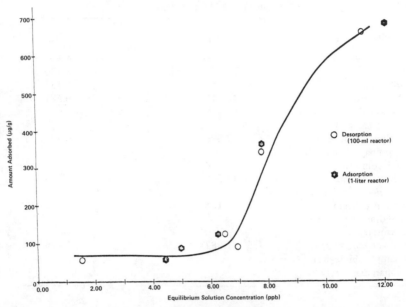

Figure 7.   Experimental adsorption and desorption data for Aroclor 1254 for Saginaw River sediment (step-spike equilibrium procedure).

within a few minutes. Additional experiments were conducted for organic-stripped Saginaw River sediment to investigate the effect of the organic layer on sorption kinetics. The results are also presented in Figure 8. The results indicate that the rate of sorption is decreased by removal of the organic layer. Further, smaller particles demonstrated increased adsorptivity, as illustrated in Figure 9.

Comparative CMB rate studies on Rouge River drainage solids were also performed using deionized-distilled water and filtered Huron River water to evaluate the effect of different background solutions. The data are presented in Figure 10. These results demonstrate the increased sorption competition due to the presence of organic and inorganic components of the river water.

SEDIMENTATION DYNAMICS SUBMODEL DEVELOPMENT

As noted earlier, suspended solids tend to stratify and accumulate in dredged portions of rivers and harbors, turning basins and nearshore regions of the Great Lakes. The suspended materials of interest with respect to

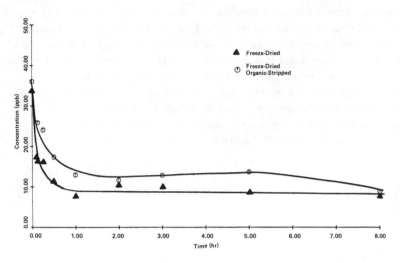

Figure 8. Rate of adsorption of Aroclor 1254 on Saginaw River sediment.

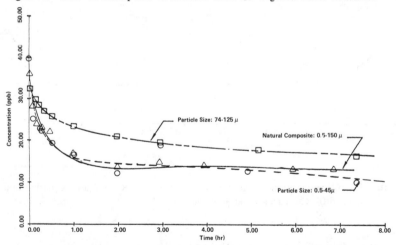

Figure 9. Rates of adsorption of PCB (Aroclor 1254) on Rouge River drainage basin solids of different particle sizes.

PCB transport do not generally act as discrete particles, the settling velocities of which can be described by Stokes' law. Rather, the solids of principal interest interact to coalesce and flocculate, thus necessitating direct settling analysis to determine empirical settling rates.

Equally important in the sedimentation submodel development is the initial spatial distribution of suspended solids in the water column. Two distributions have been considered in the present work, namely, (1) a uniform distribution at the water surface and (2) a uniform distribution throughout the entire water column. A condition of uniform distribution at the water surface most closely obtains in lakes or deep rivers to which tributaries and/or sewage treatment plants containing large amounts of solids discharge near the surface. Surface discharge of dredge spoils is another example of this type of initial distribution. A condition of uniform distribution throughout the water column is a more appropriate assumption for simulation of a river-harbor-lake system.

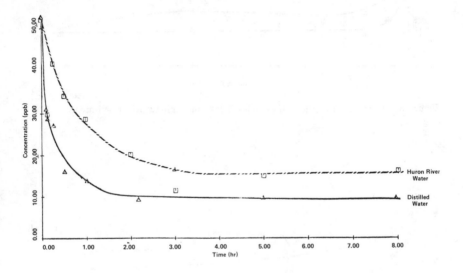

Figure 10. Rate of adsorption of Aroclor 1254 on Rouge River drainage basin solids from different background solutions in CMB reactors, 100 mg/l solids.

Since the principal burden of PCB-bearing suspended solids to the Great Lakes has its origin in river discharges, a uniform distribution of suspended solids throughout the water column was used as the initial spatial distribution throughout this phase of this investigation.

Experimental System and Procedure

Settling analyses were performed by dispersing representative solids in a 10-ft-deep, 5-in.-diam settling column. A schematic diagram of the experimental system is given in Figure 11. Samples were withdrawn at fixed time intervals and 2-ft depth intervals along the column for determination of solids concentration.

Five suspensions were analyzed:

1. montmorillonite clay in distilled water (initial concentration C_0 = 51.7 mg/l);
2. Saginaw River suspended solids in Saginaw River water (C_0 = 25.3 mg/l);
3. Saginaw River Bottom sediment in Saginaw River water (C_0 = 32.7 mg/l);
4. Rouge River drainage basin solids in distilled water (C_0 = 45.1 mg/l); and
5. kaolinite clay in distilled water (C_0 = 64.7 mg/l).

Tables II through VI summarize the experimental results.

A plot of temporal changes in suspended solids concentration for the five suspensions at two depths is presented in Figure 12. The ordinate represents the percentage of suspended solids removed, while the abscissa represents time (t) after initiation of settling.

Based on these experiments, settling behavior can be qualitatively related to the size distribution of solids as follows:

1. Submicron size particles initially settle slowly, but subsequent coalescence and flocculation can markedly increase their settling rate.
2. Particles larger than 1 μm and smaller than 100 μm settle at a relatively constant moderate rate compared to larger solids, which exhibit rapid initial settling.
3. Suspended solids concentrations increase with depth. This concentration gradient is relatively large if the particles have a large standard deviation about their mean size, and is small if the particle sizes are tightly grouped about their mean size. A concentration gradient does not exist when a majority of the particles are submicron in size.

ECOSYSTEM MODEL CONCEPTUALIZATION

Present water quality models describe the distribution of pollutants in terms of advective and dispersive transport associated with the water column alone, assuming that the pollutants move in the same manner as the elements of the water column. The transport of pollutants adsorbed on or contained in suspended solids does not adhere to this assumption because of sedimentation, differential settling velocities and stratification, all of which occur in river, harbor and lake systems. Due to the vertical settling velocities of the suspended solids and proportion of the fluid

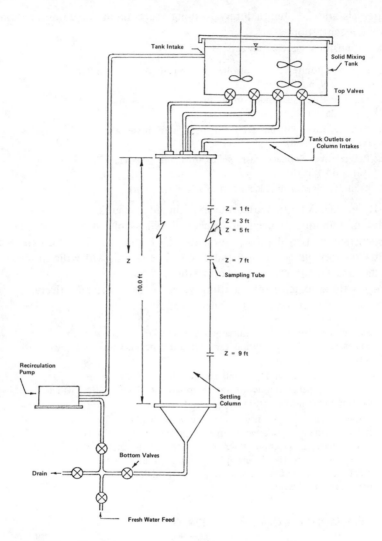

Figure 11. Schematic diagram of experimental settling column.

continuum, particle gradiation in river-harbor-lake systems occurs not only longitudinally but also vertically. To accurately describe the transport, distribution and differential accumulation of suspended solids and associated pollutants, a multidimensional approach must be taken.

The law of conservation of mass in three dimensions yields the following equation,

$$\frac{\delta c}{\delta t} + u\frac{\delta c}{\delta x} + v\frac{\delta c}{\delta y} + w\frac{\delta c}{\delta z} =$$

$$\frac{\delta}{\delta x}(D_x \frac{\delta c}{\delta x}) + \frac{\delta}{\delta y}(D_y \frac{\delta c}{\delta y}) + \frac{\delta}{\delta z}(D_z \frac{\delta c}{\delta z}) - F(c) \quad (2)$$

where:

t = time;

x,y,z = cartesian coordinates;

u,v,w = average velocities in the x,y,z directions;

c = the instantaneous concentration of the particles; and

D = the turbulent diffusion coefficient in the subscripted direction.

Table II. Spatial and Temporal Suspended Solids Concentration Distribution of Montmorillonite Clay in Distilled Water (C_0 = 51.7 mg/l).

Time from Initiation of Settling (hr)	Suspended Solids Concentration (mg/l) at Depth Z Below Water Surface				
	Z = 1 ft	Z = 3 ft	Z = 5 ft	Z = 7 ft	Z = 9 ft
0	48.48	44.95	48.99	60.02	56.06
2	45.45	43.43	45.45	45.45	45.45
24	43.43	43.94	43.43	43.94	42.93
99	34.85	36.87	37.88	37.88	38.89
194	28.79	30.81	31.82	30.81	29.29

For homogeneous, conservative substances, the term F(c) is zero, and for homogeneous, nonconservative substances, it is the reaction rate kC^n. However, since suspended solids in natural waters constitute a heterogeneous phase, this term can become a complex function of time, dispersion, sedimentation velocity, flow, and, in the case of nonconservative substances, reaction velocity. The current research effort is attempting to quantify this function by coupling the aforementioned sorption and sedimentation models. Once the function has been quantified, work on the ecosystem model can progress from the conceptualization stage to the simulation/prediction stage. To accomplish this progression, the water transport terms of Equation 2 must also be quantified. The proposed methodology will

Table III. Spatial and Temporal Suspended Solids Concentration Distribution of Saginaw River Suspended Solids in Saginaw River water (C_0 = 25.3 mg/l).

Time from Initiation of Settling (hr)	Suspended Solids Concentration (mg/l) at Depth Z Below Water Surface				
	Z = 1 ft	Z = 3 ft	Z = 5 ft	Z = 7 ft	Z = 9 ft
0	24.95	24.19	25.70	25.96	25.45
1.0	22.18	24.70	23.44	23.44	23.94
4.0	19.91	19.91	21.42	19.91	20.66
8.0	14.11	16.13	19.91	21.42	17.89
24.0	9.07	9.32	12.60	14.11	–
50.0	7.31	6.30	7.31	9.58	9.07
120.0	–	4.79	5.80	4.54	4.54
288.0	2.02	0.50	0.76	0.50	1.01

Table IV. Spatial and Temporal Suspended Solids Concentration (mg/l) Distribution of Saginaw River Bottom Sediment in Saginaw River Water (C_0 = 32.7 mg/l).

Time from Initiation of Settling (hr)	Suspended Solids Concentration (mg/l) at Depth Z Below Water Surface				
	Z = 1 ft	Z = 3 ft	Z = 5 ft	Z = 7 ft	Z = 9 ft
0	31.07	—	34.66	32.05	33.03
0.07	17.00	17.99	17.66	17.33	16.35
3.0	13.73	15.04	14.72	14.72	13.73
8.0	11.45	11.77	13.08	13.08	13.41
57.0	6.54	7.19	8.83	9.48	9.48
103.0	6.54	7.19	8.83	9.48	9.48

will include simulation of the water movement of the study system using the Navier-Stokes equations of motion. Having thus quantified appropriate water movement parameters and reaction dynamics, the progression can be attempted.

It is anticipated that the ultimate model will employ a finite element solution to Equation 2. This procedure essentially transforms the three-dimensional water body continuum into a series of finite cells, each of which can be considered to be completely mixed. Transport, other than direct inputs, into and out of each cell thus occurs at the interfaces of contiguous cells. A schematic of a two-dimensional cell is displayed in

Table V. Spatial and Temporal Suspended Solids Concentration (mg/l) Distribution of Rouge River Drainage Basin Solids in Distilled Water (C_0 = 45.1 mg/l).

Time from Initiation of Settling (hr)	Suspended Solids Concentration (mg/l) at Depth Z Below Water Surface				
	Z = 1 ft	Z = 3 ft	Z = 5 ft	Z = 7 ft	Z = 9 ft
0	45.0	49.5	43.65	40.95	46.35
0.17	34.65	—	39.15	38.70	37.35
0.6	27.90	33.75	34.65	—	36.90
1.38	23.85	30.60	31.05	34.20	36.0
3.18	18.90	24.30	25.65	29.25	34.65
5.39	12.60	17.10	19.35	24.75	27.90
34.15	—	11.70	13.50	15.75	18.45
146	—	7.65	—	8.10	8.10

Figure 13. A mass balance equation can be written about this segment such that:

$$\begin{matrix}\text{Advective}\\ \text{Transport}\\ \text{in}\end{matrix} - \begin{matrix}\text{Advective}\\ \text{Transport}\\ \text{out}\end{matrix} + \begin{matrix}\text{Dispersive}\\ \text{Transport}\\ \text{in}\end{matrix} - \begin{matrix}\text{Dispersive}\\ \text{Transport}\\ \text{out}\end{matrix} - VF(c) = \frac{Vd(c)}{dt}$$

Thomann,[14] among others, has demonstrated methods for developing this formulation into suitable water quality models. The present research anticipates a similar solution technique.

SUMMARY

The problem of simulation and prediction of suspended-solids borne-pollutants is complex. The University of Michigan's Environmental and Water Resources Engineering Laboratory is conducting a multifaceted and integrated research effort involving development of two submodels to be integrated into a comprehensive system model. Experiments have been designed and preliminary data collected to begin calibration of these submodels. Although the system model is still in the conceptual stage, it is scheduled for completion shortly after development of the two submodels. The overall research effort should represent a significant advance in developing a better understanding of the transport processes involved in the distribution of suspended solids and associated pollutants in natural water systems.

Table VI. Spatial and Temporal Suspended Solids Concentration Distribution of Kaolinite Clay in Distilled Water (C_0 = 64.7 mg/l).

Time from Initiation of Settling (hr)	Suspended Solids Concentration (mg/l) at Depth Z Below Water Surface				
	Z = 1 ft	Z = 3 ft	Z = 5 ft	Z = 7 ft	Z = 9 ft
0	83.85	66.30	57.85	58.50	57.20
0.083	35.10	--	—	--	--
0.12	—	—	41.60	—	—
0.125	--	—	—	42.90	—
0.13	—	—	—	--	46.80
0.17	36.40	--	—	--	--
0.18	—	35.10	--	—	—
0.19	—	--	37.05	—	—
0.20	--	--	—	40.30	—
0.21	—	—	—	—	42.25
0.25	29.90	—	—	--	—
0.27	—	33.80	—	--	--
0.275	—	—	33.15	--	—
0.28	—	—	—	35.10	--
0.42	25.35	—	—	—	—
0.43	—	29.25	—	—	--
0.44	—	—	30.55	—	—
0.45	—	—	—	33.15	—
0.46	—	—	—	--	34.45
1.0	--	24.05	27.95	30.55	31.85
3.0	24.70	—	26.65	—	29.25
11.0	19.50	—	23.40	--	—
24.0	--	13.65	—	20.15	--
72.0	11.70	13.65	16.90	16.90	16.90
242.0	12.35	12.35	13.65	14.30	—

ACKNOWLEDGMENT

The work described herein has been supported in part by the National Oceanic and Atmospheric Administration, U.S. Department of Commerce, through the University of Michigan Institutional Sea Grant Program.

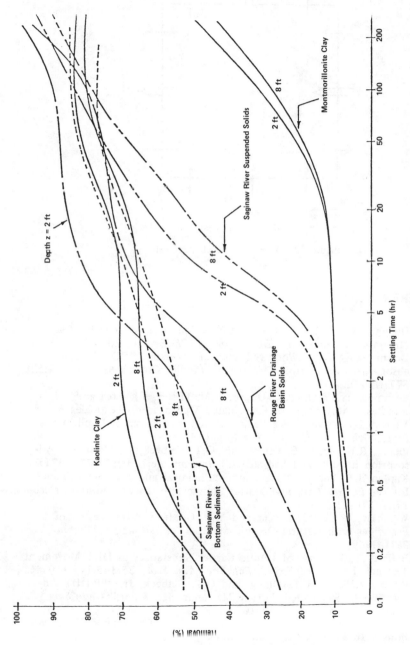

Figure 12. Comparison of suspended solids removal at 2-ft and 8-ft depths below the water surface.

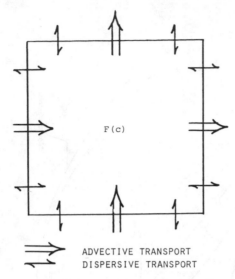

Figure 13. Typical finite element cell.

REFERENCES

1. Huang, J.C., and C.S. Liao. "Adsorption of Pesticides by Clay Minerals," *J. San. Eng. Div., Proc. ASCE* SA5 (1970).
2. Odum, W.E., G. M. Woodwell and C.F. Wurster. "DDT Residues absorbed From Organic Detritus by Fiddler Crabs," *Science* 164:576-577 (1969).
3. Leshniowsky, W.O., P.R. Dugan, R.M. Pfister, J.I. Frea and C.I. Randles. "Adsorption of Chlorinated Hydrocarbon Pesticides by Microbial Floc and Lake Sediment and its Ecological Implications," *Proc. 17th Conf. Great Lakes Res.* (1970).
4. Pierce, R.H., Jr., C.E. Olney and G.T. Felbeck, Jr. "Pesticide Adsorption in Soils and Sediments," *Environ. Lett.* 1:157-172 (1971).
5. Kaplan, I.R., and A. Nissenbaum. "Chemical and Isotopic Evidence for the *in Situ* Origin of Marine Humic Substances," *Limnol. Oceanog.* 17(4):570-82 (1972).
6. Wershaw, R. L., P.J. Burcar and M.C. Goldfrey. "Interaction of Pesticides with Natural Organic Material," *Environ. Sci. Technol.* 3(3) (1969).
7. Bellard, T.M. "Role of Humic Carrier Substances in DDT Movement Through Forest Soil," *Soil. Sci. Soc. Am. Proc.* 35:145-147 (1971).
8. Pierce, R.H., Jr., C.E. Olney and G.T. Felbeck, Jr. "PP-DDT Adsorption to Suspended Particle Matter in Sea Water," *Geo-Chem. Cosmochim. Acta* 38:1061-1073 (1974).
9. Chol, W.W , and K.Y. Chen. "Association of Chlorinated Hydrocarbons with Fine Particles and Humic Substances in Nearshore Surficial Sediments," *Environ. Sci. Technol.* 10(8) (1976).

10. Palmer, H.D., and T.O. Munson. "Sediments as Vehicles for Trace Metals and Chlorinated Hydrocarbons: Upper Chesapeake Bay," in Abstracts with programs, *Geol. Soc. America* 6:62 (1974).

11. Hague, R., D.W. Schmelding and V.H. Freed. "Aqueous Solubility, Adsorption, and Vapor Phase Behavior of A-1254," *Environ. Sci. Technol.* 8:139 (1974).

12. Goerlitz, D.F., and L.M. Law. "Determination of Chlorinated Insecticides in Suspended Sediment and Bottom Material," *J. AOAC* 57(1) (1974).

13. Royce, C.F., Jr. "An Introduction to Sediment Analysis," Arizona State University (1970).

14. Thomann, R.V. *Systems Analysis and Water Quality Management* (New York: Environmental Research Agency, 1971).

16

ROLE OF HUMIC SUBSTANCES IN PREDICTING FATE AND TRANSPORT OF POLLUTANTS IN THE ENVIRONMENT

Shahamat U. Khan

Chemistry and Biology Research Institute
Research Branch, Agriculture Canada
Ottawa, Ontario, Canada

INTRODUCTION

Toxic chemicals are being used in increased quantity for various purposes. Once the toxic chemical finds its way into the environment a major part of it comes in contact with soils and waters, which may act as a sink for many chemicals. Humic substances, the major organic constituents of soils, sediments and waters, are widely distributed in nature, occurring in almost all terrestrial and aquatic environments. They are remarkable materials in that the small amounts present in soils and waters may influence the behavior and fate of many toxic chemicals considerably. The importance and role of humic substances have been indicated in studies conducted over the past decade for a wide variety of chemicals. Numerous examples where binding, persistence, chemical and biodegradation, leachability, bioaccumulation and translocation of toxic chemicals have been shown to bear a direct relationship to humic substances content can be found in several review articles.

Of fundamental importance to any such study is a basic understanding of humic substances. The first section of this chapter will provide some background information on humic substances. This will be followed by a discussion of the role of humic substances in determining the fate and transport of toxic chemicals in the environment.

HUMIC SUBSTANCES

Humic substances are probably the most widely distributed natural products on the earth's surface, occurring in soils, lakes, rivers and the sea. Chemical investigations on humic substances go back more than 200 years. The capacity of humic substances to absorb water and plant nutrients was one of the first observations. Humic substances were thought to arise from the prolonged rotting of animal and plant bodies. Since that time several thousand scientific papers have been written on humic materials, yet much remains to be learned about their origin, synthesis, chemical structure and reactions, and their functions in terrestrial and aquatic environments.

Soils and sediments contain a large variety of organic compounds which may conveniently be grouped into nonhumic and humic substances. Nonhumic substances include those with definite chemical characteristics such as carbohydrates, proteins, amino acids, fats, waxes and low-molecular-weight organic acids. Most of these substances are relatively easily attacked by microorganisms and have a comparatively short life-span in soils and sediments. By contrast, humic substances are more stable and constitute the bulk of the organic matter in most soils. They are acidic, dark colored, predominantly aromatic, hydrophilic, chemically complex, polyelectrolyte-like materials that range in molecular weights from a few hundred to several thousand.

Based on their solubilities, humic substances are usually partitioned into three main fractions (Figure 1):

1. humic acid (HA), which is soluble in dilute alkali but is precipitated on acidification of the alkaline extract;
2. fulvic acid (FA), which is that humic fraction which remains in solution when alkaline extract is acidified, that is, it is soluble in both dilute alkali and acid; and
3. humin, which is that humic fraction that cannot be extracted from the soil or sediment by dilute base or acid.

From the analytical data published in the literature[1] it appears that structurally the three humic fractions are similar but that they differ in molecular weight, ultimate analysis and functional groups content, with FA having a lower molecular weight but higher content of oxygen-containing functional groups per unit weight. The chemical structure and properties of the humin fraction appear to be similar to those of HA. The insolubility of humin seems to arise from its being firmly adsorbed on or

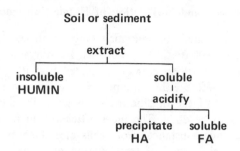

Figure 1. Fractionation of humic substances.

bonded to inorganic soil and sediment constituents.

The yellowish-brown color of the natural water, the so-called aquatic humic material, is the water-extractable fraction of the soil humic materials. It seems to be generally agreed that the differences in the physical and chemical properties between the humic substances in soil and water are relatively small, and it is not necessary, therefore, to consider these as separate groups of organic matter.

The synthesis of humic substances has been the subject of much speculation. The following four hypotheses for the formation of these materials have been listed in the literature: (1) the plant alteration hypothesis; (2) the chemical polymerization hypothesis; (3) the cell autolysis hypothesis; and (4) the microbial synthesis hypothesis.

It is difficult to decide at this time which hypothesis is the most valid. It is likely that all four processes occur simultaneously, although under certain conditions one or another could dominate. However, what all four hypotheses suggest is that the more complex, high-molecular-weight humic materials are formed first and that these are then degraded, most likely oxidatively, into lower-molecular-weight materials. Thus, the sequence of events appears to be HA → FA.

It should be added here that recent evidence shows that marine humic acid can be formed *in situ* from degradation products of plankton and is not necessarily transported from the continent.

With regard to the fresh water humic materials—the water-extractable fraction of the humus in soil—there is at present no reason to believe that this fraction is less complex and, as mentioned before, this should not be considered as a separate group of humic compounds.

Extraction, Fractionation and Purification of Humic Substances

The organic matter content of soils may range from less than 0.1% in desert soils to close to 100.0% in organic soils. In inorganic soils, organic and inorganic compounds are so closely associated that it is necessary to first separate the two before either component can be studied in greater detail. Thus, extraction of the organic matter is generally the first major operation that needs to be done. The most efficient and most widely used extractant of humic substances from soils or sediments is dilute aqueous NaOH. The classical method of fractionation of humic substances is based on differences in solubility in aqueous solutions at widely differing pH levels, in alcohol and in the presence of different electrolyte concentrations. The major humic fractions are HA, FA and humin. Fractionation of HA into hymatomelanic acid or into gray HA or brown HA is not done very often, since such preparations are not very useful.

Characterization of Humic Substances

Elementary analysis provides information on the distribution of major elements (C,H,N,S and O) in humic substances. The major oxygen-containing functional groups in humic substances are carboxyls, hydroxyls and carbonyls. Table I shows some analytical characteristics of much researched HA and FA in our laboratory. Elementary and functional group analysis of HA differ from that for FA in the following respects: (1) HA contains more C,H,N and S but less O than does FA; (2) the total acidity and COOH-content of FA are approximately twice as great as those of HA, and (3) the ratio of COOH to phenolic OH groups is about 3 for FA but only approximately 2 for HA.

Aside from elementary and functional group analysis, the methods most frequently used for the characterization of humic substances can be divided into nondegradative and degradative ones. Nondegradative methods (Table II) include spectrophotometric, spectrometric, x-ray, electron microscopy, electron diffraction, viscosity, surface tension and molecular weight measurements as well as electrometric titrations.

Humic substances, like many relatively high-molecular-weight materials, yield generally uncharacteristic spectra in the ultraviolet (UV) and visible regions. Absorption spectra of alkaline and neutral aqueous solutions of HAs and FAs and of acidic, aqueous FA solutions are featureless, showing no maxima or minima; the optical density usually decreases as the wavelength increases.

Infrared (IR) spectroscopy has been found useful in humic research. However, the assignment of absorptions to certain groupings with the aid

Table I. Analytical Characteristics of HA and FA

Element (%)	HA	FA
C	56.4	50.9
H	5.5	3.3
N	4.1	0.7
S	1.1	0.3
O	32.9	44.8
Functional groups (meq/g)		
Total Acidity	6.6	12.4
COOH	4.5	9.1
Phenolic OH	2.1	3.3
Alcohol OH	2.8	3.6
Quinonoid C=O	2.5	0.6
Ketonic C=O	1.9	2.5
OCH_3	0.3	0.1

Table II. Nondegradative Methods Used for the Characterization of Humic Substances

Spectrophotometry in the UV and visible range
Infrared (IR) spectrophotometry
Nuclear Magnetic Resonance (NMR) spectrometry
Electron Spin Resonance (ESR) spectrometry
Electron microscopy
X-Ray analysis
Viscosity measurements
Surface tension measurements
Molecular weight measurements
Electrometric titrations

of correlation chart is still fraught with considerable uncertainty. It is therefore, always advisable to corroborate spectral data with information obtained by other methods. Figure 2 shows IR spectra of the HA and FA. The absorption bands are broad because of extensive overlapping of individual absorptions. The most striking difference between the two spectra is in the intensities of the bands in the 2900-2800 cm^{-1} region and in the 1725 cm^{-1} band. The HA contains more aliphatic C-H groups than does the FA. The 1725 cm^{-1} band is very strong in the case of FA but only a shoulder for HA, thus substantiating the chemical data shown earlier that indicated that the FA contains considerably more COOH groups than does the HA. In general, IR spectra of humic substances of diverse origins

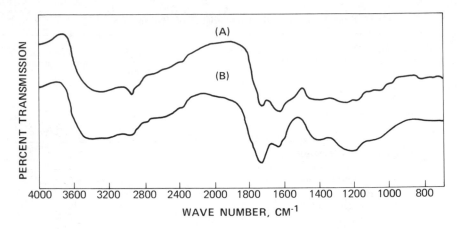

Figure 2. Infrared spectra of (A) HA and (B) FA.

are very similar, which may indicate the presence of essentially similar chemical structures differing mainly in the contents of functional groups.

Because untreated humic materials are not soluble in organic solvents such as $CHCl_3$ or CCl_4, the use of nuclear magnetic resonance (NMR) spectrometry has been confined to methylated humic fractions or to degradation products. So far, proton-NMR has provided little information on the chemical structure of humic materials.

Humic substances are known to be rich in stable free radicals which most likely play important roles in polymerization-depolymerization reactions, in reactions with other organic molecules, including pesticides and toxic pollutants, and in the physiological effects that these substances are known to exert. Electron spin resonance (ESR) spectra of aqueous HA and FA solutions usually consist of single lines devoid of hyperfine splitting with g-valves ranging from 2.0031 to 2.0045, line widths from 2.0 to 3.6 G and free radical concentration from 1.4×10^{17} to 34.4×10^{17} spins/g.[2]

X-Ray analysis has been used by several workers for elucidating the structure of soil humic substances. Naturally occurring humic substances are noncrystalline. Kodama and Schnitzer[3] concluded that the C skeleton of FA consists of a broken network of poorly condensed aromatic rings with appreciable numbers of disordered aliphatic or alicyclic chains around the edges of the aromatic layers.

Several workers have used the electron microscope for observing shapes and sizes of HA and Fa particles. A few years ago Khan[4] published electron micrographs of HAs which show a loose spongy structure with many internal spaces. Recently, Schnitzer and his co-workers[5,6] used a scanning electron microscope (SEM) to explore the effect of pH on the shape, size and degree of aggregation of HA and FA particles. Compared to the conventional transmission electron microscope (TEM), SEM offers the following advantages: (1) it yields three-dimensional pictures of samples; (2) surfaces can be directly observed; and (3) the orientation of particles with respect to each other and to other sample features can be observed.

Molecular weights reported range from a few hundred for FAs to several millions for HAs.[1] There is considerable disagreement in the literature between methods measuring the same type of molecular weight. Osmometry, cryoscopy and the diffusion and viscosity methods have given values of about 700-26,000 for humic acids and 200-300 for fulvic acids, whereas ultracentrifugation and light-scattering methods have given 30,000-80,000.[1]

Degradative Methods

Degradative methods used for the characterization of humic materials (Table III) include oxidation in alkaline and acidic media, reduction, hydrolysis, thermal, radiochemical and biological degradation. With complex materials such as HAs and FAs, degradation is often a useful approach for obtaining information on their chemical structures. The expectation here is to produce simpler compounds than can be identified and whose chemical structures can be related to those of the starting material. The method of choice should not be too drastic or lead to the formation of unwanted by-products and/or artifacts. Recent advances in the development of efficient gas chromatographic/mass spectrometric-(GC/MS)-computer systems, that make possible the separation and the qualitative and quantitative identification of microamounts of organic compounds in a complex mixture, have greatly enhanced the efficiency of chemical and possibly also of biological degradation as structural tools.

The general procedure which has been used by Schnitzer and his co-workers[1] for the degradation of humic substances and identification of degradation products is shown in Figure 3. Following degradation, the products are extracted into organic solvents, methylated and then separated by column and thin-layer chromatography. Portions of the fractions are then further separated by preparative gas chromatography into well-defined compounds which are identified by comparing their mass- and micro-IR-spectra with those of authentic specimens. Other portions of each fraction are injected directly into a GLC/MS-computer system.

Maximum yields of major oxidation products resulting from the degradation of the HA and FA are listed in Table IV.[1] The data show that HA contains about equal proportions of aliphatic and phenolic structures but

Table III. Degradative Methods Used for the Characterization of Humic Substances

Oxidation	Thermal Degradation
Reduction	Biological Degradation
Hydrolysis	

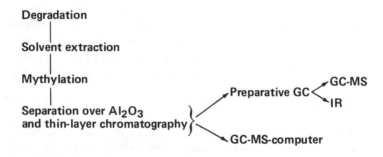

Degradation

|

Solvent extraction

|

Mythylation

|

Separation over Al$_2$O$_3$ and thin-layer chromatography → **Preparative GC** → **GC-MS** / **IR**

→ **GC-MS-computer**

Figure 3. Procedure used for the chemical degradation of humic materials and for identification of products.

a greater percentage of benzenecarboxylic structures or structures producing benzenecarboxylic acids on oxidation. By contrast, the FA contains more phenolic than benzenecarboxylic and aliphatic structures. It is interesting that the HA and FA contain approximately equal proportions of aliphatic structures. Thus, as far as chemical structure is concerned, the HA and FA are quite similar, except that the FA is richer in phenolic but poorer in benzenecarboxylic structures than the HA. The main difference between the two materials is that the HA contains more C and N and fewer COOH groups per unit weight, but it has a higher particle or molecular weight than the FA and is thus less soluble in aqueous solutions at pH < 7.0.

The Chemical Structure of Humic Substances

From the data that have been obtained over a number of years it appears that up to 50% of the aliphatic structures in HAs and FAs consist of n-fatty acids esterfied to phenolic OH groups.[1] The remaining aliphatics are made up of more "loosely" held fatty acids and alkanes which seem to be physically adsorbed on the humic materials and which are not structural humic compounds, and possibly of aliphatic chains joining

Table IV. Major Oxidation Products of HA and FA

Major Products	HA (%)	FA (%)
Aliphatic	24.0	22.2
Phenolic	20.3	30.2
Benzenecarboxylic	32.0	23.0
Total	76.3	75.4
Ratio $\dfrac{\text{benzenecarboxylic}}{\text{phenolic}}$	1.6	0.8

aromatic rings. As shown by a wide variety of chemical degradation experiments on HAs and FAs extracted from soils, which differ widely in locations and pedological histories, the major HA and FA degradation products are phenolic and benzenecarboxylic acids. These could have originated from more complex aromatic structures or could have occurred in the initial humic materials in essentially the same forms in which they were isolated but held together by relatively weak bonding. If the latter hypothesis is correct, then phenolic and benzenecarboxylic acids would be the "building blocks" of humic materials.

It is noteworthy that X-ray analysis, electron microscopy and viscosity measurements of FA point to a relatively open flexible structure perforated by voids of varying dimensions that can trap or fix organic and inorganic compounds that fit into the voids, provided the charges are complimentary. A chemical structure that is in harmony with many of the requirements discussed earlier has been proposed by Schnitzer.[1] This molecular arrangement (Figure 4) may account for a significant part of the FA structure. Each of the compounds that make up the structure has been isolated from FA without and after chemical degradation. Bonding between building blocks is by hydrogen-bonds, which makes the structure flexible, permits the building blocks to aggregate and disperse reversibly, depending on pH, ionic strength etc., and also allows the FA to react with organic and inorganic soil constituents either via oxygen-containing functional groups on the large external and internal surfaces, or by trapping them in internal voids.

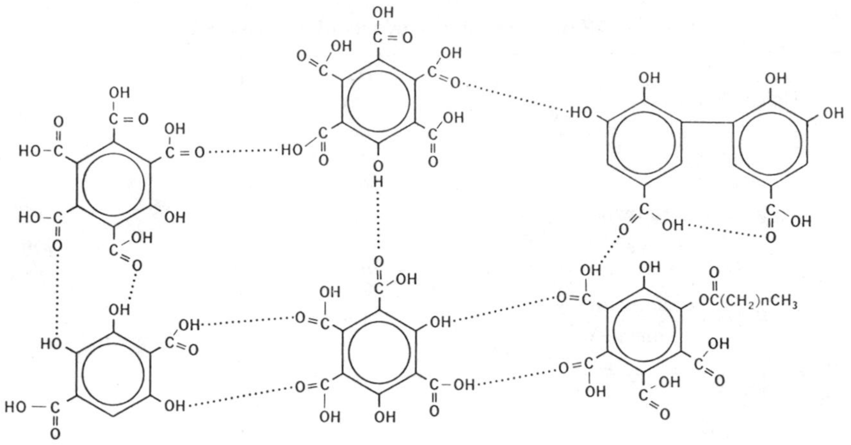

Figure 4. A partial chemical structure for FA.

HUMIC SUBSTANCES-TOXIC CHEMICALS INTERACTION

Humic materials are capable of attacking and degrading soil minerals by complexing and dissolving metals and transporting them within soils and waters.[7,8] Especially active in this regard is FA, which has a relatively low molecular weight and is a water-soluble humic substance. The dissolution of cations from chlorite by 0.2% FA solution is shown in Table V. These data show that FA can degrade clay minerals, especially those rich in Fe, in a relatively short time, thus bringing substantial amounts of Al, Fe and Mg into aqueous solution and so enhancing the mobility of these metals in soils and waters, some of which may be toxic.

The ability of humic materials to complex common heavy metals may result in reducing their immediate toxic effect, and also in increasing the possibilities for accumulation of elements in organisms with long-term consequences. Presence of humic materials in drinking water may pose a possible hazard to human organisms.[9] The fixation of certain essential elements, such as copper and iodine, on humic materials is supposed to decrease the availability of the elements to the organisms. The fixation can take place either in natural waters or inside the organisms after a long-term exposure to humic waters.

There has been considerable evidence that suggests that humic substance can solubilize organic compounds that are otherwise water-insoluble, and so modify their behavior and activity. Solubilization of some dialkyl

Table V. Dissolution of Metals from Chlorite by 0.2% Aqueous FA Solution (expressed in mg/g of sample and 1 liter of solution)

Time	Leuchtenbergite			Thuringite		
(hr)	Al	Fe	Mg	Al	Fe	Mg
3				3.50	8.63	0.84
40				11.30	32.00	3.08
48	2.00	0.40	3.50			
120	3.20	0.90	5.70			
144				19.30	53.10	5.30
192	3.20	1.00	6.90	21.90	61.10	6.03
312	4.20	1.00	8.50	23.70	65.30	6.73

phthalates by aqueous FA solution is shown in Table VI.[10] It appears that the amount of dialkyl phthalates solubilized depends on the type of phthalate. Analysis of IR spectra showed that there was no chemical interaction between the FA and the phthalate, but that the latter was firmly adsorbed, possibly by hydrogen bonding, on the FA surface. Weshaw et al.[11] have shown that the solubility of DDT in 0.5% aqueous sodium humate solution is at least twenty times greater than that in water (Figure 5). Sodium humate appreciably lowers the surface tension of water.[12] Ballard[13] has shown that humic substances can act quantitatively important carriers of DDT in the organic layer of a forest soil (Table VII). Thus, the downward movement of DDT in the organic layer can be greatly increased by application of fertilizers such as urea, which raises the pH of humus layer, thereby promoting leaching of those carriers having humic materials properties.

Several methods for the removal of humus from water have been employed.[9] The most common methods are (1) flocculation, (2) bleaching-mineralization, and (3) filtration. Bleaching-mineralization includes chlorination, ozonation and UV irradiation. From a pollution point of view it is important to know more about the ability of these treatments to mineralize humus in water and the nature of the compounds thus produced.

Recently, Chen et al.[14] investigated the effect of UV irradiation on dilute FA solution. Table VIII shows the compounds identified among products resulting from the UV irradiation of dilute FA at various pH. Major compounds identified were benzene-di-, -tri-, -tetra-, -penta-, and -hexa-carboxylic acids. Small amounts of n-C_{16} and n-C_{18} fatty acids were also identified.

Table VI.   Solubilization of Dialkyl Phtalates by Aqueous FA Solution (pH = 2.35)

Compound	Amount Solubilized (g phthalate/g FA)
Bis(2-ethylhexyl) Phtalate	1.64
Dicyclohexyl Phtalate	0.52
Dibutyl Phthalate	0.29

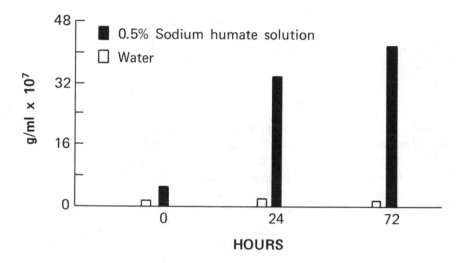

Figure 5.   Solubility of DDT in water and sodium humate solution.

Especially noteworthy is the absence of phenolic compounds among the organics identified since both benzenecarboxylic and phenolic acids are the major FA degradation products. UV irradiation belached the FA solutions until they became colorless, so that the failure to detect phenolics among irradiation products may mean that these compounds are destroyed by irradiation and are the chromophores that give FA its distinctive color. Thus, while the technical aspects of humus removal by bleaching-mineralization methods may be acceptable, it seems at present to be equally important to know the nature of the compounds that are produced in the treated water. Some of them may be potentially harmful to humans.

Substantial evidence exists to indicate that pesticides and other man-made organic chemicals can form stable complexes with humic substances and that such binding greatly increases the persistence of these compounds

Table VII. Leaching of DDT from Forest Floor Columns[13]

Treatment	Days After Treatment				
	3	11	18	25	30
DDT	0.14	0.00	0.00	0.00	----
DDT & Urea	1.06	1.26	0.83	0.72	0.53
Urea	0.00	0.00	0.00	0.00	----
Control	0.00	0.00	0.00	0.00	----

Table VIII. Compounds Identified Among Products Resulting from UV-Irradiation of 1.0 g FA

Compound	0.01% FA solon		
	pH 3.5	pH 7.0	pH 11.0
Benzenedicarboxylic Acid	3.2	----	1.2
Benzenetricarboxylic Acid	23.6	21.4	7.0[
Benzenetetracarboxylic Acid	2.0	41.0	4.0
Benzenepentacarboxylic Acid	1.9	11.3	17.8
Benzenehexacarboxylic Acid	11.3	4.0	20.1
n-C_{16} Fatty Acid	1.5	----	0.6
n-C_{18} Fatty Acid	1.3	----	0.7

in the soil and aquatic environments.[15] Binding of an organic compound with humic materials may render the molecule innocuous thereby allowing slow degradation in the buried state to products that pose short- or long-term problems. On the other hand, binding of toxic chemicals with humic substances may also represent a mechanism of immobilizing the toxicant so that other environmental processes can degrade the molecule.

It is well known that FA is present in many surface waters and imparts a yellow to brown color in natural waters. Because of its acidic functional groups it may chemically degrade a wide variety of pesticides. Recently, the catalytic effect of FA on the hydrolysis of atrazine was examined by Khan.[16] The hydrolysis of atrazine to the nonphytotoxic hydroxy derivative, 2-ethylamino-4-hydroxy-6-isopropylamino-s-triazine, was the only chemical degradation reaction observed in this study. The half-lives of atrazine calculated from the slope of the first-order plots are shown in Table IX. The half-life of atrazine was shortened with increases in the amount of FA in solution (Table IX). Furthermore, the half-life values were lowest at low pH and increased with increasing pH of the solution. For comparison,

Table IX. Half-Lives of Hydrolysis of Atrazine in Aqueous FA

Concentration of Fa (mg/ml)	pH	Half-Life (t-½ day) 25°C	40°C
0.5[a]	2.9	35	12
	4.5	174	42
	6.0	398	96
	7.0	742	199
5.0[a]	2.4	5	2
	4.5	16	6
	6.0	52	16
	7.0	87	25
HA (0.02 g/ml)[b]	4.0	2	0.7
Acetate Buffer (0.08 M)[b]	4.0	71	18
Distilled Water[b]	4.0	244	——
Potassium Acid Phthalate[c]	4.0	9	3

[a]From reference 16.
[b]From reference 17.
[c]From reference 18.

other data for HA and buffer solutions have been also shown in Table IX. Note the half-life of atrazine in water at 25°C is about 244 days.[17]

Humic materials have the potential for promoting the nonbiological degradation of many pesticides. Because surface waters and soil solutions contain humic substances, which can strongly absorb UV light, it has been suggested that these materials can perhaps act as photosensitizers for other nonabsorbing pesticides. Recently, we investigated the photodecomposition of a widely used herbicide, atrazine in dilute FA solutions, a system that is likely to simulate conditions that prevail when some light strikes wet soil and water surfaces to which pesticide has been applied. The pathway of the photolytic degradation of atrazine in the absence and presence of FA is shown in Figure 6.[19] The photolysis of atrazine (I) in water in the absence of FA produces only one product, hydroxyatrazine (II). However, photolysis under the same conditions in the presence of FA also yielded N-dealkylated compounds III and IV, demonstrating N-dealkylation in addition to hydrolysis. Further photochemical dealkylation of III and IV gave rise to a totally N-dealkylated analog, namely, 2-hydroxy-4,6-diamino-s-atriazine (V). In another experiment it was observed that photolysis of hydroxyatrazine (II) in the presence of FA yielded compounds III, IV and V, thereby indicating that either FA or its photoproducts, or both, assisted successive N-dealkylation.

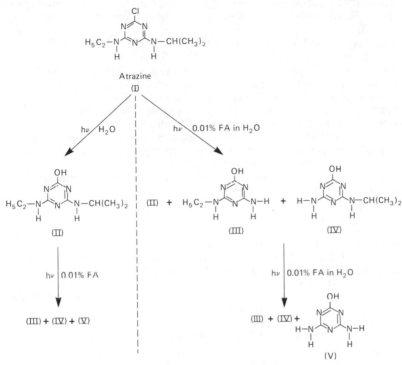

Figure. 6. Pathway of photolytic degradation of atrazine in aqueous FA solution.[19]

CONCLUSIONS

The data presented herein serve to emphasize that the behavior and fate of pesticides in soil and waters are influenced by humic substances. However, much remains to be learned about the interactions of these substances with other pesticides. With the modern analytical techniques presently available, there should be rapid progress in this area in the near future. Significant advances may then be expected in our understanding of the many roles that humic substances play in influencing pesticide behavior in soil and aquatic environments.

REFERENCES

1. Schnitzer, M., and S.U. Khan. *Humic Substances in the Environment* New York: Marcel Dekker, Inc., 1972).

2. Riffaldi, R., and M. Schnitzer. "Electron Spin Resonance Spectrometry of Humic Substances," *Soil Sci. Soc. Am. Proc.* 36:301-305 (1972).
3. Kodama, H., and M. Schnitzer. "X-Ray Studies of Fulvic Acid, a Soil Humic Compound," *Fuel* 46:87-94 (1967).
4. Khan, S.U. "Distribution and Characteristics of Organic Matter Extracted from the Black Solometric and Black Chemozemic Soils of Alberta: The Humic Acid Fraction. *Soil Sci.* 112:401-409 (1971).
5. Schnitzer, M., and H. Kodama. "An Electron Microscopic Examination of Fulvic Acid," *Geoderma* 13:279-287 (1975).
6. Chen, Y., and M. Schnitzer. "Scanning Electron Microscopy of a Humic Acid and of a Fulvic Acid and its Metal and Clay Complexes," *Soil Sci. Soc. Am. J.* 40:682-686 (1976).
7. Kodama, H., and M. Schnitzer. "Dissolution of Chlorite Minerals by Fulvic Acid," *Can. J. Soil Sci.* 53:240-243 (1973).
8. Schnitzer, M., and M. Kodama. "The Dissolution of Micas by Fulvic Acid," *Geoderma* 15:381-391 (1976).
9. Gjessing, E.T. *Physical and Chemical Characteristics of Aquatic Humus* (Ann Arbor, MI: Ann Arbor Science Publishers, Inc., 1976).
10. Matsuda, K., and M. Schnitzer. "Reaction Between Fulvic Acid, a Soil Humic Material, and Dialkyl Phthalates," *Bull. Environ. Contam. Toxicol.* 6:200-203 (1971).
11. Wershaw, R.L., P.J. Burcar and M.C. Goldberg. "Interaction of Pesticides with Natural Organic Materials," *Environ. Sci. Technol.* 3:271-273 (1969).
12. Visser, S.A. "Oxidation-Reduction Potentials and Capillary Activities of Humic Acids," *Nature* 204:581 (1964).
13. Ballard, T.M. "Role of Humic Carrier Substances in DDT Movement Through Forest Soil," *Soil Sci. Soc. Am. Proc.* 25:145-147 (1971).
14. Chen, Y., S.U. Khan and M. Schnitzer. "Ultraviolet Irradiation of Dilute Fulvic Acid Solutions," *Soil Sci. Soc. Am. J.* 42:292-296 (1978).
15. Khan, S.U. In: *Fate of Pollutants in the Air and Water Environments,* I.H. Suffet, Ed. (New York: John Wiley & Sons, 1977), p. 367.
16. Khan, S.U. "Kinetics of Hydrolysis of Atrazine in Aqueous Fulvic Acid Solution," *Pestic. Sci.* 9:39-43 (1978).
17. Li, G., and G.T. Felbeck, Jr. "Atrazine Hydrolysis as Catalyzed by Humic Acids," *Soil Sci.* 114:201-209 (1972).
18. Brown, N.P.H., C.G.H.L. Furmidge and B.T. Grayson. "Hydrolysis of the Triazine Herbicide, Cyanazine," *Pestic. Sci.* 3:669-678 (1972).
19. Khan, S.U., and M. Schnitzer. "UV Irradiation of Atrazine in Aqueous Fulvic Acid Solution," *J. Environ. Sci. Health.* B13:299-310 (1978).

CHAPTER 17

TERRESTRIAL MICROCOSM TECHNOLOGY IN ASSESSING FATE, TRANSPORT AND EFFECTS OF TOXIC CHEMICALS

James W. Gillett

Corvallis Environmental Research Laboratory
U.S. Environmental Protection Agency
200 S. W. 35th
Corvallis, Oregon 97330

INTRODUCTION

In a search for more rapid, yet informative tests of the fate, transport and effects of pesticides, toxic substances and pollutants in general, a wide variety of laboratory-model ecosystems or microcosms have been developed. This book presents the opportunity to review and examine how such systems have been used and what may lie ahead.

The past two decades have seen an enormous increase in public and scientific awareness of the potential and often very real problems caused by the distribution of chemicals in the environment—into wildlife, the food chain and even humans. The result has been legislation and regulation based on specific tests for which theory and experience have provided predictions of hazard or safety. These tests are considered reliable when performed in a certain manner, and application factors could be established to set the rate of exposure. With the enactment of the amended Federal Insecticide, Fungicide, and Rodenticide Act[1] and the Toxic Substances Control Act,[2] the requirements have become increasingly detailed. In essence, they provide for "cradle to grave" regulation of the manufacture, transport, use, storage and disposal of all potentially toxic substances, whether their use is intentionally dispersive (as for pesticides) or simply provides distribution as an adventitious result.

Faced with the staggering prospect of the evaluation of some 70,000 industrial chemicals and the reassessment of over 600 pesticidal agents, and with approximately 1000 new chemicals introduced each year,[3] a number of researchers have turned to various laboratory systems which might yield an integrated response and afford means of evaluating at least classes of chemicals. Thus, in 1971 Metcalf and co-workers[4] introduced the "farm pond" model laboratory ecosystem (Figure 1), which has subsequently been used to test the environmental fate of over 100 substances.[5]

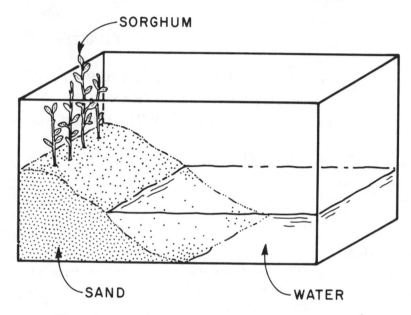

Figure 1. Aquatic/terrestrial model ecosystem of Metcalf *et al.*[4]

During the same period biologists, and particularly ecologists, sought to free their experiments from the vagaries of seasonal changes and uncontrolled forces by bringing pieces of the environment literally into the laboratory. These studies were focused on *interrelationships* and *processes*, instead of on single species responses. Through such devices as the soil core microcosm[6] (V, Figure 2), ecologists have been able to assess such critical phenomena as the relationship of ecosystem complexity to stability (resistence and resilience) in the face of chemical insult.[7]

This emerging terrestrial microcosm technology has been reviewed by Gillett and Witt[8] with the able assistance of many colleagues, with regard to the state-of-the-art, our ability to set forth a protocol or series of protocols which might be used in evaluation or regulation, and the research needed to improve or better define such protocols. Although many of the practical and theoretical problems of these systems are shared with aquatic

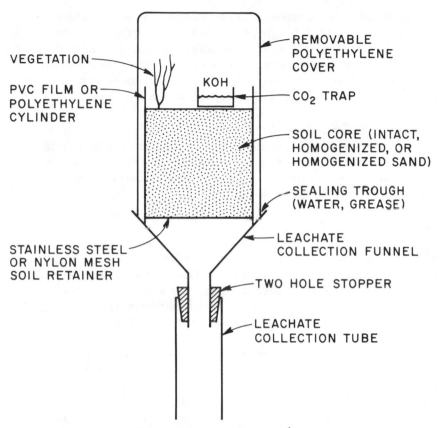

VEGETATION

PVC FILM OR
POLYETHYLENE
CYLINDER

KOH

STAINLESS STEEL
OR NYLON MESH
SOIL RETAINER

REMOVABLE
POLYETHYLENE
COVER

CO_2 TRAP

SOIL CORE (INTACT,
HOMOGENIZED, OR
HOMOGENIZED SAND)

SEALING TROUGH
(WATER, GREASE)

LEACHATE
COLLECTION FUNNEL

TWO HOLE STOPPER

LEACHATE
COLLECTION TUBE

Figure 2. Soil core microcosm.[6]

systems, only those based principally on soil will be considered herein. From the evaluation of these terrestrial systems, it will be clear that microcosms are not a panacea, but are rather part of a logical framework that may offer a reasonable alternative to accomplishment of a seemingly impossible task—the individual testing of each chemical in all manner of tests and assays.

The Systems

Irrespective of the researchers' design or interest, microcosms share certain characteristics, whether terrestrial or aquatic. Essentially, a microcosm is a controlled, reproducible laboratory system which attempts to simulate the processes and interactions of a natural ecosystem. It has a boundary and is under the control of the investigator regarding environmental variables; the components can be described or defined and are relevant to real world situations. Microcosms are not expected to be self-sustaining for long periods of time, hence the processes under study must fit within the scale and scope of the system. The critical element in the definition of microcosms is that they are *process-oriented*, whether addressing soil chemodynamics of a chemical or the soil community response to it.

Another significant facet of microcosm technology is that it stands intermediate between the laboratory, with its great investigator control yet lack of reality, and the field conversely beyond the control of the investigator and all too real. Thus microcosms offer several distinct advantages over conventional single-species testing at the laboratory bench, on one hand, and field trials, on the other. Table I compares these advantages and some of the disadvantages.

Table II shows the principal characteristics of the several terrestrial microcosm systems that were reviewed at the Workshop on Terrestrial Microcosms.[8] The plant/soil system (I) of Lichtenstein and co-workers[9] (Figure 3) has subsequently been modified[10] to a modular plant/soil/ water system (Figure 4). The physical model ecosystem (II) or terrestrial monoculture system of Cole, Metcalf and Sanborn[11] (Figure 5) may also be operated so as to have an aquatic testing phase following removal of terrestrial species.[12] The two largest constructed systems are the micro-agroecosystem[13,16] (III, Figure 6) of Nash and co-workers and the CERL Terrestrial Microcosm Chamber[15,16] (IV, Figure 7). The latter has since been modified extensively (based on suggestions at the Workshop) and the extent systems are being replaced by a prototype chamber under evaluation. The "Mark II" TMC is 1.0 m x 0.75 m x 1.5 m (high) and has air flows in excess of 700 liters/min, a soil lysimeter system for regulation of soil moisture and collection of groundwater, and a soil capacity of 200-250 kg (maximum 50-cm depth). Photochemical capacity is being built in by using a quartz glass top and xenon/metal halide lighting.

The soil core microcosm[6] (V, Figure 2) is the smallest of the systems derived from intact materials removed from the natural environment. Related systems are the grassland microcosm of Van Voris and co-workers[7] and the "treecosm" system [a 1-m² woodlands block containing a young red maple (*Acer rubrens*)] of Ausmus and co-workers.[17] The soil/litter ecosystem respirometer[18,19] (VI, Figure 8) is totally enclosed for respirometric (O_2, CO_2 changes) and calorimetric measurements of soil/litter decomposition processes. Of all these systems, only V has been tested in more than one laboratory.

Table I. Comparison of Microcosms to Laboratory and Field Studies for Screening Toxic Substances

Problem	Laboratory	Microcosm	Field
Risk of irreversible harm to populations or the environment	Low	Low	High or unacceptable
Cost-effectiveness for screening	Good, once interactions are known, for 10^4–10^5 chemicals	Good for 10^3–10^4 chemicals and reveals interactions	Poor, probably less than 10^2 chemicals can be assayed
Step in verification process	Initial, hence cannot verify predictions	Intermediate, verifying lab predictions	Final, but limited to specific sites
Leads to fundamental analysis of process or problems	Seldom	Basic requirement especially for classes of compounds and types of environments	Usually after the fact
Degree of complexity in tests	Does not deal with complex processes or systems	Process focus on systems of moderate complexity	Usually too complex for disaggregation of effects
Provides broad indices of chemical behavior and effects	Seldom	Usually, thus helping to indicate where and what to look for in real world	With difficulty and at the risk of mistakes and delays
Provides chemical and ecological data simultaneously	No, except in large and cumbersome experiments	Generally, although the numbers of processes are limited	Generally
Utility of information in guiding use, safety	Weak, requires great extrapolation	Directly applicable or evident requiring only some extrapolation	Direct, especially for environment tested
Lifetime; ability to be self-sustaining	Short, not sustained except by investigator	Moderate (months) with care and input by investigator; not self-sustaining for long period	Subjected to uncontrollable outside forces but long and self-sustaining
Ability to determine simple properties of chemical or process	Excellent for properties; generally impossible for processes	Fair to weak on properties; good for processes	Poor on properties; excellent for some processes, poor on most

Table II. Chief Characteristics of Terrestrial Microcosms used to Study Fate and Effects of Chemicals in the Environment[8]

Characteristic	I Plant/Soil[9]	II Terrestrial Monoculture System[11,12]	III Microagro-ecosystem[13,14]	IV Terrestrial Microcosm Chamber[15,16]	V Soil Core[6,17]	VI Soil/Litter Ecosystem Respirometer[18,19]
Unit size	1.0 liter	19 liter	863 liter (1.5 x 0.5 x 1.15 m)	458 liter (1.0 x 0.75 x 0.6 m)	40 cc	500 cc
Mass of soil (kg)	0.7	0.4 (vermiculite) 3.0 (Drummer)	165	150	80 g	100 g
Type of soil(s)	Silty loam; sandy; quartz sand	Vermiculite; silty clay loam	Sandy loam	Synthetic potting mix; silty clay loam	Various forest and grassland types	(Douglas fir, red alder litter)
Temperature (°C)	28/20	26/19	Ambient[a]	30/19	26/20	19
Light/Dark (hr)	12/12	12/12	Ambient	16/8	12/12	0/24
Air flow rate (liter/min)	Ambient	Ambient	2500	10, 50	Ambient	O_2 by demand
Water inputs	Addition to weight; percolation	Addition to weight; postterrestrial aquatic study	"Rain" to set humidity or excess for percolation	"Rain" to set humidity; "spring"	Fixed volume on weekly basis for percolation	None
Plants	Corn	Corn; soybeans	Corn; cotton; tomatoes; tobacco; cereals and grasses	Alfalfa/ryegrass; Douglas fir/red alder/ryegrass	Endemic plant types (e.g., fescue meadow plants)	None
Invertebrates added	None	Caterpillar, slugs, pillbugs, earthworms	None	Tenebrio larvae, snails, pillbugs, earthworms, crickets, Collembola spp., nematode spp.	None (endemic fauna)	None (endemic fauna)
Vertebrates	None	Prairie vole	None	Gray-tailed vole	None	None
Microbiota	Ambient	Ambient	Ambient	Ambient	Ambient	Ambient
Operating time (per experiment)	Up to 200 days	30 days as terrestrial; 27 days as aquatic	60-90 days	60-90 days	6 to 9 weeks (including pre-equilibration time of 5 to 7 weeks)	60 days (including pre-treatment equilibration of about 30 days)

[a]Greenhouse with supplemental lighting available.

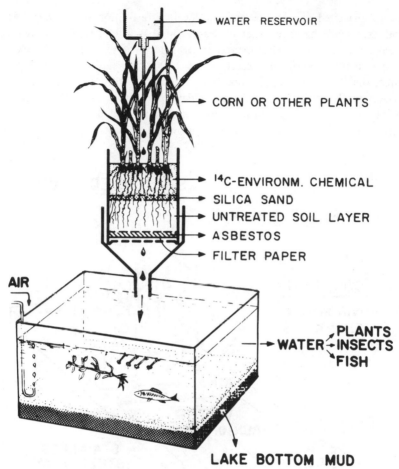

Figure 3. Plant/soil microcosm of Lichtenstien *et al.*[9]

STATE-OF-THE-ART REVIEW

Capabilities

The reported applications of these several systems for the study of chemical fate, transport and effects have been evaluated[8] and the suggested capabilities are shown in Tables III and IV. The constructed systems have largely been used to examine pesticides, while the emphasis in the natural or intact systems has been with respect to heavy metals.

The cost of achieving reported data has been estimated for each of the systems (Table V), but these figures must be interpreted cautiously with regard to projected capabilities and the level of resolution of the data. The general experience has been that of Nash,[8] in that attempts to achieve significant increases in environmental control, level of resolution of data and experimental complexity result in 100-1000% increases in costs. For example, the "Mark II" TMC (modified system IV) will cost about three times system IV per unit in order to achieve photolytic capability, a more than tenfold increase in air flow, a 50% increase in soil depth, and 2.5 to 3.5 times as much head space for plant growth.

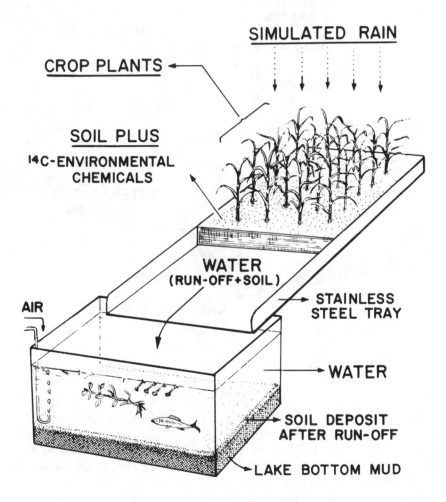

Figure 4. Modified plant/soil/water modular microcosm.[10]

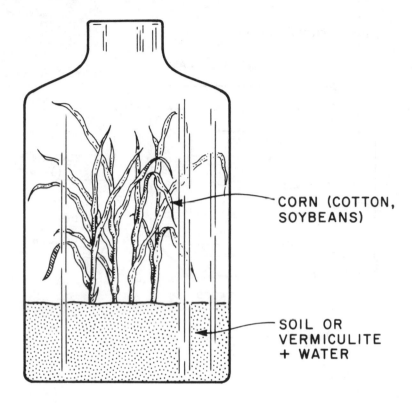

Figure 5. Physical model ecosystem of Cole *et al.*[11]

Chemical Fate

Microcosm technology has been applied to the three main processes of chemical disposition: transport—the movement of chemicals and their alteration products between environment compartments; transformation—the nature and rate of formation of chemical and biological transformation products in various compartments of the environment; and bioaccumulation—accumulation of residues in the biota or biomagnification between trophic levels.

Transport

Because the several systems appear to have difficulty in achieving accurate representations of environmental conditions and media flux, aspects of transport are not necessarily simulated accurately. This led to recommendations to increase soil depth, increase air flows, and provide

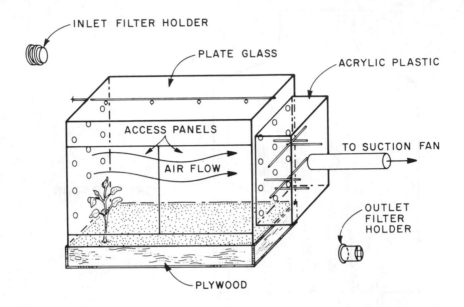

Figure 6. Microagroecosystem of PDL/USDA.[14]

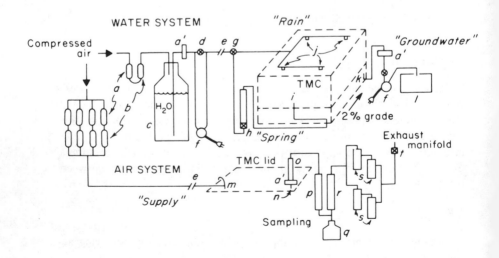

Figure 7. CERL Terrestrial Microcosm Chamber schematic diagram.[15]

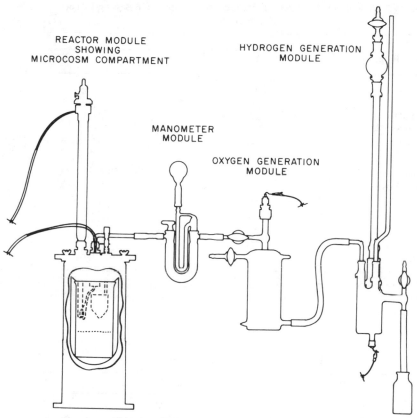

REACTOR MODULE
SHOWING
MICROCOSM COMPARTMENT

HYDROGEN GENERATION
MODULE

MANOMETER
MODULE

OXYGEN GENERATION
MODULE

Figure 8. Soil/litter ecosystem respirometer.[19]

Table III. Capabilities of Terrestrial Microcosms for Assessing Fate and Transport of Chemicals[8]

Process or Property	System[a]					
	I	II	III	IV	V	VI
Mass Balance[b]	–	–	(X)	X	–	(X)
Leaching, Soil Binding	X(*)	X	X(*)	X(*)	X*	–
Transformations	X*	X	X*	X*	(X*)	(X*)
Volatility	–	X(*)	X*	X*	(X*)	–
Bioaccumulation in Animals	–	X	–	X	–	–

[a]X indicates reported determinations, * indicates measured with respect to time; parentheses indicate authors' or reviewers' projection of capabilities with modification of previous experiments.

[b]Requires use of radiolabled chemicals not used in system III or VI to date.

Table IV. Capabilities of Terrestrial Microcosm Systems for Assessing Effects of Chemicals[8]

Process or Property	System[a]					
	I	II	III	IV	V	VI
Toxicity to:						
Plants	(X)	X	X	(X)	–	–
Insects and Invertebrates	–	(X)	–	X*	–	(X)
Vertebrates	–	X	–	X*	–	–
Soil						
Respiration	–	–	–	–	X*	X*
Nutrient Loss	(X*)	–	(X*)	(X*)	X*	(X)
Calorimetry	–	–	–	–	–	X*
Interactions:						
Plant/Soil	X	X	(X)	(X)	(X)	–
Plant/Animal	–	(X)	–	(X)	–	–
Predator/Prey	–	–	–	X*	–	–

[a]X indicates reported determinations, * indicates measured with respect to time; parentheses indicate authors' or reviewers' projection of capabilities with modification of previous experiments.

Table V. Microcosm System Costs[8]

Item	System					
	I	II	III	IV	V	VI
Unit ($100)[a]	0.25	0.5	15	15	0.002	0.5
Cost/Compound ($1,000)[b]	3-5	15	100	50-250	2-4	25
Time/Compound (man-mo)	6	6	18	20-40	2	6

[a]Cost for construction and assembly of each unit exclusive of facilities, environmental controls, monitoring equipment, etc.

[b]Through-put expense only for items not in parentheses in Tables III and IV.

closer approximations to temperature and moisture changes in real systems. Although many reviewers felt that most observations in terrestrial microcosms could be easily projected from simple tests, they noted that the opportunity to carry out chemical mass balance studies could provide more complete information on the multimedia fate of chemicals.

Transformation

Most of the work in systems I-IV has been devoted to aspects of metabolism and residues in the various biota, so that we know more about realization of capabilities in this area than others. Both advantages and disadvantages of microcosms have been demonstrated in relation to laboratory tests of chemical and biochemical transformations in single species or components. The dynamic state of transformations are easily considered simultaneously with effects. However, since metabolite patterns are only quantitatively different in multispecies system and not qualitatively different from results in two or three species (plant, animal, soil) exposed individually, it may be more cost-effective to determine patterns in one or two species. None of the systems to date have been able to demonstrate photochemical products (*e.g.,* photoaldrin in II or photodieldrin in IV) at a level of 1% or more in several runs. Some of these difficulties may be resolved soon, further buttressing this area as among the strongest points of demonstrable accuracy of microcosm systems. Research is also underway to establish transformation rates in intact soil core microcosms. These studies will not only clarify the utility of system V but also provide scaling information between systems II, III, and IV and system V.

Bioaccumulation

The potential hazard of a chemical or its transformation products to a species is often judged by the relative extent of residue accumulation, so that indices of relative persistence and relative bioaccumulation have been a major factor in terrestrial microcosm applications from the beginning.[4,11,12] Such tests may be performed outside a microcosm or field as in the modular food chain approach of Reinerts.[20] However, the multimedia exposure in microcosms containing predator and prey should give a much more realistic indication of hazard in natural environments. At the same time, effects can be evaluated in light of residue dynamics.

Some[8] have criticized the Bioaccumulation Index (BI) values from terrestrial microcosms as being less useful than those from aquatic systems. BI (ratio of concentration in whole animal or plant to concentration in soil[11]) in the vole, as shown in Table VI, is much less than the ratio in guppy/water (approximately 1.7×10^4).[21] However, when BI (vole/soil) is multiplied by the concentration ratio of dieldrin in soil to that in soil water (greater than 3.8×10^3),[16] it can be seen that the resultant BI may be even more sensitive in the terrestrial microcosm. The results in the voles were obtained in 2 to 14 days,[11,16,22] whereas the guppy exposure was 2 months.[21] Care must be taken, however, in

Table VI. Bioaccumulation Index of Dieldrin in *Microtus* spp

Rate (kg/ha)	Route	Soil Type	Days Exposure	Bioaccumulation Index Soil[a]	Water[b]	Reference
1.12	Soil	Silty clay loam	5	1.2	4.6×10^3	14
1.12	Soil	Vermiculite	5	4.1	15.6×10^3	14
1.12	Foliar	Sand/clay/ peat moss 45/45/10	1.5-8	$59.5(0.38)^c$	2.3×10^5	15
0.64	Seed Coating	Sand/clay/ peat moss 45/45/10	11	2.1	8.0×10^3	22
0.64	Foliar	Silty loam	5,10	14^c	5.8×10^3	22
0.25	Foliar	Silty loam	14	10	3.8×10^3	22

Microtus ochragaster in reference 14; *M. canicaudus* in reference 15, 22.

[a] Ppm of HEOD in vole (fresh wgt)/ppm in soil (wet wt.) at termination

[b] BI(soil) x 3800 (see text)

[c] Coefficient of variation (n = 6) in parentheses; all voles dead at termination.

trying to use these values very exactly, since considerable difference results from route and rate of application and the nature of the substrate (Table VI).

Effects

Demonstration and measurement of biological effects in microcosms has been one of the prime goals of this emerging technology.[17] Single species testing for toxic effects can be advantageously supplemented and expanded, but not replaced by microcosm studies. The opportunity to assess effects simultaneously with transport, transformation and residue accumulation, particularly in subchronic exposures, is an important aspect of microcosm operations. Furthermore, complex effects may only be observed in more complex systems, for which the microcosm may be an adequate approximation of the natural system. Except in the isolated intact systems, however, more attention has been devoted to chemical fate than effects. Table VII shows some of the biological effects demonstrated to date with various toxic substances in terrestrial microcosms.

The system currently most useful in screening for biological effect and judged[8] to be most highly developed is system V, the soil core microcosm[6,17,22] The results focus primarily on soil community processes (nutrient export/retention, respiration) and further work is needed to assess transport and effects determined simultaneously. System V and

Table VII. Biological Effects Observed or Measured in Terrestrial Microcosms in
in Response to Toxic Chemicals

Effect	Reference
Chronic and acute phytotoxicity	11, 12, 14, 16
Acute and subacute insecticidal action	4, 5, 11, 12, 16, 22
Vertebrate toxicity	5, 11, 12, 16, 22
Bioaccumulation of residues	5, 11, 12, 16, 22
Fetal transfer of residues	16, 22
Altered soil respiration	6, 7, 17, 18, 19
Interference in nutrient retention	6, 7, 17
"Reentry" (safe period post application) for vole	16
Interference with predation (vole-cricket)	16, 22
Community disruption in relation to complexity (diversity)	7

its successors[7,17] meet a number of criteria for acceptability in screening
for toxic substances.[17] First, they can be site-specific to target or nontarget
ecosystems. Second, dose-response information can be obtained at relatively
low cost. Third, the error terms in the various measurements are similar
to those for the same processes in field studies. The relative simplicity
of the test and the degree of low-level technology required in support lend
themselves to rapid screening procedures in the first tiers of testing.

System V generally relies heavily on measurement of elutrative loss of
soil nutrients (nitrogen forms, metal ions, dissolved organic carbon, phosphate, sulfate) to show disruption of the cohesiveness or intactness of the
soil community. For example, Van Voris and co-workers[7] were able to
demonstrate that the resistance or resilience of ecosystems (ecosystem
stability), as measured by loss of calcium ion, nitrate, etc., is related to
ecosystem complexity, as measured in the richness of cycles of soil respiration. Although the mechanisms involved are not yet known, the
system appears applicable to clinical evaluation of ecosystems and their
vulnerability to chemicals.

Mathematical Modeling

One of the critical deficiencies in microcosm systems to date has been
the paucity of mathematical models as part of the systems. Such models
would not only aid in interpretation of data and lead to improvements in
experimental design, but they would also provide a means of validating
models of chemical fate, based upon physical chemical and biological

laboratory data. Validation studies of microcosm systems (for reproducibility) will assist in providing experimental detail of the level of resolution to be included in the model. Effects of scale (size) and scope (complexity) must be related to functions of real systems, and this seems likely to occur only through careful mathematical modeling. Finally, models of microcosms must be related to models of biomes or fields and the degree of correspondence improved as much as possible. Figure 9 illustrates the relationships of microcosms and single species/component testing to fate and effects of chemicals in the real world.[8] By iteration of the interactions, the models support scale-up of microcosm experiments and ultimately better predictions from benchmark data.

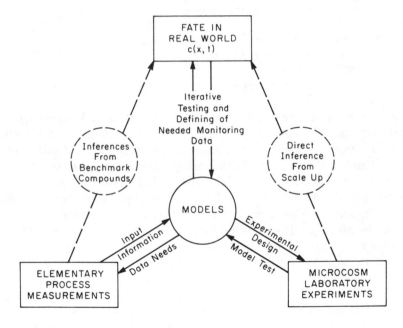

Figure 9. Relationship of modeling to laboratory tests, microcosms and monitoring.[8]

RESEARCH NEEDS

Standardization

There are two divergent thrusts to microcosm research: standardization for screening purposes, and development of an understanding of microcosms and ecosystems generally. The first requires that a very large number of compounds be tested in one system under essentially similar circumstances, while the second requires that a few chemicals be thoroughly understood in a great variety of conditions for a related set of systems. The demands of the regulator can be met by the former, but "the fundamental basis of scientific understanding"[2] can only truly come with the latter. Both points of view, however, are reflected in recommendations in four important areas:

1. validation—interlaboratory testing to determine how well a given system can be used and what statistical precision may be attributed to the test;
2. verification—specific studies to determine the accuracy of predictions from microcosm studies to field exposures;
3. mathematical modeling—application of *a priori* modeling techniques to microcosm studies of both chemical fate and effect;
4. design criteria and performance standards—studies to determine how different systems can be related and how validity and accuracy may be maintained.

To achieve these, the following suggestions can be made to improve standardization:

1. Use reference chemicals in comparative tests of new chemicals.
2. Test at the anticipated environmental level (and multiples/ fractions thereof).
3. Provide kinetic data.
4. Use trophic levels consistent and equivalent to the environment modeled.
5. Use standard organisms in constructed microcosms, with at least two soils of diverse properties.
6. Whenever possible, determine the mass balance of the chemical, which should be applied in a formulation consistent with how it might reach the environment tested.
7. To the extent possible obtain from the system data on the transformation products, including fate and biological effects.

To enhance likelihood of the systems being accurate and precise, greater attention must be paid to the following problems:

1. photochemical transformations,
2. better simulation of airflows and environmental temperatures throughout the system,
3. scaling of systems and determination of edge effects or errors inherent in system design,
4. use of nondestructive tests of plant and plant-soil interaction,
5. criteria for fauna appropriate to the scale of the system and test objectives, and
6. parameters of reproducibility which can serve as operating criteria.

Specific studies must be undertaken in the field in relation to several microcosm systems, and these should be joined to both analysis and predictive mathematical modeling.

CONCLUSIONS

Terrestrial microcosm technology is a relatively new approach to assessment of the fate and effects of chemicals in the environment. While much has been achieved, even more work must be performed to establish the validity and accuracy of these systems. Each microcosm system used must be designed to test specific hypotheses about chemical fate and its effects as a toxicant. Currently, microcosms serve best to resolve issues raised at the laboratory bench and to clarify for further study in the field those questions which cannot be answered under laboratory conditions. Ultimately, microcosms may be most useful in hierarchical testing schemes employing mathematical models from benchmark data and structure/activity relationships.

REFERENCES

1. The Federal Insecticide, Fungicide and Rodenticide Act, PL 94-140 (1975).
2. The Toxic Substances Control Act, PL 94-469 (1976).
3. Costle, D. M. "Dealing with the Chemical Revolution," *EPA Journal* 4(8):2-3 (1978).
4. Metcalf, R. L., G. K. Sangha, and I. P. Kapoor. "Model Ecosystem for the Evaluation of Pesticide Degradability and Ecological Magnification," *Environ. Sci. Techol.* 5:709-13 (1971).
5. Metcalf, R. L., and J. R. Sanborn. "Pesticides and Environmental Quality," Bull. Ill. Nat. Hist. Surv. *31:381-436 (1975).*
6. Draggan, S. "The Microcosm as a Tool for Estimation of Environmental Transport of Toxic Materials," *Int. J. Environ. Studies* 10: 65-70 (1976).
7. Van Voris, P., R. V. O'Neill, H. H. Shugart and W. R. Emanual. "Functional Complexity and Ecosystem Stability: An Experimental Approach." Environmental Sciences Division Publication N. 1123, ORNL/TM-6199 (1978).
8. Gillett, J. W., and J. M Witt, Eds. *Terrestrial Microcosms* (Washington, DC: National Science Foundation, in press).
9. Lichtenstein, E. P., T. W. Furhemann and K. R. Schulz. "Translocation and Metabolism of ^{14}C-Phorate as Affected by Percolating in a Model Soil-plant Ecosystem," *J. Agric. Food Chem.* 22:991-6 (1974).
10. Lichtenstein, E. P., T. T. Liang and T. W. Fuhremann. "A Compartmentalized Microcosm for Studying the Fate of Chemicals in the Environment," *J. Agric. Food Chem.* 26:948-53 (1978).
11. Cole, L. K., R. L. Metcalf and J. R. Sanborn. "Environmental Fate of Insecticides in a Terrestrial Model Ecosystem," *Int. J. Environ. Studies* 10:7-14 (1976).

12. Cole, L. K., and R. L. Metcalf. "Terrestrial/Aquatic Laboratory Model Ecosystem for Pesticide Studies." In: *Terrestrial Microcosms and Environmental Chemistry,* J. M. Witt and J. W. Gillett, Eds. (Washington, DC: National Science Foundation, in press.)

13. Beall, M. L., Jr., R. G. Nash and P. C. Kearney. "Agroecosystem—a Laboratory Model Ecosystem to Simulate Agricultural Field Conditions for Monitoring Pesticides." In: *Environmental Modeling and Simulation,* W. Ott, Ed. U. S. Environmental Protection Agency EPA-600/9-76-016, U.S. Government Printing Office (1976), pp. 790-793.

14. Nash, R. G., M. L. Beall, Jr., and P. C. Kearney. "A Microagroecosystem to Monitor Environmental Fate of Pesticides," In: *Terrestrial Microcosms And Environmental Chemistry,* J. M. Witt and J. W. Gillett, Eds. (Washington, DC: National Science Foundation, in press).

15. Gillett, J. W., and J. D. Gile. "Pesticide Fate in Terrestrial Laboratory Ecosystems," *Int. J. Environ. Studies* 10:15-22 (1976).

16. Gile, J. D., and J. W. Gillett, "Terrestrial Microcosm Chamber Evaluations of Substitute Chemicals," In: *Terrestrial Microcosms and Environmental Chemistry,* J. M. Witt and J. W. Gillett, Eds. (Washington, DC: National Science Foundation, in press).

17. Ausmus, B. S., D. R. Jackson and P. Van Voris, "The Accuracy of Screening Techniques." In: *Terrestrial Microcosms and Environmental Chemistry,* J. M. Witt and J. W. Gillett, Eds. (Washington, DC: National Science Foundation, in press).

18. Bond, J., B. Lighthart, R. Shimabuka and L. Russell. "Some effects of Cadmuim and Coniferous Forest Soil/Litter Microcosms," U.S. Environmental Protection Agency, EPA-600/3-75-036, U.S. Government Printing Office (1975).

19. Lighthart, B., J. Bond, and M. Richard. "Trace Elements Research using Coniferous Soil/Litter Microcosms," U.S. Environmental Protection Agency EPA-600/3-33-091, U.S. Government Printing Office (1977).

20. Reinerts, R. T. "Accumulation of Dieldrin in an Alga (*Scendesmus obligans*), *Daphnia magna,* and the Guppy (*Poecilia Reticulata*)," *J. Fish. Res. Board* 29:413-9 (1972).

21. Sanborn, J. R., and C. C. Yu. "The Fate of Dieldrin in a Model Ecosystem," *Bull. Environ. Contam. Toxicol.* 10:340-6 (1973).

22. Gile, J. D., J. Collins and J. W. Gillett. Unpublished data.

18

FATE OF 3,3′-DICHLOROBENZIDINE
IN THE AQUATIC ENVIRONMENT

Henry T. Appleton, Sujit Banerjee and Harish C. Sikka

Syracuse Research Corporation
Merrill Lane
Syracuse, New York 13210

INTRODUCTION

3,3′-Dichlorobenzidine (3,3′-dichloro 4,4′-diaminobiphenyl), hereinafter referred to as DCB, is widely used as an intermediate in the manufacture of azo pigments. It is of considerable commercial importance; total DCB production in the United States in 1972 was about 4.6 million pounds.[1] With current work practices, effluents containing this chemical are discharged directly into receiving waters. Moreover, the discharge of dichlorobenzidine-pigment wastes into receiving waters constitutes an additional source of DCB contamination in the environment since free, unreacted DCB is reported to be present in these pigments.

Dichlorobenzidine is strongly carcinogenic in animals[2] and is regarded by the Occupational Health and Safety Administration (OSHA) as being carcinogenic to man.[3] The discharge of DCB into the aquatic environment is of great concern to human health because of possible exposure to the chemical through drinking-water supplies. Also, DCB and its metabolites may accumulate in fish and could pose a health hazard if fish from contaminated waters were to be used as human food. The potential hazard of DCB may be compounded by its biological or nonbiological conversion to compounds of even greater toxicity and/or persistence than the parent chemical. For instance, it is believed that in the case of carcinogenic aromatic amines, it is the metabolites of the chemicals that produce the carcinogenic response.[4] Furthermore, potential degradation products of DCB, such as benzidine, may constitute an even greater carcinogenic hazard. Therefore, in order to fully evaluate the hazards associated with the release of DCB into the environment, it becomes important to study the environmental fate of the chemical because its persistence, disappearance or partial transformation will determine the degree of its hazard.

Several physical, chemical and biological factors determine the fate of a chemical in the aquatic environment. These include adsorption to sediment, chemical hydrolysis, photodegradation, microbial degradation and uptake and metabolism by aquatic organisms. Currently nothing is known about the effect of these factors on the persistence and transformation of DCB in the aquatic environment. This study was undertaken to assess the role of some of the processes that may determine the environmental behavior of the chemical. The overall objective of this investigation was to obtain information needed for establishing effluent guidelines for DCB, examining the sorption/desorption of DCB in sediment, biodegradation of DCB by naturally occurring aquatic microbial communities, photodegradation of DCB in aqueous solution and the uptake and metabolism of DCB by fish.

ADSORPTION/DESORPTION OF DCB TO AQUATIC SEDIMENTS

Adsorption of pollutant chemicals to aquatic sediments can affect the availability of the chemical for both microbial biodegradation and bioconcentration by aquatic species, and can also determine the mobility of environmental residues. Therefore, the adsorption of DCB to several aquatic sediments was assessed.

Procedures

Bottom sediments from USDA pond, Hickory Hill pond and Doe Run pond of the Athens, Georgia, area were provided by the Athens Environmental Research Laboratory. The composition of these sediments is listed below in Table I.

To determine sorption, a solution of ^{14}C-DCB.2HCl (obtained from California Bionuclear Corporation, Sun Valley, California) radiochemical purity greater than 99%) was mixed with sediment to give a water:sediment (dry-weight equivalent) ratio of 100:1 in Erlenmeyer flasks shaken

Table I. Composition of Bottom Sediments

Pond	pH	Organic Carbon (%)	Clay (%)	Silt (%)	Sand (%)
USDA	6.4	0.8			
Doe Run	6.1	1 4	<1	44	56
Hickory Hill	6.3	2.4	<1	45	55

at 250 rpm on a rotary shaker. All samples were protected from light and maintained at 22°C. At the appropriate time and at equilibrium, samples were centrifuged at 15,000 rpm for 10 min, and ^{14}C content in the supernatant was determined by scintillation counting. The amount of ^{14}C-DCB disappearing from the solution was assumed to be sorbed by the sediment. Adsorption of DCB to glass was not detected over 24 hr in Erlenmeyer flasks containing a solution of ^{14}C-DCB in distilled water. The effect of pH on sorption was studied in appropriately buffered solutions: pH 5 (0.01 M acetate buffer), pH 7 (0.01 M phosphate buffer), pH 9 (0.01 M borate buffer). All other sorption studies were done in distilled water-sediment suspensions. Sorption solutions contained an initial DCB.2 HCl concentration of 2.0 ppm.

Desorption was studied in sediments that were suspended in 2 ppm DCB.2 HCl for either 24 hr or 7 days, at which time sorption was measured and desorption initiated. After centrifugation, the supernatant solution was removed and its ^{14}C content determined. The sediment pellet was washed once by suspension in distilled H_2O to remove DCB from the solution that may have been trapped in interstitial spaces of the pellet. After ^{14}C determination, the wash was discarded, an appropriate volume of distilled water was added (maintaining a liquid/sediment ratio of 100), a 0-time sample was taken, and the samples were then shaken vigorously for up to 72 hr. During this period, the suspension was centrifuged at appropriate intervals and the supernatant was counted for ^{14}C to determine the amount of desorbed DCB. Extraction of DCB from sediment by organic solvents and other agents was performed by similar procedures.

Results

The rate of DCB adsorption to all sediments studied was initially very rapid. The end-point of the sorption was generally achieved within the first 24 hr of the experiments. The comparative distribution coefficients (K_d = 100 x μg DCB sorbed/μg DCB in solution) listed in Table II indicate the high affinity of DCB for a variety of sediments. In addition,

Table II. Distribution Coefficients (K_d) of Sediment Sorption of DCB

Sediment	$K_d{}^a$
USDA Pond	2850 (666)
Hickory Hill Pond	3210 (328)
Doe Run Pond	2670 (310)

[a]Values are the average of three replications. Figures in parentheses are standard deviations.

sorption was reduced significantly at pH 9, compared to values obtained at pH values of 5 or 7 (Table III).

Investigation of the desorption of DCB by water and other agents confirmed the high affinity for sediment (Table IV). These studies also showed that DCB desorption was related to the age of the sample at the time when desorption was initiated. That is, a higher degree of desorption was obtained from samples in which DCB had been in contact with sediment for one day than with those where contact had been maintained for several days.

Table III. Effect of pH on Sorption of DCB by Sediment

Sediment	K_d [a]		
	pH 5	pH 7	pH 9
Doe Run Pond	3325 (655)	2417 (295)	206 (18)
Hickory Hill Pond	2225 (252)	2122 (545)	269 (98)
USDA Pond	3783 (1254)	3083 (445)	95 (2)

[a]The values are the average of three replications. Values in parentheses are standard deviations.

Table IV. Desorption of DCB From Sediment by Various Agents
As a Function of Sample Age

	Percent ^{14}C extracted[a]	
	Period of Sediment exposure to ^{14}C-DCB	
	24 hr	7 Days
Distilled H_2O	1.2	0.3
5 N NH$_4$Cl	2.1	0.7
1 N HCl	9.0	3.1
1 N NaOH	31.3	16.7
Methanol	36.2[b]	21.9

[a]Values are the average of three replicates.
[b]Extraction of these samples with a second portion of methanol liberated an additional 3.8% of adsorbed ^{14}C.

Discussion

The inability of 5 N NH_4Cl to appreciably enhance desorption of DCB from the sediments studied would indicate that cation-exchange mechanisms may play a minimal role, if any, in sorption of DCB under normal environmental conditions.[5] DCB is predominantly in a neutral state under our experimental conditions, minimizing any potential ionic interactions between DCB and sediment. For the same reason, it is unlikely that the decrease in sorption seen under alkaline conditions in this study is caused by repression of ionic mechanisms. Alkaline conditions may change the sorptive properties of the sediments, since treatment of sediments with NaOH releases humic and fulvic acides from soil or sediment.[6]

A high degree of difficulty was encountered in extracting more than a fraction of adsorbed DCB from sediment and the efficiency of desorption and extraction decreased according to the length of time DCB had been in contact with the sediment. These observations are consistent with the occurrence of chemical reaction between DCB and sediment constituents. Hsu and Bartha[7,8] have demonstrated that chloroanilines react with organic constituents of soil, forming covalent complexes, by condensation of the amine moiety with carbonyl and other functions in the soil. This process would probably lead to formation of humic-like materials that presumably would be nonhazardous.[6] Thus, although there is little doubt that the rapid decrease in free DCB from aqueous suspensions of sediment is predominantly the result of a physical adsorptive process, it seems likely that, once absorbed, DCB may covalently bind to the sediment and become increasingly resistant to desorption and extraction as this process occurs over time.

BIODEGRADATION OF DCB BY AQUATIC MICROORGANISMS

Metabolic conversion by aquatic microbial communities can represent one route for the ultimate degradation and elimination of DCB residues. However, metabolites of aromatic amines may be more toxic than the parent compound, thereby compounding the potential hazard. The capacity of microorganisms of lake water or activated sludge to degrade or otherwise alter DCB was therefore examined.

Procedures

The degradation of DCB by aquatic microorganisms was studied using samples of lake water and activated sludge.

Samples of water were obtained from Oneida Lake, a large mesotrophic lake—and from Jamesville Reservoir, a small eutrophic lake. Prior to starting the biodegradation studies, microbial counts in the water samples were made using standard serial dilution and plating techniques.[9]

To determine the rate of biodegradation of DCB, DCB was added to the water in Erlenmeyer flasks at a concentration of 2 ppm. The flasks were stoppered with foam plugs and incubated on a gyrotary shaker in the dark at 21 ± 1°C. Sterilized water samples containing DCB were included as controls to account for any nonbiological degradation. At appropriate intervals, the contents of the flasks were analyzed for DCB by high performance liquid chromatography (HPLC) using a Waters Associates (Milford, Massachusetts) liquid chromatograph (model M6000A) equipped with a UV detector (Schoeffel Instrument Corp., Westwood, New Jersey, model GM770). A 4-mm (i.d.) 30-cm column of μ Bondapak C18 medium and a solvent system of acetonitrile:5% acetic acid (70:30 v/v) was used. Samples were diluted with acetonitrile prior to HPLC analysis not only to prevent filter retention by DCB in the preanalytical sample filtration, but also to reverse the association of DCB with materials present in turbid lake-water samples.

Carbon-14-labeled DCB was used to characterize the products resulting from the biodegradation of the chemical. Carbon 14-DCB was added to the water. samples at a concentration of 2 ppm. At periodic intervals, the water samples were adjusted to pH 11 and shaken with ether, and the ^{14}C distribution between the aqueous and organic phases was determined. The ether extracts were analyzed by thin-layer chromatography (TLC). Additional search for metabolic products was performed by HPLC coupled with scintillation counting of the eluted fractions.

The evolution of $^{14}CO_2$ was measured using the biometer flask described by Bartha and Pramer.[10] The water samples were incubated with 2 ppm of ^{14}C-DCB in the biometer flask and 0.1 N KOH was used as the CO_2-trapping solution in the side arm. The solution in the side arm was removed and replaced at appropriate intervals. The amount of ^{14}C in the CO_2-trapping solution was measured by liquid scintillation counting. The radioactivity collected in the KOH trap was verified as $^{14}CO_2$ by acidifying with HCl.

The biodegradation of DCB was also conducted using activated sludge as a microbial inoculum[11] to present extensive adsorption of DCB to solid sludge particles. In this test, 10 ml of settled activated-sludge supernatant was added as a source of microbial inoculum to 90 ml of a mineral medium[11] containing 50 or 100 mg of yeast extract and 2 ppm of DCB. After an incubation period of 7 days, 10 ml of the contents from the inoculated flask were withdrawn and transferred to a fresh medium containing the DCB, and the procedure was repeated for three consecutive weeks. After incubation for 7 days, each subculture was analyzed for DCB.

Results

In order to assess the degradation of DCB under conditions of high microbial activity, a study was initiated in mid-July when microbial assay of Jamesville Reservoir water showed approximately 5×10^6 cells/ml and Oneida Lake water 900 cells/ml.

As shown in Table V (first experiment), the DCB assayed decreased with time in both lake-water sources. No degradation products were detected by either HPLC or TLC of the samples, and ^{14}C trapped in biometer flasks was less than 0.1% over 14 days in both lake water and sterile samples.

Table V. Persistence of DCB in Lake Water

Time (days)	DCB Concentration (ppm)[a]	
First Experiment	Jamesville Reservoir	Oneida Lake
0	2.07	2.26
7	1.81	2.22
21	1.55	1.66
28	1.47	1.68
Second Experiment	Sterile Control	Jamesville Reservoir
0	2.07	1.96
10	2.02	1.68
20	2.03	1.56
30	2.04	1.52

[a]As analyzed by HPLC. Values are the average of three replicates.

During the course of incubation, the samples became increasingly turbid, probably as a result of microbial growth and the accumulation of metabolic by-products (lipid, protein) and dead microorganisms. This accumulated material might have provided a surface for DCB adsorption. To verify this, Jamesville Reservoir water (5×10^6 cells/ml) was incubated with ^{14}C-DCB.2 HCl of higher specific activity to facilitate determination of the distribution of ^{14}C in the incubation medium. As in the previous study, a progressive loss of DCB was observed by HPLC over time (Table

V, second experiment), but no degradation products were detected. Also, as before, sample turbidity increased over time.

Portions of the 30-day samples prior to acetonitrile treatment were centrifuged, with 64.4% of original ^{14}C detected in the supernatant. In contrast, the distribution seen in acetonitrile-treated samples was 94.7% in the supernatant and 3.3% in the precipitate, indicating that a large portion of the ^{14}C was associated with the organic matter.

Portions of the 30-day samples (diluted with acetonitrile) containing known amounts of ^{14}C were applied to the HPLC column, eluted fractions collected, and the ^{14}C content of the fractions determined. The eluted fraction containing DCB showed 85.6 ± 3.8% of the applied radioactivity. Carbon-14 in the fractions before and after DCB, although present, was too low to quantify. Therefore, a portion of ^{14}C-material in the lake-water samples does not chromatograph distinctly with the HPLC systems employed. In addition, portions of the 30-day samples were treated with acetonitrile and centrifuged. The supernatant was then basified and partitioned with ether. Only 0.9% of the original supernatant ^{14}C remained in the aqueous phase after ether partition. The extracts were analyzed by TLC. No distinct materials other than DCB were seen, under conditions capable of detecting components that represented 5% or more of the material applied to the plate.

The fate of DCB in activated sludge was also studied because waste treatment systems may represent an important site of entry of DCB into the environment and because the sludge microbial community may have a higher biodegradation capacity than lake-water communities.

To minimize the loss of DCB from adsorption on the sludge solids and microorganisms and to determine whether activated sludge microorganisms would adapt to degrading DCB, the biodegradation of DCB was examined in a system that utilized settled activated sludge as a source of microbes and included repeated weekly subcultures into fresh medium.

The results shown in Table VI show that the subculture enrichment method through weekly subcultures into fresh medium did not increase the biodegradation of DCB. The decrease of DCB seen in this system was incomparable to that seen in lake water. As in the previous studies, a considerable amount of the ^{14}C unaccounted for in the analyzed water samples was detected in the organic material formed during incubation. For example, ^{14}C in the isolated organic material of the third culture accounted for 12 to 17% of original ^{14}C in the incubations. Again, no DCB metabolites were detected by HPLC analysis.

Table VI. Degradation of DCB by Activated Sludge
(Weekly Subculture Enrichment)

| | Percent DCB Remaining after 7 Days of Incubation DCB (% 0 time) | | |
| | Subculture[a] | | |
Treatment	0	1	2
Sterile Control	100.6	101.2	99.6
Activated Sludge + 100 mg/l Yeast Extract	84.6	87.3	90.5
Activated Sludge + 50 mg/l Yeast Extract	˙87.3	73.6	82.5

[a]Sampling and culturing procedures are explained in Section 4. The data represent one experiment per treatment.

Discussion

Our findings indicate that DCB is not readily degraded by microorganisms under the experimental conditions. There was some indication of biodegradation in lakewater, although no distinct metabolites were detected. It is speculated that the loss of the chemical noticed was primarily the result of adsorption and/or accumulation by microorganisms in the water, since our results show that, during the course of incubation, a portion of DCB becomes associated with organic matter in the water. The addition of acetonitrile releases or solubilizes the majority of this bound ^{14}C-material, although a portion remains associated with the organic debris. This suggests that the progressive decrease in DCB detected during the incubation period is probably related to the increase of organic material formed during incubation, resulting in a DCB/organic matter complex that cannot be analyzed by chromatographic techniques. No evidence was seen to indicate that sludge microbial populations may be induced to metabolize DCB.

Even in the event that DCB loss was the result of active microbial degradation, however, the rate of this process is slow. The rate of loss of detected DCB in the lakewater system tested indicates half-lives greater than 40 days, based on the initial rate of DCB loss. In contrast, the half-life of DCB loss in the sediment sorption and photodegradation experiments was measured in minutes. Also, the stability of DCB in sterile samples protected from light indicates that hydrolytic, oxidative or other chemical mechanisms do not significantly degrade DCB in aqueous solution

PHOTODEGRADATION

The intense absorption characteristics of DCB makes it a likely candidate for photodegradation, and a number of studies were conducted to determine the photochemical fate of the compound under a variety of conditions.

Procedures

Quantification of the benzidines was carried out by HPLC, and the identity of the photoproducts was confirmed by GC using an HP 5730A FID instrument and a 10% UCW 932 on Chromosorb W column maintained at 225°, and by mass spectrometry (MS).

Preparative photolyses were conducted in a 1-liter immersion well type photoreactor equipped with a 450 W high-pressure Hanovia lamp fitted with a Pyrex filter to exclude light of wavelength less than 280 nm. Quantum yields were determined in a Rayonet RMR-400 merry-go-round photoreactor at 2537 Å or at 3000 Å.

Results

Irradiation of aqueous solutions of DCB under preparative conditions results in a hypsochromic shift of the UV absorption band of the substrate, and is accompanied by about a 50% decrease in intensity. Extraction of the photolyzed solution by ether after basification, and analysis of the ether concentrate by HPLC, GC and MS revealed the presence of benzidine and 3-chlorobenzidine (MCB). In a separate experiment, aliquots were withdrawn periodically and analyzed for DCB, MCB and benzidine by HPLC. These results are listed in Table VII and suggest that DCB is, in part, degraded sequentially to MCB and benzidine. In addition, DCB is also photodegraded to a number of relatively water-insoluble products which adhere to the walls of the photoreactor. These materials are ether soluble, and TLC (Silica gel, ether:hexane-2:1) of an ethereal concentrate resolved the products into more than five brightly colored components. The extent of conversion to these water-insoluble materials was determined through the photolysis of ^{14}C-labeled DCB. Aliquots of the aqueous solution were counted periodically, and after 30 min, 63% of the counts were observed to be adsorbed on the surface of the reactor. It is evident that very little DCB, MCB or benzidine remains after 15 min, and the residual spectral absorption probably derives from partial solution of the unidentified products deposited on the reactor walls. Furthermore, less than 40% of the initial substrate gives rise to this absorption, and, consequently, the extinction coefficient is >25,000. Hence, it is probable that these products are dimers or higher analogs of DCB and/or its photoproducts.

Table VII. Photolysis of DCB[a]

Irradiation (min)	DCB · 2 HCl[b] (ppm)	MCB · 2 HCl[b] (ppm)	Benzidine · 2 HCl[b] (ppm)
0	0.64	0	
1	0.60	0.034	
2	0.56	0.060	
3	0.42	0.068	
4	0.24	0.114	
5	0.078	0.044	0.044
10	–	0.044	0.072
15	–	–	0.086
45	–	0.034	0.072

[a]Irradiated with light from a Hanovia 450 W high-pressure lamp filtered through Pyrex.

[b]Average of 2 determinations. The "2 HCl" represents the dihydrochloride.

In addition, 5-ml aliquots of an aqueous solution of DCB were exposed to noonday sunlight in identical quartz tubes. Tubes were withdrawn every 30 sec, covered in foil and analyzed by HPLC. The results illustrated in Figure 1 confirm the expected photolability of DCB. No benzidine was observed in this experiment, but its presence was detected in prolonged exposures (>20 min) of DCB to sunlight.

The disappearance quantum yields of benzidine and MCB were measured with respect to a ferrioxalate actionometer[12] at 2357 Å and 3000 Å. The wavelength dependence of the quantum yields was minimal, and the results obtained at 3000 Å are reported in Table VIII. The limited solubility of DCB in water made direct determination of the quantum yield at 3000 Å difficult, and, consequently, the measurement was made with respect to MCB at 2537 Å. The high quantum yields for DCB and MCB confirm the photolability of these substrates, whereas the relatively low value for benzidine is in keeping with Metcalf's finding that the disappearance half-life

Table VIII. Disappearance Quantum Yields for DCB, MCB and Benzidine

	ϕ	Wavelength (Å)	pH
Benzidine	0.012	2537, 3000	7.0, 8.2
MCB	0.70	2537, 3000	7.0, 8.2
DCB	0.43	2537	6.7, 8.1

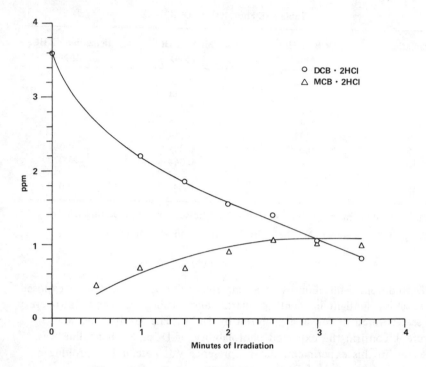

Figure 1. Irradiation of DCB in natural sunlight.

of benzidine in methanol at 254 nm is about 2 hr.[13] A number of experiments were also conducted to determine the pH-dependence of the quantum yield for DCB. These results, presented in Table IX, indicate that the rate of degradation has a slight dependency on pH. The appearance quantum yields for MCB, although less accurate due to the reactivity of this material, also exhibit the same dependency.

Table IX. pH Dependence of the Rate of Disappearance of DCB

pH	ϕ Rel (DCB)	ϕ Rel (MCB)[a]
1.96	1.0	1.0
3.96	0.71	0.73
6.01	0.42	0.49
8.28	0.50	0.56

[a]Appearance quantum yield.

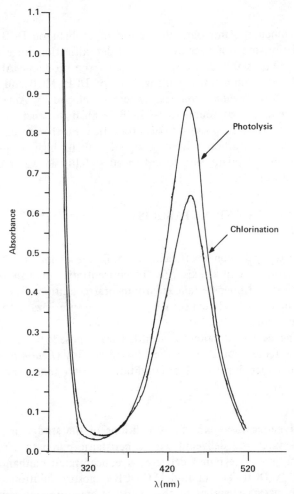

Figure 2. Spectra of transients generated from the photolysis and chlorination of DCB.

To determine the photoreactivity of DCB in solvents other than water, a number of preliminary experiments were carried out in hexane, isopropanol and methanol. Irradiation of these solutions, under approximately the same conditions under which disappearance of DCB in aqueous solution is complete, showed that virtually no degradation had occurred. For example, irradiation of a 15-ppm solution of DCB.2HCl in isopropanol at 2537 Å for 1 hr led to 28% degradation. Under similar conditions, a 4-ppm aqueous solution of DCB is decomposed to an extent of 42% in 0.3 min. It is, therefore, evident that photodegradation of DCB in water proceeds through a mechanism different from that in organic solvents.

Discussion

From an environmental standpoint, the action of sunlight on DCB will lead to its degradation but not necessarily to its detoxification since benzidine, a relatively photostable carcinogen, is one of the products. Also, in a separate study, the comparative mutagenicity of DCB, MCB and benzidine in the *Salmonella typhimurium* test system developed by Ames,[14] the order of mutagenicity was found to be DCB > MCB > benzidine, indicating that MCB may also possess toxic properties. Furthermore, the inertness of DCB in hydrocarbon solvents coupled with its high octanol/water partition coefficient might lead to enhanced stability in water contaminated with hydrocarbons.

UPTAKE, ELIMINATION AND METABOLISM OF DCB IN FISH

The n-octanol:water partition coefficient of DCB free amine is 3,210, indicating that a strong potential exists for bioconcentration in fish and other aquatic organisms. Under certain environmental conditions, therefore, DCB might accumulate to levels constituting a carcinogenic hazard to humans consuming contaminated fish.

To determine the degree of potential health hazard posed, the uptake, elimination and metabolism of DCB in the bluegill sunfish (*Lepomis macrochirus*) was studied under dynamic flow conditions at two concentrations.

Procedures

To create the dynamic flow exposure conditions of the study, uncontaminated dilution water was delivered with a peristaltic pump to a mixing chamber into which a syringe pump delivered a concentrated methanol stock solution of ^{14}C-DCB. The diluted ^{14}C-DCB exposure solution was conveyed from the mixing chamber to the exposure aquarium by gravity flow. The exposure tank consisted of a 15-liter glass aquarium fitted with a standpipe drain. The tank was mildly aerated to ensure proper circulation of incoming exposure solution.

The concentrations of ^{14}C-DCB used in the studies were 5 ppb and 0.1 ppm (nominal), and were maintained by adjusting the concentration of the stock solution and the rate of flow of the syringe pump. In no case did the concentration of methanol in the exposure solution exceed 0.02%. The flow through the exposure tanks was approximately 8 tank volumes per 24 hr.

The bioconcentration experiments were initiated by allowing the DCB delivery system to operate for 48 hr prior to addition of fish to ensure maintenance of a constant chemical concentration. At that time, seventy bluegills (*ca.* 1 in. long) were introduced to the system.

The fish were removed from the treated water at appropriate intervals, rinsed with clean water and sacrificed. The content and distribution of ^{14}C-DCB and derived materials in exposed fish was measured in the following manner. After sacrifice, the fish were dissected into two portions prior to analysis: the head plus viscera fraction (consisting of the head, gills and internal organs), and the edible flesh fraction (flesh and skin). After weighing, the tissue fractions were homogenized with methanol (5 ml/g fresh weight) in a blender. The slurry was centrifuged and the supernatant was decanted. The residue was reextracted with methanol. After centrifugation, the two extracts were combined and the amount of ^{14}C in the pooled extract was determined by liquid scintillation counting. The amount of ^{14}C in the tissue residue was determined by solubilizing it in the NCS tissue solubilizer (Amersham Searle Corp.) as described by Sikka *et al.*[15] The radioactivity in the methanol extract and in the tissue residue was combined to calculate the ^{14}C concentration in the fish. The whole fish ^{14}C residues were calculated from the combined values of the edible flesh and head plus viscera fractions. To study the metabolism of DCB by the fish, the methanol extracts of the tissues were analyzed by partition extraction, TLC and HPLC.

Results

The results of the uptake studies are given in Tables X and XI. From these results, it is evident that DCB is rapidly and significantly bioconcentrated by bluegills. The equilibrium bioconcentration factors achieved in the 5-ppb (nominal) study were 114, 856 and 495 for edible, head and viscera, and whole fish, respectively, based on total ^{14}C residue. In comparison, total ^{14}C residue bioconcentration factors for the 0.1 ppm-

Table X. ^{14}C Distribution in Bluegills Exposed to 5 ppb ^{14}C-DCB · 2 HCl ppm-DCB Equivalent[a]

Exposure Time (hr)	Exposure Water	Edible Flesh	Head and Viscera	Whole Fish
24	0.0038	0.062 ± 0.010	3.796 ± 0.027	1.902 ± 0.049
48	0.0027	0.394 ± 0.036	3.246 ± 0.362	1.815 ± 0.086
72	0.0056	0.493 ± 0.038	3.150 ± 0.546	1.902 ± 0.200
96	0.0034	0.356 ± 0.007	3.285 ± 0.001	1.867 ± 0.327
120	0.0052	0.578 ± 0.015	3.719 ± 0.465	2.180 ± 0.090

[a] The values are the average of duplicate samples per exposure interval, with three fish per sample.

Table XI. [14]C Distribution in Bluegills Exposed to 0.1 ppm ^{14}C-DCB · 2 HCl ppm-DCB Equivalent[a]

Exposure Time (hr)	Exposure Water	Edible Flesh	Head and Viscera	Whole Fish
24	0.093	2.390 ± 0.052	28.771 ± 4.198	14.822 ± 1.097
48	0.105	5.333 ± 0.586	54.388 ± 7.780	29.536 ± 1.686
72	0.094	6.944 ± 0.696	47.737 ± 3.097	29.154 ± 0.640
96	0.105	17.726 ± 1.616	82.830 ± 7.430	48.174 ± 2.074
120	0.112	16.142 ± 1.886	81.235 ± 3.352	56.438 ± 0.987
168	0.105	18.360 ± 0.622	85.776 ± 7.413	50.978 ± 2.198

[a]The values are the average of two samples per exposure interval, with three fish per sample.

(nominal) study were 170, 814 and 507 for edible flesh, head and viscera, and whole fish, respectively. The higher concentration of ^{14}C residues in the head and viscera fraction may be due to a higher concentration of nonpolar materials in that fraction, compared to the edible flesh.

Some of the fish exposed to 0.1 ppm and 5.0 ppb DCB.2 HCl were transferred to water free of the chemical after equilibration to determine the rate of ^{14}C-residue elimination (depuration). The fish were placed in fresh water, flowing at a rate to give 12 complete turnovers of water per 24-hr period. During this period, the fish were periodically sampled and analyzed for ^{14}C content.

The depuration of DCB and derived materials is shown in Tables XII and XIII. Although the rate of elimination is rapid initially, DCB levels in the fish appear to be relatively constant in samples taken in the later phase

Table XII. Elimination of [14]C From Bluegills Exposed to 5 ppb ^{14}C-DCB·2 HCl

Depuration Time (hr)	Percent of Initial [14]C Remaining[a]		
	Edible FleshMM	Head and Viscera	Whole Fish
24	74.2	86.5	80.7
72	43.1	14.9	16.7
168	30.3	5.3	18.7
336	5.2	7.2	6.9

[a]Initial ^{14}C residues were 0.58 ppm, 3.72 ppm and 2.18 ppm for edible flesh, head and viscera, and whole fish, respectively. Depuration data are the average of duplicate samples per time point with three fish per sample.

Table XIII. Elimination of [14]C From Bluegills Exposed to 0.1 ppm ^{14}C-DCB·2 HCl

Depuration Time (hr)	Percent of Initial [14]C Remaining[a]		
	Edible Flesh	Head and Viscera	Whole Fish
24	69.2	72.1	75.0
72	42.9	63.3	35.4
144	29.2	30.4	32.0
216	18.3	8.0	10.7
288	18.4	6.5	8.9

[a]Initial ^{14}C residues were 18.36 ppm, 85.78 ppm and 50.98 ppm for edible flesh, head and viscera, and whole fish, respectively. Depuration data are the average of duplicate samples per time point with three fish per sample.

of depuration, particularly in the 0.1-ppm study. Other lipid-soluble chemicals such as lindane and methoprene are bioconcentrated to a large degree but are progressively eliminated after removal to noncontaminated water.[16,17] In this investigation, however, a rapid initial rate of elimination, followed by a low or negligible rate, with appreciable residues remaining even after 14 days of exposure to fresh water, suggests the possibility of an enterohepatic circulation of DCB and metabolite or that a portion of the residues are tightly bound to lipoproteins or other substances and resistant to elimination.

Preliminary evidence of metabolism of DCB by fish was obtained by TLC of methanol extracts of the edible flesh or head and viscera fractions. Chromatography in two solvent systems showed the presence of at least two materials. One radioactive peak cochromatographed with DCB whereas another remained at the origin, indicating that it was of a polar nature. To confirm this possibility, the fish extracts were fractionated according to partitioning behavior. The ^{14}C material that partitioned into ether from basic (pH 11) water (designated fraction A) was found to contain only DCB as shown by TLC. The water-soluble material remaining (fraction B) did not partition into ether when the aqueous medium was acidified to pH 1 with HCl. Because authentic DCB is almost quantitatively recovered into ether from basic water in a one-step partition, the water-solubility of the metabolic derivative might be attributed to a zwitterionic nature (such as hydroxylated products), or the conjugation of DCB to a material that retains a free acidic functional group and liberates free DCB upon exposure to mildly acidic conditions. The latter possibility was investigated by rebasification and partitioning of the acid-treated aqueous phase with ether. The ^{14}C was quantitatively recovered into ether and cochromatographed with DCB in two thin-layer chromatographic systems, indicating that a highly acid-labile conjugate of DCB is formed in the fish. The conjugate was

Table XV. Relative Abundance of DCB and Metabolite in Bluegills Exposed to 0.1 ppm ^{14}C-DCB·2 HCl

| Exposure (hr) | Percent Abundance[a] | | | |
| | Edible Flesh | | Head and Viscera | |
	Fraction A (DCB)	Fraction B (Metabolite)	Fraction A (DCB)	Fraction B (Metabolite)
24	72.0	28.0	25.6	74.4
48	66.2	33.8	22.5	77.5
72	70.2	29.8	25.6	74.4
96	77.7	22.3	16.4	83.6
120	58.7	41.3	20.7	79.3
168	66.1	33.9	21.6	78.4
Depuration (hr)				
24	21.4	78.6	16.9	73.1
72	18.9	71.0	19.7	80.3
144	44.3	55.7	35.6	64.4
216	49.5	50.5	38.9	61.1
288	14.5	85.5	57.3	42.7

[a]Values are the average of duplicate analyses of 6 fish pooled at each time interval. The data were obtained from the experiment listed in Table XIII.

which was stable in pH 10 media, exhibiting a λ max. of 285 nm (as opposed to 282 nm for DCB under identical conditions). When the pH was adjusted to 2 with HCl, a maximum absorbance at 248 nm (corresponding to DCB) rapidly appeared. The acid-lability of the synthesized product, producing free DCB, was confirmed by both TLC and HPLC. Without the acid treatment, both the synthetic and biological materials appeared to elute with the column void volume. Bovine liver β-glucuronidase towards N-glucuronide conjugates of aromatic amines has been noted previously.[20]

The results of the separation of the fish extracts from the uptake/elimination studies into Fraction A (DCB) and Fraction B (metabolite) are given in Tables XIV and XV. Based on the portion of ^{14}C residues which are free DCB, the bioconcentration factor for the edible portions in the 5-ppb study was 27.8 and 86.2 for the head and viscera fraction. In the 0.1-ppm study the bioconcentration value for edible flesh was 114.9 and 159.3 for nonedible tissue. These differences probably reflect the concentration-dependence of metabolism, as well as distribution and storage of DCB and its metabolite.

Table XIV. Relative Abundance of DCB and Metabolite in Bluegills Exposed to 5 ppb ^{14}C-DCB·2 HCl

| | Percent Abundance[a] | | | |
| | Edible Flesh | | Head and Viscera | |
Exposure (hr)	Fraction A (DCB)	Fraction B (Metabolite)	Fraction A (DCB)	Fraction B (Metabolite)
24	75.3	24.7	12.1	87.9
48	41.6	58.4	10.4	89.6
72	26.9	73.1	8.8	91.2
96	26.6	73.4	11.9	88.1
120	22.1	77.9	8.2	91.8
Depuration (hr)				
24	34.3	65.7	17.8	82.2
72	36.7	63.3	38.8	61.1
168	37.1	62.9	51.6	48.3
336	62.7	37.3	46.8	53.2

[a]Values are the average of duplicate analyses of six fish pooled at each time interval. The data were obtained from the experiment listed in Table XIII.

hydrolyzed to form free DCB in as little as 30 sec at pH 1, but was stable at pH 9 or higher. The identity of the fraction A material and the material produced by acid treatment of fraction B was further confirmed to be DCB, both qualitatively and quantitatively, by HPLC. In addition to identical retention times, UV absorbance of the eluted unknown materials was scanned with the chromatograph UV detector. The UV spectra of both samples was virtually the same as authentic DCB, including identical λ max. = 282 nm. Therefore, the degree of metabolism seen in the uptake and elimination studies is reported as the distribution of extracted materials into the fraction A (DCB), and fraction B (metabolite) as determined radiometrically (Tables XIV and XV)..

The partition behavior of the metabolite in basic aqueous media indicates that the DCB nucleus was modified by addition of at least one ionizable acidic group. Mono- or *bis*-N sulfate or N-glucuronide conjugates of DCB would be expected to behave in such a manner. Glucuronidation is well documented as an important metabolic pathway in vertebrates, including fish.[18] Furthermore, previous studies of aromatic amine metabolism indicates the occurrence of N-glucuronidation.[19] These conjugates are highly labile under mildly acidic conditions[20] and may be formed in the absence of the enzymes required for synthesis of N-glucuronide conjugates.[21] When DCB was mixed with glucuronic acid, a crystalline product was obtained

Discussion

Bioconcentration factors in fish of greater than 10^4 have been recorded for such chemicals as DDT[22] and Arochlor 1254[23] under equilibrium conditions. In comparison, the bioconcentration factors for DCB are rather low. However, the presence of even relatively low concentrations of DCB must be regarded as significant in view of the carcinogenic hazard posed by the chemical. Also, in view of the ease of hydrolysis of the DCB conjugate present in the fish, the conjugated residues should be viewed as the toxicological equivalent of DCB, since rapid hydrolysis to free DCB is likely to occur in the acid medium of the digestive tract after consumption.

Based on the results of our studies it appears that photodegradation, sorption to suspended sediment, and bioaccumulation and biotransformation are among the processes which are more likely to control the transport and fate of DCB. These studies have only identified the likely pathways of DCB dissipation. The contribution of each of these processes will vary with the ecosystems, however, since the rates of these processes are controlled by environmental conditions which are characteristic of an individual aquatic system. For instance, photolysis of DCB can be expected to be a dominant process in clear waters. On the contrary, sorption by suspended sediments may be more important in turbid waters. While the photolytic half-life of DCB in clear distilled water may be as short as five minutes, this rate may not necessarily be representative of an entire natural aquatic system. In most cases transparency depth, degree of mixing and season would be critical factors in determining the rate at which DCB is degraded by sunlight.

Also, physical, chemical and biological processes compete with each other in natural ecosystems. Our studies have shown that DCB is rapidly photodegraded in aqueous solution, is readily bioaccumulated and is extensively sorbed by the sediment. It is quite likely that sorption of DCB by sediment may decrease its availability to aquatic organisms. It may also affect the rate of photodegradation. Therefore, to determine the relative importance of various mechanisms controlling DCB fate, further studies should be done in a system in which the processes are acting simultaneously.

ACKNOWLEDGMENTS

We wish to thank the staff of the Environmental Protection Agency, Athens, Georgia and, in particular, Dr. W. C. Steen (project monitor) for suggestions and information that were useful in completion of this work. We also acknowledge the technical assistance of Mr. E. Pack and Mr. R. Gray.

The work was supported by EPA Grant No. R 804-584-010.

REFERENCES

1. U.S. Tariff Commission. "Synthetic Organic Chemicals, U.S. Production and Sales, 1972," T.C. Publication 681, U.S. Government Printing Office (1974).
2. Pliss, G.B. "Some Regular Relations Between Carcinogenicity of Aminodiphenyl Derivatives and the Structure of These Substances," *Acta Unio. Int. Contra Concrum.* 19:499-501 (1963).
3. *Federal Register,* 38 (85) (1973).
4. Boyland, E. *Biochemical Mechanism of Induction of Bladder Cancer* (Boston: Little, Brown, and Co., 1959).
5. Riley, D., W. Wilkinson and B. V. Tucker. "Biological Unavailability of Bound Paraquat Residues," In: *Bound and Conjugated Pesticide Residues,* D. D. Kaufman, G. G. Still, G. D. Paulson and S. K. Bandal, Eds. (Washington, DC: American Chemical Society, 1976).
6. Stevenson, F. J. "Organic Matter Reactions Involving Pesticides in Soil," In: *Bound and Conjugated Pesticide Residues,* D. D. Kaufmann, G. G. Still, G. D. Paulson and S. K. Bandal, Eds. (Washington, DC: American Chemical Society, 1976).
7. Hsu, T., and R. Bartha. "Interaction of Pesticide-Derived Chloroaniline Residues with Soil Organic Matter," *Soil Sci.* 116:444-452 (1974).
8. Hsu, T., and R. Bartha. "Hydrolyzable and Nonhydrolyzable 3,4-DichloroanilinepHumus Complexes and Their Respective Rates of Biodegradation," *J. Agric. Food Chem.* 24:118-122 (1976).
9. Johnson, L. F., and E. A. Curl. *Methods for Research on the Ecology of Soil-Borne Plant Pathogens* (Minneapolis, MN: Burgess Publishing Co.).
10. Bartha, R., and D. Pramer. "Features of a Flask and Method for Measuring the Persistence and Biological Effects of Pesticides in Soil," *Soil Sci.* 100:68-70 (1965).
11. Bunch, R. L., and C. W. Chambers. "A Biodegradability Test for Organic Compounds," *J. Water Poll. Control Fed.* 39:181-187 (1967).
12. Calvert, J. G., and J. N. Pitts, Jr. *Photochemistry* (New York: Wiley Interscience, 1966).
13. Lu, P.-Y., R. L. Metcalf, N. Plummer and D. Mandel. "The Environmental Fate of Three Carcinogens: Benzo-(α)-Pyrene, Benzidine, and Vinyl Chloride in Laboratory Model Ecosystems," *Arch. Environ. Contam. Toxicol.* 6:129 (1977).
14. Ames, B. N., W. E. Durston, E. Yamasaki and F. D. Lee. "Carcinogens are Mutagens. Simple Homogenates for Activation and Bacteria for Detection," *Proc. Nat. Acad. Sci.* (USA) 70:2281-2285 (1973).
15. Sikka, H. C., D. Ford and R. S. Lynch. "Uptake, Distribution and Metabolism of Endothall in Fish," *J. Agric. Food Chem.* 23:849-851 (1975).
16. Schimmel, S. C., J. M. Patrick, Jr. and J. Forester. "Haptachlor: Uptake, Depuration, Retention and Metabolism by Spot, Leiostomus Xanthurus," *J. Toxicol. Environ. Health* 2:169-178 (1976).

17. Quistad, G. B., D. A. Schooley, L. E. Staiger, B. J. Bergot, B. H. Sleight and K. J. Macek. "Environmental Degradation of the Insect Growth Regulator Methoprene. IX. Metabolism by Bluegill Fish," *Pestic. Biochem. Physiol.* 6:523-529 (1976).

18. Chambers, J. E., and J. D. Yarbrough. "Xenobiotic Transformation Systems in Fishes," *Comp. Biochem. Physiol.* 55C:77-84 (1976).

19. Shuster, L. "Metabolism of Drugs and Toxic Substances," *Ann Rev. Biochem.* 33:571-596 (1964).

20. Axelrod, J., J. D. Inscoe and G. M. Tomkins. "Enzymatic Synthesis of N-Glucosyluronic Acid Conjugates," *J. Biol. Chem.* 232:835-841 (1958).

21. Bridges, J. W., and R. T. Williams. "N-glucuronide Formation *in Vivo* and *in Vitro*," *Biochem. J.* 83:27P (1962).

22. Hansen, D. J., and A. J. Wilson, Jr. "Significance of DDT Residues From the Estuary Near Pensacola, Fla.," *Pestic. Monit. J.* 4:51-56 (1970).

23. Hansen, D. J., P. R. Parrish, J. L. Lowe, A. J. Wilson, Jr. and P. D. Wilson. "Chronic Toxicity, Uptake, and Retention of Arachlor 1254 in Two Estuarine Fishes," *Bull. Environ. Contam. Toxicol.* 6:1130119 (1971).

19

THE ROLE OF THE PARTITION COEFFICIENT
IN ENVIRONMENTAL TOXICITY

Corwin Hansch

Department of Chemistry
Pomona College
Claremont, California   91711

INTRODUCTION

Experimental evidence for the role of the oil/water partition coefficient (P) of organic compounds in nonspecific toxicity was first pointed out about the turn of the century independently by Meyer[1] and Overton.[2] Both workers showed that there was a relationship between olive oil/water partition coefficients and the narcotic action of organic compounds on various organisms. Overton in particular measured the concentration in water of many simple alcohols, esters, ketones, ethers, etc., necessary to stop motion of tadpoles. The partition coefficients of many of these compounds were not determined until the early 1960s, at which time it was possible to show a mathematical relationship between Overton's data and log P for octanol/water partition coefficients:[3]

$$\log 1/C = 0.94 \log P + 0.87 \tag{1}$$
$$n = 51; r = 0.971; s = 0.280$$

where: C is the molar concentration of compound necessary to produce narcosis,
n is the number of data points
r is the correlation coefficient and
s is the standard deviation.

This equation indicates that the higher the partition coefficient is, the greater the toxicity will be; of course, such a linear relationship cannot go on forever and, as discussed below, we eventually find that as log P increases, activity begins to fall off and even decreases.

Equation 1 can be compared with Equation 2 for the bioconcentration of various chemicals in trout.[4]

$$\text{log bioconcentration factor} = 0.54 \log P + 0.12 \qquad (2)$$
$$n = 8; \ r = 0.948; \ s = 0.342$$

Equation 2 is not as good a correlation as Equation 1, which is due at least in part to greater experimental difficulty. Three of the eight compounds on which Equation 4 is based have log P values > 5.5, which makes their determination extremely difficult.

CALCULATION OF LOG P

Equations 1 and 2 show how increasing the hydrophobicity of chemicals can produce increased toxicity in an aqueous environment. The interaction of all kinds of systems with organic compounds in an aqueous environment can be related to the partitioning process. In order to analyze such interactions, the hydrophobic parameters log P, π and f are being studied. An important discovery is that log P is an additive-constitutive parameter;[5] this means that one can calculate log P values for many compounds without actually measuring them. This can be illustrated with the parameter π which is defined as: $\pi_X = \log P_X - \log P_H$. P_H refers to the partition coefficient of the parent compound and P_X to that of a derivative; for example, π for Cl is calculated as follows:

$$\pi_{Cl} = \log P_{C_6H_5-Cl} - \log P_{C_6H_6}$$
$$\pi_{Cl} = 2.84 - 2.13 = 0.71$$

The values of π vary from system to system, but they can be assumed to be constant for similar systems.[6]

The partition coefficient for 2-acetylaminofluorene can be calculated as follows:

$$\begin{array}{ccccc} 4.18 & - & 0.97 & = & 3.21 \end{array}$$

The π_{NHCOCH_3} value has been determined[7] in the benzene system as illustrated above for Cl. Using this technique, an enormous number of log P values for various monosubstituted aromatic hydrocarbons may be accurately calculated. However, when two or more substituents, at least one of which has a strong electronic effect, are placed on the same ring, the above approach may be in error by as much as 0.5 to 1.0 log unit.

The parameter π does not lend itself easily to the calculation of log P values from scratch. Rekker[8] has devised a fragment constant, f, which is now being developed in several laboratories for the computerized calculation of large numbers of log P values. Leo et al.[9] have been developing f along slightly different lines from Rekker.

The calculation of log P for aflatoxin B_1 can be illustrated using fragment constants:

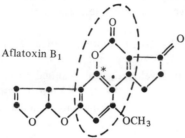

Aflatoxin B_1

Step 1. Two of fused rings are aromatic; one fusion is $\overset{*}{\underset{.}{C}}$; other $\underset{.}{C}$.

Step 2. (n–1) Bonds in each fused set of alicyclic rings counted separately: (8–1) = 7 + (4–1) = 10 total

Step 3. σ Enhancement of $-C=O$ ortho to $-\overset{o}{C}-$ in ring; use $f_1{}^X$

Step 4. f_o^{vinyl} = -0.61 Correction for oxygen attached to vinyl group.

$5f_{\underline{C}}$ + $f_{\underline{\dot{C}}}$ + $f_{\underline{\overset{*}{C}}}$ + f_{CH} + $4f_{CH_2}$ + $F_{(=)}$ + $2f_{CH}$

5(0.13) + 0.225 + 0.44 + 0.35$_5$ + 4(0.66) – 0.55 + 2(0.43)

+ f_{CH_3} + $3f_o^{\phi}$ + $10F_b$ + F_b + F_{P-1}^{o-p} + f_{vo}^{ϕ}

+ 0.89 – 3(0.61) – 10(0.09) – 0.12 + 0.32(2.61) – 1.40

+ f_v^{X1} = 0.82 Calcd
                1.18 Obsd

= 0.82

The "aromatic" part of aflatoxin B_1 is circled. In calculating bond f constants,[10] these bonds are not included. In the above calculations, $F_{(=)}$ is the fragment constant for an isolated double bond, f_o^{ϕ} is for oxygen (2 in the furan moieties and 1 in OCH_3), $F_{\underline{b}}$ is the bond factor for a ring bond (nonaromatic), and F_b is for normal bonds. F_b is the number of normal single bonds less 1. In the present case, only OCH_3 is outside the ring system and $n-1 = 1$ for methoxy. $F_{P-1}^{o \to o}$ is an interaction factor for the two furan oxygen atoms, $f_{\underline{vo}}$ is a constant for a fused C=O in a ring, and f_v^{X1} is the fragment value which accounts for the electron-withdrawal effect of the carbonyl group in the 5-membered ring on the cyclic ester group.

The observed value for aflatoxin B_1 was determined by high-pressure liquid chromatography rather than the usual shake flask method. The agreement between observed and calculated values is satisfactory for many purposes.

The calculation of log P of such complex molecules as aflatoxin B_1 from fragments is still in the developmental stage. Once an internally self-consistent system has been developed, it is planned to computerize the calculations. There are so many fragments to keep in mind for large molecules that, unless one is constantly making such calculations, it is all too easy to make a mistake.

HYDROPHOBIC INTERACTIONS IN SIMPLE MODEL SYSTEMS

Hydrophobic interactions are important at the simplest levels of molecular complexity for systems in an aqueous environment. This can be illustrated by the recent work of Murakami et al.[11] on the paracyclophane-assisted hydrolysis of $RCOOC_6H_4$-p-NO_2 congeners:

The above paracyclophane with the attached imidazole moiety can be viewed as a kind of "mini-enzyme". Its rate enhancement of the aqueous hydrolysis of a set of 11 acyl nitrophenols is correlated[12] by Equation 3:

$$\log P = 0.45\pi - 0.53 \qquad (3)$$
$$n = 11; \; r = 0.968; \; s = 0.260$$

The rate of hydrolysis depends on the hydrophobic character of R. The dependence is much the same as that of Equation 2. Even for rather small molecules in aqueous solution, the hydrophobic interaction is important.

No doubt more complex synthetic mini-enzymes than the above cyclophane will be made and allowed to react with more complex substrates. The hydrophobic parameters π and f, along with steric and electronic parameters, will be very useful in the study and development of such systems.

Another example of the use of model systems to study hydrophobic interactions is illustrated[13] in Equations 4-7.

Disaggregation of 0.2-mm Silanized Glass Beads by ROH

$$\log 1/C = 0.98 \log P - 0.80 \qquad (4)$$
$$n = 4; \; r = 0.995; \; s = 0.077$$

Change in Resistance of Black Lipid Membrane by ROH

$$\log 1/C = 1.16 \log P - 0.51 \qquad (5)$$
$$n = 7; \; r = 0.985; \; s = 0.262$$

Hemolysis of Red Cells

$$\log 1/C = 0.93 \log P - 0.09 \qquad (6)$$

-10 μv Change in Rest Potential of Lobster Axon by ROH

$$\log 1/C = 0.87 \log P - 0.24 \qquad (7)$$
$$n = 5; \; r = 0.993; \; s = 0.100$$

Equation 4 correlates the concentration of alcohol necessary to disaggregate glass beads covered with an oily film. This is the simplest possible model of a membrane. The dependence of disruption of this membrane on log P is essentially the same (same slope) as for the more complex membranes (Equations 6 and 7). Equation 5 correlates a much more complex system. The black lipid membrane (BLM) is made by dissolving lipid of sheep red cell ghosts in a hydrocarbon and brushing this solution over a small hole; this gives an extremely thin membrane. The change in resistance from 10^8 to 10^6 ohms/cm^2 was used as the end point for finding log $1/C$. Even the intercepts of Equations 4 and 5 are close; the higher intercept for Equation 5 shows that the BLM is more easily disrupted by isolipophilic molecules. Twice the concentration of ROH is needed to produce the separation of the glass beads that is needed to change the resistance of the BLM.

Equation 6 is an average of 15 different linear equations for the hemolysis of red blood cells by a wide variety of organic compounds. Since it is an average equation, r and s have not been calculated. Comparison of the intercepts of Equations 5 and 6 shows that rupture of the red blood cell membrane is about 2.5 times (in terms of molar concentration) as easy as changing the resistance of the BLM.

Finally, perturbation of the lobster axon (Equation 7) by alcohols can be compared with the simpler processes of Equations 4-6. The slopes and intercepts of Equations 6 and 7 for the two living systems are close. Disruption of these living systems by lipophilic compounds which can be compared via log P appear to be very similar processes. This disruptive action of lipophilic organic compounds on membranes can have lethal effects on larger organisms, as the following correlation shows:

LD_{100} of ROH for cats

$$\log 1/C = 1.06 \log P + 1.37 \qquad (8)$$
$$n = 8; \; r = 0.986; \; s = 0.134$$

Killing cats has the same dependency on the hydrophobic character of alcohols as putting tadpoles to sleep (Equation 1) or disrupting aggregated oil-coated glass beads.

The lesson from Equations 4-8 and many others like them[13] is that organic compounds are toxic to living systems in proportion to their lipophilic character, *other factors being equal.*

The kind of toxicity considered in the above correlations is that of simple neutral compounds without any special toxic groups. The effect of introducing a group with a potent pharmacophoric function can be seen in Equation 9.[13]

$$R-\overset{+}{N}\diagdown\diagup I^-$$

$$\log 1/C = 0.88 \log P + 2.35 \qquad (9)$$
$$n = 5; \ r = 0.988; \ s = 0.252$$

Comparing Equations 6 and 9, we find the same dependence on hydrophobicity for the two systems; however, the differences in intercepts indicates that the pyridinium function is about 200 times as potent at membrane perturbation as nonspecific functions such as OH, OR, COOR and COR. There is now a large amount of experience with OSAR which shows that one can alter the activity of almost any pharmacophore by modulating its lipophilic character.

ROLE OF HYDROPHOBICITY IN THE SEQUESTERING OF CHEMICALS BY PROTEINS AND OTHER MATERIALS

Membrane perturbation by lipophilic organic compounds is an extremely important mechanism of environmental toxicity. Another mechanism for toxic activity is the interaction of hydrophobic compounds with proteins rather than the lipids of membranes. Equations 10 and 11 illustrate the point.

Binding of Miscellaneous Compounds by Bovine Serum Albumin[14]

$$\log 1/C = 0.67 \log P + 2.60 \qquad (10)$$
$$n = 25; \ r = 0.945; \ s = 0.242$$

Binding of Miscellaneous Compounds by Bovine Hemoglobin[15]

$$\log 1/C = 0.67 \log P + 1.96 \qquad (11)$$
$$n = 17; \ r = 0.941; \ s = 0.163$$

C in these two equations is the molar concentration of ligand producing a 1:1 ligand-protein complex. These results show that two different proteins bind a wide variety of molecules in proportion to their octanol/water partition coefficients. This means that once organic compounds onto living organisms, they are bound by various proteins.

The hydrophobic property of organic compounds causes them to bind to all kinds of material,[13,16] as shown by Equations 12 and 13.

Adsorption of $X\text{-}C_6H_4OCH_2COOH$ by Carbon[16]

$$\log \% \text{ ad.} = 0.81 \log P - 2.13 \tag{12}$$
$$n = 16; \ r = 0.952; \ s = 0.213$$

Binding of Acetanilides by Nylon[16]

$$\log k = 0.70 \log P - 1.16 \tag{13}$$
$$n = 7; \ r = 0.965; \ s = 0.198$$

Many other such examples show that the way organic compounds accumulate in the environment is dependent on log P.

Knowing that hydrophobic compounds are bound by simple proteins, it is not surprising that lipophilic character plays an important role in the interaction of all kinds of inhibitors with enzymes.[17] Exactly how enzyme inhibitors in the environment might affect our lives is largely unknown. Of course, in the case of a certain few inhibitors such as those for acetylcholinesterase, there is no doubt about their great toxicity. While lipophilic character plays an important role in the inhibition of isolated enzymes, what the role is in whole animals has unfortunately received little systematic study.

NONLINEAR DEPENDENCE OF TOXICITY ON LOG P

Up to this point in the chapter the correlation equations relating toxicity and hydrophobicity have shown a linear relationship between these two parameters. That such a relationship cannot go on forever has long been appreciated; however, the nature of the break in linearity is still not entirely understood. The first general advance in our understanding resulted from the postulate that log l/C was, in the general sense, parabolically related to the partition coefficient:[18,19]

$$\log 1/C = a \log P - b(\log P)^2 + c \tag{14}$$

Since the advent of the "parabolic model", a number of other mathematical relationships have been developed to explain the nonlinearity between biological activity and hydrophobicity[20-27]; of these models, only that of Kubinyi[25,26] has received extensive application. The Kubinyi bilinear model is defined as:

$$\log 1/C = a \log P + b \log(\beta P + 1) + c \qquad (15)$$

In Equation 14, a, b and c are evaluated by the method of least squares. An additional parameter, β, must be evaluated for Equation 15. Since Equation 15 is a nonlinear equation, the simple least squares method cannot be used; instead, an iterative procedure must be employed.

Equations 16 and 17 compare the two models using the data on the toxicity of aliphatic hydrocarbons to mice.

$$LD_{100} \text{ in Mice of } C_nH_{2n+2}$$
$$\log 1/C = 0.94 \log P - 0.11(\log P)^2 + 0.20 \qquad (16)$$
$$n = 11; \ r = 0.930; \ s = 0.148; \text{ ideal } \log P = 4.37$$

$$\log 1/C = 0.96 \log P - 1.31 \log(\beta P + 1) - 0.65 \qquad (17)$$
$$n = 11; \ r = 0.996; \ s = 0.039; \ \log \beta = -3.52; \text{ ideal } \log P = 3.96$$

In correlating the toxicity of hydrocarbons to mice, the bilinear model gives a correlation much superior to that of the parabolic model. As usual, the ideal log P (value for most potent congener) for the bilinear model is a little lower than that for the parabolic model.

It has been found from an analysis of many data sets that the bilinear model usually gives sharper correlations for a homologous series of congeners. The parabolic model is apt to give better correlations in more complex sets of congeners where many variations in heteroatoms occur.

Attention in this report is focused on toxicity and hydrophobic character. Of course, many other parameters such as electronic and steric have an important role in toxicity. When these effects are encountered, equations such as Equations 16 and 17 must be expanded with new terms to account for these effects.

MUTAGENICITY

One of the most exciting events in toxicology in recent years is the development of the Ames test for mutagenicity.[28-30] This test is based on strains of mutant bacteria which require histidine for growth. Treatment with mutagens causes some of the bacteria to revert back to histidine sufficient organisms. By placing the bacteria on the proper media and observing the development of colonies, it is possible to count the number of mutations produced by a given concentration of a mutagenic compound.

For an initial study of the Ames test in a structure-activity analysis we selected the triazenes:

The triazenes were selected because it is known they are both mutagenic[31] and carcinogenic.[32] Also, they are used in the treatment of cancer, and correlation equations have been formulated[33] for their activity against leukemia in mice, as well as their toxicity (LD_{50}) to mice.[34]

Some of the triazenes are very toxic to the bacteria (TA92) used in our study. In order not to kill the bacteria so that mutations could be observed, it was necessary to work at low concentrations. As the concentration of triazene was increased, observed mutagenicity increased up to a point and then fell off rapidly as toxicity set in. Dose response curves were run for each triazene, after which it was decided to use as an end point the concentration of triazene which, under standard conditions, produced 30 mutations above background (about 50 mutations) in 10^8 bacteria Equation 18 was formulated from these results and it shows that mutagenic potency depends on lipophilic character and electron withdrawal by substituents.

$$\log 1/C = 1.09 \log P - 1.63\sigma^+ + 5.58 \tag{18}$$
$$n = 17; \; r = 0.974; \; s = 0.315$$

It was found that σ^+ gave considerably better correlation than σ which brings out the importance of through resonance; this was also observed in the antileukemia activity of the triazenes.

To observe the mutagenicity of many compounds, including the triazenes, it is necessary to include liver microsomes along with the bacteria in the test medium. The microsomes apparently oxidize aromatic compounds to forms which produce the mutagenic action. It has been shown via QSAR that microsomal oxidation is highly dependent[35,36] on log P; therefore, it is not surprising to find the high dependence of mutagenicity on lipophilicity which is brought out by Equation 18. It is interesting that although a wide range of hydrophobicity is covered (log P varied from 0.98 to 4.40), the dependence on log P is still linear. It seems unlikely that this sitation would hold for action in animals; here one would expect equations of the form of 16 or 17 to hold.

Many years ago we showed[19] via QSAR that the carcinogenicity of polycyclic aromatic hydrocarbons and acridines to mice was highly dependent on log P. It is quite clear that these undesirable properties of molecules can be varied greatly simply by varying hydrophobic character. For example, changing log P by 4 changes mutagenicity according to Equation 4 by a factor of 10,000. Changes in σ^+ could result in another 3 log units of activity. In other words, by proper variation of substituents, it should be possible to make a set of triazenes with a mutagenicity range of 10 million (7 log units).

Equation 18 shows how one might greatly reduce the mutagenic character of pesticides, for example. It is hoped that this can be done without destroying the valuable pesticide activity.

It is now fully apparent that the hydrophobic properties of organic compounds are of enormous importance in their interaction with all biochemical and biological systems. Research in toxicology must undertake a more systematic and vigorous study of this universally important property of the many organic chemicals, both natural and unnatural, which make up our environment. Correlation equations can greatly simplify our job; for example, Equation 18, rationalizing the mutagenicity of triazenes, was derived using 17 congeners. These congeners had log P values ranging from 0.98 to 4.40 and σ^+ values ranging from -0.84 to 1.12. Enough experience is now in hand about the predictive value of correlation equations[37] that we can be fairly confident that Equation 18 will accurately predict mutagenicity of new triazenes having log P and σ^+ values within these ranges. One can expect moderately accurate predictions even outside these ranges if one does not venture too far beyond the explored log P and σ^+ ranges. It is not enough to know that triazenes are mutagenic. As has been shown above, very large changes in potency can be effected simply by making changes in substituents; for example, Equation 18 predicts that replacing a 4-CN group in the phenyl triazenes with a $4\text{-}OC_4H_9$ group would increase the mutagenic potency by a factor of 10 million. The $4\text{-}OC_4H_9$ analog is expected to be more mutagenic than aflatoxin B_1.

With the many tens of thousands of synthetic chemicals already in commercial use, and with the countless number of natural products of possible toxic character, we cannot afford to test each individual compound in scores or possibly hundreds of biochemical systems to develop toxicological profiles. QSAR, in its broadest sense, can be of enormous help in deciding how toxic a chemical may be—even before it is synthesized.

REFERENCES

1. Meyer, H. "Zur Theorie der Alkoholnarkose," *Arch. Exp. Pathol. Pharmakol.* 42:109-118 (1899).
2. Overton, E. *Studien uber die Narkose* (Jena, Germany: Fischer, 1901).
3. Hansch, C. In: *Drug Design*, Vol. 1, E. J. Ariëns, Ed. (New York: Academic Press, 1971), pp. 271-337.
4. Neely, W. B., D. R. Branson and G. E. Blau. "Partition Coefficient to Measure Bioconcentration Potential of Organic Chemicals in Fish," *Environ. Sci. Technol.* 8:1113-1115 (1974).
5. Leo, A., C. Hansch and D. Elkins. "Partition Coefficients and Their Uses," *Chem. Rev.* 71:525-616 (1971).
6. Fujita, T., J. Iwasa and C. Hansch. "A New Substituent Constant, π, Derived from Partition Coefficients," *J. Am. Chem. Soc.* 86: 5175-5180 (1964).
7. Hansch, C., A. Leo, S. H. Unger, K. H. Kim, D. Nikaitani and E. J. Lien. "Aromatic Substituent Constants for Structure Correlations," *J. Med. Chem.* 16:1207-1216 (1973).
8. Rekker, R. F. *The Hydrophobic Fragmental Constant* (Amsterdam, Holland: Elsevier, 1977).
9. Leo, A., P. Y. C. Jow, C. Silipo and C. Hansch. "Calculation of the Hydrophobic Constant (Log P) from π and f Constants," *J. Med. Chem.* 18:865-868 (1975).
10. Leo, A. Private communication.
11. Murakami, Y., Y. Aoyama, M. Kida and A. Nakano. "A Macrocyclic Enzyme Model System. Deacylation of *p*-Nitrophenyl Carboxylates as Effected by a [20]Paracyclophane Bearing an Imidazole Group," *Bull. Chem. Soc. Jap.* 50:3365-3371 (1977).
12. Hansch, C. Unpublished results.
13. Hansch, C., and W. J. Dunn III. "Linear Relationships Between Lipophilic Character and Biological Activity of Drugs," *J. Pharm. Sci.* 61:1-19 (1972).
14. Vandenbelt, J. M., C. Hansch and C. Church. "Binding of Apolar Molecules by Serum Albumin," *J. Med. Chem.* 15:787-789 (1972).
15. Kiehs, K., C. Hansch and L. Moore. "The Role of Hydrophobic Bonding in the Binding of Organic Compounds by Bovine Hemoglobin," *Biochemistry* 5:2602-2605 (1966).
16. Dunn, W. J. III, and C. Hansch. "Chemicobiological Interactions and the Use of Partition Coefficients in Their Correlation," *Chem.-Biol. Interact.* 9:75-85 (1974).
17. Yoshimoto, M., and C. Hansch. "Correlation Analysis of Baker's Studies on Enzyme Inhibition. 2. Chymotrypsin, Trypsin, Thymidine Phosphorylase, Uridine Phosphorylase, Thymidylate Synthetase, Cytosine Nucleoside Deaminase, Dihydrofolate Reductase, Malate Dehydrogenase, Glutamate Dehydrogenase, Lactate Dehydrogenase, and

Glyceraldehyde-phosphate Dehydrogenase," *J. Med. Chem.* 19:71-98 (1976).

18. Hansch, C., P. P. Maloney, T. Fujita and R. M. Muir. "Correlation of Biological Activity of Phenoxyacetic Acids with Hammett Substituent Constants and Partition Coefficients," *Nature* 194:178-180 (1962).

19. Hansch, C., and T. Fujita. "*ρ-σ-π* Analysis. A Method for the Correlation of Biological Activity and Chemical Structure," *J. Am. Chem. Soc.* 86:1616-1626 (1964).

20. McFarland, J. W. "On the Parabolic Relationship between Drug Potency and Hydrophobicity," *J. Med. Chem.* 13:1192-1196 (1970).

21. Higuchi, T., and S. S. Davis. "Thermodynamic Analysis of Structure-Activity Relationships of Drugs: Prediction of Optimal Structure," *J. Pharmacol. Sci.* 59:1376-1383 (1970).

22. Hyde, R. M. "Relationship Between the Biological and Physiochemical Properties of Series of Compounds," *J. Med. Chem.* 18:231-233 (1975).

23. Dearden, J. C., and M. S. Townend. "A Theoretical Approach to Structure-Activity Relationships. Some Implications for the Concept of Optimal Lipophilicity," *J. Pharm. Pharmacol.* 28S:13P (1976).

24. Martin, Y. C., and J. J. Hackbarth. "Theoretical Model-Base Equations for the Linear Free-Energy Relationships of the Biological Activity of Ionizable Substances. 1. Equilibrium-Controlled Potency," *J. Med. Chem.* 19:1033-1039 (1976).

25. Kubinyi, H. "Quantitative Structure-Activity Relationships. IV. Nonlinear Dependence of Biological Activity on Hydrophobic Character: A New Model," *Arzneim.-Forsch.* 28:1991-1997 (1976).

26. Kubinyi, H. "Quantitative Structure-activity Relationships. 7. The Bilinear Mode, A New Model for Nonlinear Dependence of Biological Activity on Hydrophobic Character," *J. Med. Chem.* 20:625-629 (1977).

27. Kubinyi, H. "Quantitative Structure-Activity Relationships. VI. Nonlinear Dependence of Biological Activity on Hydrophobic Character: Calculation Procedures for the Bilinear Model," *Arzneim.-Forsch.* 28:598-601 (1978).

28. Ames, B. N., J. McCann and E. Yamasaki. "Methods for Detecting Carcinogens and Mutagens with *Salmonella*/Mammalian-Microsome Mutagenicity," *Mut. Res.* 31:347-364 (1975).

29. Ames, B. N., and J. McCann. "Detection of Carcinogens as Mutagens in the *Salmonella*/Microsome Test: As say of 300 Chemicals: Discussion," *Proc. Nat. Acad. Sci. U.S.* 73:950-954 (1976).

30. Ames, B. N., and C. Yanofsky. In: *Chemical Mutagens*, A. Hollaender, Ed. (New York: Plenum Publishing Corporation, 1971), p. 267.

31. Malaveille, C., G. F. Kolar, and H. Bartsch, "Rat and Mouse Tissue–Mediated Mutagenicity of Ring Substituted 3,3-Dimethyl-1-Phenyl triazenes in *Salmonella typhimurium*," *Mut. Res.* 36:1-10 (1976).

32. Preussmann, R., S. Ivankovic, C. Landschuetz, J. Gimmy, E. Flohr

and O. Griesbach. "Carcinogene Wirkung von 1,3-Aryldialkyltriazenen an BD-Ratten," *Z. Krebsforsch. Klin. Onkol.* 81:285-310 (1974).

33.   Hatheway, G. J., C. Hansch, K. H. Kim, S. R. Milstein, C. L. Schmidt, R. N. Smith and F. R. Quinn, "Antitumor 1-(X-Aryl)-3,3-di-Alkyltriazenes. 1. Quantitative Structure-Activity Relationships vs L1210 Leukemia in Mice," *J. Med. Chem.* 21:563-573 (1978).

34.   Hansch, C., G. J. Hatheway, F. R. Quinn and N. Greenberg, "Antitumor 1-(X-Aryl)-3,3-di-Alkyltriazenes. 2. On the Role of Correlation Analysis in Decision Making in Drug Modification. Toxicity Quantitative Structure-Activity Relationships of 1-(X-Phenyl)-3,3-dialkyltriazenes in Mice," *J. Med. Chem.* 21:574-577 (1978).

35.   Martin, Y. C., and C. Hansch. "Influence of Hydrophobic Character on the Relative Rate of Oxidation of Drugs by Rat Liver Microsomes," *J. Med. Chem.* 14:777-779 (1971).

36.   Hansch, C., "Quantitative Relationships between Lipophilic Character and Drug Metabolism," *Drug Met. Rev.* 1:1-13 (1972).

37.   Hansch, C. In: *Biological Activity and Chemical Structure* (Amsterdam, Holland: Elsevier, 1977), p. 47.

20

A METHOD FOR SELECTING THE
MOST APPROPRIATE ENVIRONMENTAL
EXPERIMENTS ON A NEW CHEMICAL

W. Brock Neely

>Environmental Sciences Research
>The Dow Chemical Company
>Midland, Michigan 48640

INTRODUCTION

The chemical industry has long been concerned with the health and environmental properties of the products that they manufacture and distribute. The effort that is expended in this area has grown exponentially in the past few years due to growing understanding of the environment. This increased awareness of potential problems is requiring better predictive techniques for making early decisions on what tests are needed (see, for example, Howard et al.[1]). Of necessity, such predictions must be based on laboratory findings, since it is not feasible to use the environment as a testing ground and, in addition, the newly enacted Toxic Substances Control Act (TSCA) requires a company to submit information to the EPA prior to manufacture and distribution.

Section 5 of TSCA, dealing with premanufacture notification, has generated interest in defining the tests that predict the environmental impact of a chemical. One of the concepts that is emerging is based on tier testing.[2] The objective of this approach is to enable the studies to proceed in a logical manner and to optimize the amount of information in a cost-effective manner.

The basic process in any hazard evaluation involving the environmental effects of chemicals is to make predictions of the expected environmental concentration (EEC) and to match this with the experimentally determined no-effect level for appropriate environmental organisms. Once the data demonstrate that the EEC is below the no-effect level, the product should be considered acceptable from an environmental point of view. Estimating environmental exposure is difficult. It may be accomplished for a localized situation where the source inputs and the ecosystem such as a river or lake can be identified. Atmospheric exposures can also be estimated for volatile compounds. However, in most other systems reliance is made on the benchmark approach.[3] In such an approach the properties of a new chemical are matched with similar chemicals of known environmental distribution, *e.g.*, DDT-like materials will behave like DDT.

This chapter will present a technique that estimates the distribution of the chemical in the air, water and soil. By comparing this profile with the intended use pattern, decisions can be made on what further action is required. This model is designed for assessing environmental as opposed to human health hazard. A different approach will be required for the latter decision.

The discussion will conclude with the presentation of several case studies using existing products.

THE ENVIRONMENTAL PROFILE

The proposed technique is the extension of several previous studies on compartmental analysis.[4-6] The output from this analysis is a ranking of the environmental distribution to be expected in the three main compartments: air, water and soil. While the results are given in percentages, the numbers are not meant to be absolute but are designed to yield a relative rank of importance. By matching this profile against the use pattern of the chemical it becomes easier to decide what future tests may be required.

A scenario is used for generating the profile where the chemical is added to a water compartment (Figure 1) at a fixed rate of 0.15 g/hr for a 30-day period, followed by a 30-day clearance phase.[6] The half-life for clearance from the fish biomass is estimated and the percentages of the total material found at 30 days in the air, water and soil compartments are calculated.

The estimated half-life ($t_{1/2}$) for clearance from fish is that which would be observed in this ecosystem, which depends on the system parameters (water depth, etc.) and is not to be confused with the clearance rate of a chemical from fish in pure water.

Using a series of common chemicals ranging from toluene to DDT exhibiting a wide range of solubilities and vapor pressures, four regression equations were found to describe the results in a statistically significant manner. These equations are shown below:

$$\text{\% of chemical in air} = -0.247\,(1/H) + 7.9 \log S + 100.6$$
$$\text{\% of chemical in water} = 0.054\,(1/H) + 1.32$$

$$\% \text{ of chemical in soil} = 0.194 \ (1/H) - 7.65 \log S - 1.93$$
$$\log (t_{1/2}) = 0.0027 \ (1/H) - 0.282 \log S + 1.08$$

where $H = \dfrac{\text{vapor pressure x molecular weight}}{\text{solubility (ppm)}} (\text{m}M \text{ Hg m}^3 \text{ mole}^{-1})$

$S = \dfrac{\text{solubility (ppm)}}{\text{molecular weight}} (\text{m}M/\text{liter})$

$t_{1/2} =$ half-life for clearance from fish in this ecosystem (hr)

The chemicals along with the relevant data are shown in Table I. Table II shows the results of the computer simulation and the prediction by means of these regression equations.

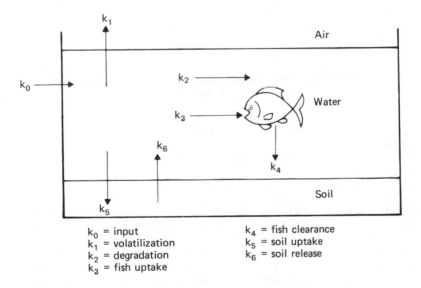

$k_0 =$ input
$k_1 =$ volatilization
$k_2 =$ degradation
$k_3 =$ fish uptake

$k_4 =$ fish clearance
$k_5 =$ soil uptake
$k_6 =$ soil release

Figure 1. Compartmental model showing the movement and distribution of a chemical in an aquatic ecosystem.

Fish Clearance

If $t_{1/2}$ is greater than 100, a potential problem of bioconcentration is indicated. This is an arbitrary decision and is based on the results of Table II. Using benchmark concept,[3] the chemicals in Table II with a $t_{1/2}$ greater than 100 are known to have bioconcentration problems; consequently, if the chemical screened has this high a number, it should be examined experimentally for degradability and possibly bioconcentration in aqueous systems.

Table I. Properties of a Series of Chemicals Tested in the
Simulated Aquatic Ecosystem

Chemical	Molecular Weight	Vapor Pressure (mM Hg)	Water Solubility (ppm)
Toluene	92	30	470
p-Dichlorobenzene	147	1	79
Trichlorobenzene	180	0.5	30
Hexachlorobenzene	285	10^{-5}	0.035
Diphenyl	154	9.7×10^{-3}	7.5
Trichlorobiphenyl	256	1.5×10^{-3}	0.05
Tetrachlorobiphenyl	291	4.9×10^{-4}	0.05
Pentachlorobiphenyl	325	7.7×10^{-5}	0.01
DDT	350	10^{-7}	1.2×10^{-3}
Perchloroethylene	166	14	150

Table II. Distribution of the Chemicals Shown in Table I in the Various
Compartments of the Simulated Ecosystem

Chemical	Water (%)	Soil (%)	Air (%)	t$\frac{1}{2}$ from Fish[a] (hr)
Toluene	0.9 (1.33)[b]	0.4 ($\sim$0)	98.6 ($\sim$100)	10 (7.6)
p-Dichlorobenzene	1.24 (1.31)	1.28 (0.24)	97.5 (98)	15 (14)
Trichlorobenzene	1.33 (1.34)	2.06 (4.09)	96 (94)	17 (20)
Hexachlorobenzene	3.57 (1.98)	39.4 (31)	56 (68)	162 (164)
Diphenyl	2.27 (1.59)	5.4 (9)	92.2 (89)	27 (29)
Trichlorobiphenyl	1.38 (1.33)	15.2 (26)	83 (71)	96 (134)
Tetrachlorobiphenyl	1.5 (1.34)	17 (27)	81 (71)	104 (139)
Pentachlorobiphenyl	1.5 (1.34)	21 (33)	77 (65)	229 (226)
DDT	1.26 (3.17)	67.5 (46.5)	28 (49)	915 (517)
Perchloroethylene	1 (1.32)	1 ($\sim$0)	98 (100)	14 (12)

[a]This is the time for clearance from the fish in the simulated aquatic ecosystem once addition of chemical was terminated.
[b]The numbers in parenthesis were estimated from the regression equations.

Soil

Again, using the benchmark approach the chemicals in Table II suggest that 4% is a reasonable cut-off point. In other words, if the amount of chemical in the soil compartment is greater than 4%, degradation in soil needs to be investigated.

Water

In a similar manner if the amount of chemical in the water compartment is greater than 2%, degradation studies are required.

This first cut is designed to give some indication as to where further testing is needed. Every case will be slightly different, and attempting to formulate a decision adequate to cover the many possibilities would be a wasted exercise. The only firm conclusion is that testing should be continued until enough is known about degradation, distribution and toxicity of the compound to ensure that the expected environmental concentration resulting from the use is below the no-effect level. Once this is demonstrated, manufacture and distribution should be allowed.

If in a particular application the concentration reflecting no adverse biological effect is close to the expected environmental level, then more refined measurements on the ecosystem will be required. For example, the actual receiving body of water will need characterization. Some typical properties are shown in Table III. Simultaneously, an improved estimate of the input function will be needed. Such a function should describe the rate and amount at which the product is anticipated to enter the particular ecosystem.

**Table III. Typical Properties of the Aquatic Environment
Needed to Predict the Concentration of a Chemical in that Environment**

Surface Area		% Carbon in Sediment
Depth		Temperature
pH		Salinity
Flow/Turbulence		Suspended Sediment Concentration
	Trophic Status	

CASE STUDIES

Kepone

This is a chemical that has received a great deal of attention (see for example Kryielski et al.[7] and Dawson et al.[8]). Produced primarily for use as a pesticide, it was accidentally discharged into the James River from the manufacturing site at Hopewell, Virginia. The physical properties are listed in Table IV. Performing the profile analysis, the results in Table V are generated. This profile immediately suggests the types of problems that can be associated with the distribution of such a chemical in an aquatic system. These may be listed as follows:

Table IV. Properties of Chemicals Examined for Potential Environmental Hazard

Chemical	Molecular Weight	Vapor Pressure (mM Hg)	Water Solubility (mg/l)
Kepone[a]	491	2.5×10^{-5}	3 at pH 7.0
Mirex[a]	546	6×10^{-6}	0.005
Chlorpyrifos	350	1.9×10^{-5}	2

[a]Values obtained from G. Dawson, Battelle Pacific Northwest Laboratories, Richmond, Washington.

Table V. The Partitioning Pattern Generated from the Regression Equations[a]

Chemical	Amount in			$t_{1/2}$ for Clearance from Fish (hr)
	Soil (%)	Air (%)	Water (%)	
Kepone	62	23	14	231
Mirex	37	60	1.4	320
Chlorpyrifos	74	8.5	18	335

[a]This partitioning is based on physical properties and it does not include any type of degradation mechanism.

1. The potential for bioconcentration is evident by the half-life for clearance (greater than 100 hr from the simulated ecosystem).
2. The great affinity for the soil and water suggests a major problem in these compartments with the continued release of Kepone into an aquatic environment.

This analysis indicates the need for further testing on possible degradative mechanisms. Such tests have been performed and indicated that Kepone is persistent in the environment, *i.e.*, it resits photo- and biological degradation, and it does in fact bioconcentrate.[8] These results confirm the conclusions from the preliminary analysis. Furthermore, these conclusions reflect the type of problems that were created by the discharge of the chemical from the manufacturing plant at Hopewell, Virginia.[8] Levels ranging up to 10 ppm were found in the James River sediment and high concentrations were found in Chesapeake Bay.[7] Even the ambient air near the plant contained detectable levels of Kepone,[7] affirming the predicted release to the atmosphere from the results in Table II. Dawson *et al.*[8] estimated that up to 200,000 pounds of Kepone were released from the Virginia site; furthermore it is estimated that up to 25% of this amount currently resides in the sediments of the river. Thus, it is seen that the actual field observations agree with the profile generated by the equations and shown in Table V.

The examination of the Kepone incident indicates that the proposed regression equations do have the capability of quickly focusing on the key areas for further testing. It also serves as an alert system of what precautions are necessary in both the manufacture and distribution of the product.

Mirex

In 1969 a large-scale, federally coordianted program was implemented to eradicate the imported fire ant in the southeastern United States. The agent chosen for this work was an insecticide known as mirex. While some early warnings over the widespread use of this close relative to Kepone were registered, it was not until mirex was found in fishes from Lake Ontario and in seals from Europe[9] that the concern over the environmental impact became important. More intensive investigations soon demonstrated that the Lake Ontario ecosystem was badly contaminated. By sampling the bottom sediments of the lake, two distinct sources were apparent: one off the mouth of the Niagara River and the other in the area of Oswego, New York. Since a chemical company on the Niagara River produced mirex, the manufacturing plant was implicated as one of the major sources. The Oswego source was traced to a plant in Volney, New York.

As the second case study, it is interesting to evaluate mirex by generating the environmental profile. Using the phsyical properties of mirex listed in Table IV, the profile of this chlorinated hydrocarbon was determined and is shown in Table V. The potential problems associated with mirex become quite evident. Its tendency to bioconcentrate in fish is indicated by the long half-life for clearance, while its association with the soil compartment is high. Such a high affinity for sediment suggests that once an aquatic ecosystem becomes contaminated the mirex in the sediment will act as a source for further contamination of the food chain long after the direct source has been terminated. A similar situation has been postulated for the PCB (polychlorinated biphenyl) contamination of Lake Michigan.[10] While there are many similarities between mirex and Kepone, there is one important difference (Table V). In the case of mirex there is a greater tendency for the chemical to escape into the atmosphere. In many ways mirex more closely resembles DDT. Due to the relatively high volatility rate, both are capable of being circulated around the globe. Fortunately, the production of mirex was much smaler than that of DDT (50 million pounds of DDT annually at the peak as compared to 50 thousand pounds for mirex), so that detectable levels in species far removed from the source such as penguins have not been observed.

However, there is no question that Lake Ontario has become contaminated with mirex. What is important in this discussion is that the simple profile presented in Table V combined with further testing showing persistence[7] has the ability to predict what actually occurred. If such a profile had been generated on a new chemical, the next steps would be to confirm the magnitude of the bioconcentration effect, determine the biodegradation rate in water and soil, and determine the acute and chronic effects on various target organisms. Armed with such information the producer would be alerted to the dangers of excessive discharges from the manufacturing site. This would allow time to build proper safeguards into the process in order to prevent such an incident from occurring. However, given a proper plant design and trained pesticide operators there appears to be no environmental reason why such a material cannot be used for the intended purpose of controlling the imported fire ant. In the case of mirex the human health problems may preclude the safe use of the pesticide.[9]

Chlorpyrifos

The third case study involves chlorpyrifos (0,0-diethyl, 3,5,6-trichloro-2-pyridyl phosphorothioate). The key properties are shown in Table IV and the profile resulting from the application of the equations is given in Table V. Without any further data, the profile suggests problems similar to Kepone. Obviously, before such an insecticide can be widely distributed degradation studies are needed. Such experiments were performed and indicated a rapid hydrolysis in water,[11] a significant rate of metabolism by fish[12] and a rapid destruction by photodegradation in both air and water.[13] When all of these rate constants were included in the computer simulation,[5] a much faster fish clearance time (less than 100 hr) was observed. In addition, the major portion of the added insecticide ended up as hydrolysis products.[5] Prior experimentation on the fate of the pyridinol entity led to the conclusion that the aquatic plants and microbial population converted this intermediate to CO_2, NH_3 and H_2O.[14] Such a situation implies that there is no persistence of chlorpyrifos in an aquatic ecosystem. The only precaution that must be observed is that when the pesticide is distributed into water for insect control, the application rate must be adjusted in order that the initial level is below the acute toxicity level for the fish species that might be present. By knowing the physical characteristics of the receiving body of water (see Table III), the application rate can be adjusted via a computer simulation to achieve this safe level.[5]

CONCLUSION

These three case studies indicate that it is possible to focus quickly on the key environmental questions that might be associated with a new product. Using the chemical and physical properties, it is possible to visualize where in the environment the chemical will reside. Based on this information the relevant biological testing can be performed. By incorporating the additional data into the model, a more refined estimate of exposure can be made. Such cycling needs to be performed until the investigator is satisfied that the expected concentration is below the no-effect level. When this is reached no further testing is required.

REFERENCES

1. Howard, P. H., J. Saxena and H. Sikka. "Determining the Fate of Chemicals," *Environ. Sci. Technol.* 12:398-407 (1978).
2. Duthie, J. R. "The Importance of Sequential Assessment in Test Programs for Estimating Hazard to Aquatic Life." In: *Aquatic Toxicology and Hazard Evaluation,* F. L. Mayer and J. L. Hamelink, Eds. ASTM STP 634, American Society for Testing Materials (1977), pp. 17-35.
3. Goring, C. A. I. "Agricultural Chemicals in the Environment: a Quantitative Viewpoint. In: *Organic Chemicals in the Soil Environment,* C. A. I. Goring and J. W. Hamaker, Eds. (New York: Marcel Dekker, Inc., 1972), pp. 793-863.
4. Blau, G. E., and W. B. Neely. "Mathematical Model Building with an Application to Determine the Distribution of DURSBAN® Insecticide Added to a Simulated Ecosystem," *Adv. Ecol. Res.* 9:113-163 (1975).
5. Neely, W. B., and G. E. Blau. "The Use of Laboratory Data to Predict the Distribution of Chlorpyrifos in a Fish Pond." In: *Pesticides in the Aquatic Environment,* M. A. Q. Khan, Ed. (New York: Plenum Publishing Corporation) 1977, pp. 145-163.
6. Neely, W. B. "A Preliminary Assessment of the Environmental Exposure to be Expected from the Addition of a Chemical to a Simulated Aquatic Ecosystem," *Int. J. Environ. Studies* 13:101-108 (1979).
7. Kryielski, D. J., and G. F. Bennett. "Intergovernmental Considerations in the Disposal of Kepone—a Municipal Perspective," *Proc. of 1978 National Conference on Control of Hazardous Materials Spills* (1978), pp. 274-280.
8. Dawson, G. W., J. A. McNeese and D. C. Christensen. "An Evaluation of Alternatives for the Removal/Destruction of Kepone Residuals in the Environment," *Proc. of 1978 National Conference on Control of Hazardous Materials Spills* (1978), pp. 244-249.
9. Kaiser, K. L. E. "The Rise and Fall of Mirex," *Environ. Sci. Technol.* 12:520-527 (1978).
10. Neely, W. B. "A Material Balance Study of Polychlorinated Biphenyls in Lake Michigan," *Sci. Total Environ.* 7:117-129 (1977).
11. Shaeffer, C. H., and E. F. Dupras. "Factors Affecting the Stability of DURSBAN® in Polluted Waters," *J. Econ. Entomol.* 63:701 (1970).
12. Smith, G. N., B. S. Watson and F. S. Fischer. "The Metabolism of ^{14}C-O, O-Diethyl O-(3,5,6-trichloro-2-pyridyl) Phosphorothioate (DURSBAN®) in Fish," *J. Econ. Entomol.* 59:1464-1475 (1966).

13. Smith, G. N. "Ultraviolet Light Decomposition Studies with DURSBAN® and 3,5,6-trichloro-2-pyridinol," *J. Econ. Entomol.* 61:793-799 (1968).
14. Smith, G. N. "Basic Studies on DURSBAN® Insecticide," *Down to Earth* 22:3-7 (1966).

21

DATA NEEDED TO PREDICT THE ENVIRONMENTAL
FATE OF ORGANIC CHEMICALS

Theodore Mill

SRI International
Menlo Park, California 94025

INTRODUCTION

Intensive investigation of the effects on plant and animal populations of exposure to synthetic chemicals present in the air, soil or water has culminated in passage of the Toxic Substances Control Act of 1976 (PL 94-469). One of the provisions of this law is that chemicals that may present an unreasonable risk of injury to health or the environment must be tested before manufacture and distribution so that EPA can assess the probable fate and effects of the chemicals if they are released into the environment. This information, combined with data on population exposure and release rates, should enable EPA to evaluate the probable hazard associated with commercial use of a specific chemical.

At present, evaluation of probable fate and exposure of chemicals is largely a matter of individual or collective judgment based on whatever laboratory or field data may be available. This procedure is necessarily fragmentary and often leads to inconsistent or nonobjective use of information.

Many environmental scientists now believe that the key to rapid, objective and reliable hazard assessment is the use of laboratory tests for fate and effects.[1,2] Predictive tests for chemical fate or pathways can provide information on the concentration and location of a chemical and its products in a specific part of the environment as a function of time. Toxicological test methods give information on concentrations required to have some effect on a sensitive species in the same environment. Thus, together these data provide the basis for hazard assessment.

Test methods for toxicological effects have been well developed and are now an accepted part of the hazard assessment procedure for many new chemicals, particularly pesticides.[3] Test methods for fate estimation have until now not been developed and validated in any systematic way.

The key to using kinetic data from laboratory tests to evaluate dynamic environmental processes lies in establishing quantitative relationships that couple kinetic properties of chemicals with kinetic properties of the environment and its biological populations. Well-conceived laboratory tests can provide many of the data needed for hazard assessment at a fraction of the cost of field tests. Moreover, the data can be applied to processes in a wide variety of specific environments including streams, soil plots or air columns, provided that the environmental and biological properties needed to characterize these systems are also available or can be measured.

BACKGROUND

Relationship Between Chemical Fate and Effects

The environmental impact or hazard of a chemical in a specific setting depends on the concentration of the chemical and its toxicity to the organisms at risk. The concentration of a chemical depends on how rapidly it is applied from some external source (which could be another part of the environment) and on how rapidly it is removed by all the processes that contribute to its loss. The nature of the effects of a chemical on organisms exposed to it depends on both the concentration of the chemical and the length of exposure. Movement of the chemical from one part of the environment to another determines the kinds and numbers of organisms affected. Thus both the population at risk and the effect of the chemical are determined partly by the detailed fate of the chemical.

This simple conceptual model of chemical exposure hazard is useful for exploring the most effective ways of coupling the fate and effects information derived from laboratory tests for hazard assessment. Figure 1 shows how test data on fate and effects can be used in a predictive way for hazard assessment in a specific environmental location. Fate tests can be used to predict how rapidly and where a chemical will move in the environment, how rapidly and to what it will be transformed, and—coupled with information on the amount entering the environment—how long the chemical will persist and at what concentrations. Effects tests can be used to predict whether, at the concentrations predicted from fate tests, organisms at risk will be affected, and how and whether the chemical will move up the food chain.

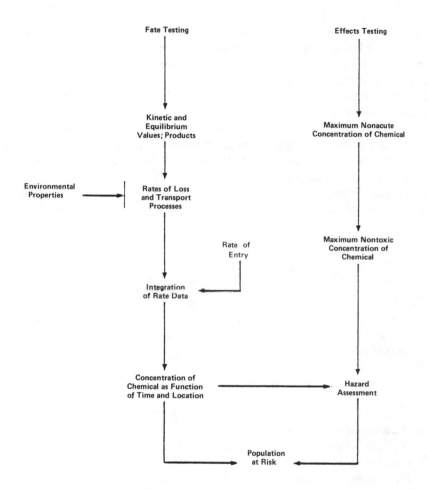

Figure 1. Hazard assessment: relation of chemical fate and effects.

Predictive Methodology for Assessing Fate

Chemical fate assessment is generally used to mean a description of all major pathways for movement or transformation* of a chemical in a selected environmental setting. This description should include concentration as a function of time and location and all major products produced by all major transformation processes. Since concentrations often change with time, kinetic and equilibrium rate terms are an essential part of a quantitative fate assessment.

---

*Transformation is the term preferred to describe any process in which a change takes place in molecular structure. Terms such as photolysis, degradation or oxidation refer to specific transformation processes and are used as appropriate.

The greatest effort at predicting fate for complex mixtures of chemicals has been expended in development of air chemistry models based on combinations of basic kinetic data from laboratory measurements,[4] large air chamber studies,[5] meteorological models[6] and a few comprehensive field studies.[7] Laboratory studies have led to some reactivity relationships for chemicals;[4] modeling has led to some success in predicting levels of nitrogen oxides and ozone in urban areas.[6,7] Extension of the principles of atmospheric models to aquatic and terrestrial systems is one important basis for development of test methods in predictive modeling and holds great promise for predictive testing of a wide variety of chemicals.

Although the environment is both complex and dynamic, the conceptual basis for predictive and evaluative methods lies in two simplifying assumptions.

1. The concentrations of chemicals and their loss rates at any location and time are determined by combinations of independent rate or distribution processes.
2. For each process a quantitative relationship exists between chemical and environmental properties.

For example, the net rate of change in concentration of chemical C is assumed to be the sum of all equilibrium and kinetic processes that affect the chemical in a specific environmental location, including input from movement or anthropogenic sources.

$$R_T = \Sigma k_n [C] [P]_n \tag{1}$$

where R_T is the net rate of change,
k_n is the rate constant for the nth loss or input process,
$[C]$ is the concentration of chemical
$[P]$ is the environmental property for the nth process

Those environmental processes now recognized as playing significant roles in transport or transformation under some conditions are summarized in Table I.

Reliable prediction of chemical fate requires:

1. accurate data on both chemical and environmental properties for important processes,
2. equations that properly represent the relationships between properties, and
3. appropriate methods to combine the equations correctly to represent the environmental location of interest.

The first requirement means that test methods are needed to produce kinetic data that are relevant to a specific environmental process (see Table I) and are in a form that can be used in some quantitative relationship such as

$$R_n = k_n [C] [P]_n \qquad (2)$$

where R_n is the rate of process n,

$\quad k_n$ is the rate constant which characterizes that process,

$\quad [C]$ is the concentration of chemical, and

$\quad [P]_n$ is the environmental property for the nth process expressed in concentration units compatible with k.

The second requirement means that detailed knowledge of the environmental locations is available or can be measured. The major environmental fate processes and their corresponding properties are summarized in Table II. The third requirement is for a simple or complex model that can combine properly process equations to give information on concentrations of chemicals as a function of time and location.

Table I. Environmental Compartments and Processes

Air	Water
Meteorological Transport	Sorption
Photolysis	Bio-uptake
Oxidation	Volatilization
Fallout	Photolysis
	Hydrolysis
Soil	Oxidation
Sorption and Sediments	Biodegradation
Bio-uptake	
Runoff	
Volatilization	
Leaching	
Transformations $\begin{cases} \text{Hydrolysis} \\ \text{Oxidation} \\ \text{Photolysis} \\ \text{Reduction} \end{cases}$	
Biodegradation	

The predictive methodology for assessing fate, which includes development, processing and validation of data, is shown in Figure 2. Data needed for the process and effects equations come from two sources: laboratory test methods and environmental measurements in selected field sites. Integration of the data is accomplished by a suitable model. As a final step, the whole estimation procedure and its constituent parts are evaluated for accuracy in predicting concentrations as a function of time and location in a field site. The most useful validation procedure is a field test in which the chemical is applied at known rates to a simple aquatic, soil or air system for which all of the important environmental properties are known or can be measured during the exposure

Table II. Environmental Processes and Properties

Process	Property[a]
Physical Transport	
Meteorological transport	Wind velocity
Bio-uptake	Biomass
Sorption	Organic content of soil or sediments, mass loading of aquatic systems
Volatilization	Turbulence, evaporation rate, reaeration coefficients, soil organic content
Runoff	Precipitation rate
Leaching	Adsorption coefficient
Fall out	Particulate concentration, wind velocity
Chemical	
Photolysis	Solar irradiance, transmissivity of water or air
Oxidation	Concentrations of oxidants and retarders
Hydrolysis	pH, Sediment or soil basicity or acidity
Reduction	Oxygen concentration, ferrous ion concentration and complexation state
Biological	
Biotransformation	Microorganism population and acclimation level

[a]At constant temperature.

period. The concentration of chemical is then monitored as a function of time and location within and without the test plot. If desired, the input rate can be altered and the effect on concentration with time and location followed. The field test gives information mostly about the net change in concentration of chemical; only limited information is available as to the relative contribution of individual processes. In effect the field test integrates with time the individual fate processes. Field data are readily compared with comparable concentration data from the laboratory tests procedure to determine how well prediction agrees with observation.

Failure to secure close agreement between predicted and measured values might arise from any one of several sources, but the most likely source is failure to reproduce field conditions in the laboratory tests. Others include errors in test methods, data integration and field sample procedures.

Kinetic and Equilibrium Rate Processes

Because the processes controlling concentrations of chemicals in the environment are dynamic, kinetic- and equilibrium-rate constants are the most

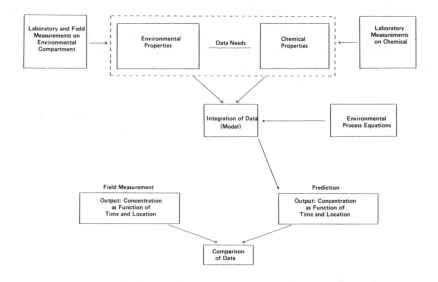

Figure 2. Predictive methodology, field test and validation scheme for fate.

useful quantitative descriptors for evaluating the contribution of various processes to the net change in concentration and to the rate of change in concentration.

Almost all processes in the environment which control* rates and concentrations involve interaction of the chemical with some environmental agent or intermediate, for convenience referred to as property (P), such that

$$C + P \rightarrow \text{Product.}$$

Examples of environmental properties which can govern rates are listed in Table II. Properties are expressed in concentration units, the values of which are characteristic of specific environments. Since chemicals are usually present in the environment at very low concentrations, the rate of reaction depends usually on the first power of concentration, and the kinetic-rate law takes the form shown in Equation 2.

$$R_n = d[C]/dt = k_n[C][P_n] \tag{2}$$

The total rate of loss of C will be given by Equation 1, the summation of individual loss or input rates

---

*Transformation of C to products may involve several steps but usually only one, the slowest, step in the sequence will control the rate of loss of C.

$$R_T = \Sigma k_n [C] [P_n] \qquad (1)$$

The kinetic representation of a series of parallel second-order processes is greatly simplified if we can assume that the environmental property, P_n, is constant during the time interval of the calculation. This is a reasonable assumption as long as $[\Delta P]$ is small compared to $[P]$, a condition usually met in low-level polluted systems.

The relatively minor loss of reality associated with this assumption is more than off-set by the simplicity and flexibility introduced by use of the simpler relation

$$R_T = \Sigma k_n' [C] = k_T [C] \qquad (3)$$

where k_n', is the pseudo-first-order constant for the nth process and k_T is the sum of these constants.

The set of simple first-order processes can be combined to quickly evaluate the contribution of each rate process to the total rate of loss of chemical and thus serve as a quick screening tool for preliminary assessment and decisions concerning the need for detailed testing. The equations can be used in a multi-compartment computer model to take into account mixing and equilibrium sorption to suspended or bottom sediments and to soil columns. Their simple form makes numerical integration unnecessary, greatly simplifying the program and cost of running the computer.

These relations are also useful for estimates of persistence. Two limiting conditions may be considered. In situations where the chemical is applied to the compartment in a single application, the time required for loss of half of the original amount is simply

$$t_{1/2} = \ln2/k_T \qquad (4)$$

However, in the situation where a chemical is added continuously to a compartment, often from a point source, a more meaningful measure of presistence is the steady-state concentration resulting from a balance of input and loss processes

$$d[C]/dt = O = R_i - R_L \qquad (5)$$

$$R_i = R_L \qquad (6)$$

where R_i is the rate of introduction and R_L is total rate of loss by transport, dilution and transformation. Since

$$R_i = [C] \Sigma k_n = k_L [C] \qquad (7)$$

$$[C] = R_i/k_L \qquad (8)$$

This simple kinetic picture of environmental fate processes must also include one or more terms for reversible sorption of chemicals to sorbates (S) such as sediments, soils, biomass or airborne particulate.

$$C + S \underset{k_{-s}}{\overset{k_s}{\rightleftharpoons}} CS \quad K = \frac{k_s}{k_{-s}} \tag{9}$$

$$C \overset{k_L}{\rightarrow} \text{products by n loss processes (1st order)} \tag{10}$$

$$C + CS = C_T \text{ (mass balance)} \tag{11}$$

The loss of C is then

$$-R_L = k_L [C_T]/(K_s[S] + 1) \tag{12}$$

The net effect of sorption on the rate of loss is to slow down the process by the factor $(1/K_s[S] + 1)$. Another way of viewing the process is that sorption provides a buffering effect in the system which maintains a higher concentration of chemical in the unsorbed state than would otherwise be present in its absence. Sorption leads to longer half-lives or higher steady concentrations; substitution in Equations 4 and 6 gives

$$t_{1/2} = (K_s[S] + 1) \ln2/k_T \tag{13}$$

$$[C]_{ss} = R_i(K_s[S] + 1)/k_L \tag{14}$$

Specific Fate Processes and Data Needs for Fate Assessment

If we accept the premise that only a limited number of discrete chemical, physical and biological processes control concentrations of chemicals in different environmental settings (Table I), then it becomes appropriate to enquire how laboratory tests might be used to measure the rate or equilibrium constant and major products for a specific chemical in each process to provide a basis for predicting the fate of that chemical in selected environments. Equations 4, 8 and 12-14, from the last section, provide the necessary set of relations to predict concentration as a function of time and space using appropriate kinetic, equilibrium and environmental parameters. The following sections discuss briefly each of the processes believed to be important in controlling fate in aquatic, atmospheric and terrestrial environments, the kinetic data needed to describe the processes, and some methods available for predicting kinetic data. Less is known about the processes in terrestrial systems than about those in water or air, although our knowledge of heterogeneous processes in water and air is also meagre.

PHOTOCHEMISTRY

Measurement Methods

The cutoff for the solar spectrum by the upper atmosphere is at about 290 nm; only absorption of photons by a chemical at this or longer wavelengths can result in direct photochemical transformations. Direct absorption of sunlight may result in cleavage of bonds, dimerization, oxidation, hydrolysis or rearrangement. No simple selection rules are available to predict the specific chemical process that may occur, although some useful generalizations have been found.[8-10]

Quantitative aspects of direct photolysis in water, on soil surfaces or in the atmosphere have the same general kinetic relationships. The rate of absorption of light, I_A (rate constant k_a), by a chemical at one wavelength is determined by ϵ_λ, the molar absorbance; I_λ, the intensity of the incident light at wavelength λ; and [C], the concentration of chemical. At low concentrations of C where only a small percentage of the light is absorbed

$$I_{A(\lambda)} = \epsilon I_\lambda [C] = k_{a(\lambda)} [C] \qquad (15)$$

The rate of direct photolysis of a chemical at wavelength λ is obtained by multiplying $k_{a(\lambda)}$ by the quantum yield ϕ_λ, which is the efficiency for converting the adsorbed light into chemical reaction, measured as the ratio of moles of substrate transformed to einsteins of photons absorbed. Thus at a single wavelength, λ

$$k_{a(\lambda)}\phi_\lambda [C] = k_{p(\lambda)} [C] \qquad (16)$$

where

$$k_{p(\lambda)} = k_{a(\lambda)}\phi_\lambda \qquad (17)$$

Figure 3 illustrates relationships between spectral properties and solar irradiance.

The simplest and most direct method of using laboratory experiments to estimate environmental photolysis rates in the field is to expose an aqueous solution, a vapor phase sample or a thin surface layer of a chemical to outdoor sunlight and monitor its rate of disappearance. At the same time, photolysis of another chemical having a well-characterized quantum yield and similar absorption spectrum should be carried out. This method will take into account variations in sunlight intensity but avoids the need for determining the detailed spectrum or the quantum yield for the chemical.

Another method for estimating environmental photolysis rates is based on laboratory measurements of ϕ at a single wavelength and ϵ_λ; average sunlight

$$k_p(sec^{-1}) \propto \phi \sum_\lambda \epsilon_\lambda Z_\lambda$$

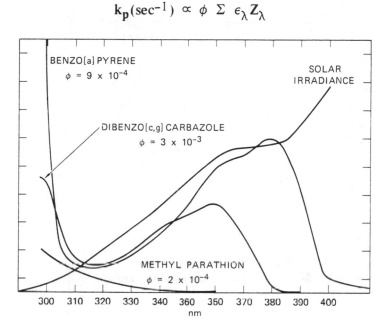

Figure 3. Absorption spectra and quantum yields for representative organic chemicals plotted on the same coordinate with solar irradiance.

intensity (I_λ) data are available in the literature as a function of time of day, season and latitude.

Since k_p is equal to the product of k_a and the quantum yield ϕ and, generally, ϕ does not vary significantly with wavelength, the rate constant in sunlight $k_{p(S)}$ is

$$k_{p(S)} = \phi \sum_\lambda \epsilon_\lambda \qquad (18)$$

and the half-life in sunlight is

$$(t_{\frac{1}{2}})_{(S)} = \frac{\ln 2}{k_{p(S)}} \qquad (19)$$

Both a computer and hand methods are available to sum the products of $\epsilon_\lambda I_\lambda$ over a wavelength range and give a plot of the half-life of the chemical toward photolysis in water or air as a function of the month of the year and latitude.[11,12]

Comparisons made at SRI[1] between measured and calculated half-lives for direct photolysis in sunlight of eight chemicals dissolved in water using procedures described above gave excellent agreement, usually within a factor of two, as shown in Figure 4.

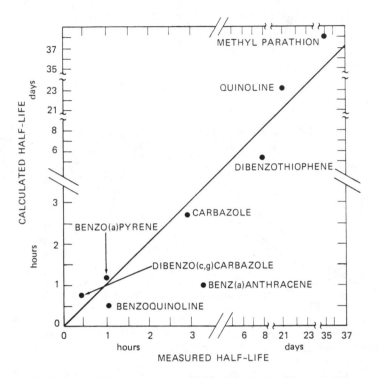

Figure 4. Comparison of measured and calculated half-lives for direct photolysis.

Predictive Methodology

Equation 18 may be used to calculate an upper limit for k_p by assuming $\phi = 1$. This method is recommended as a screening tool, since if the rate constant is small compared to rate constants for other competing environmental processes no additional photolysis measurements are needed.

Prediction of ϕ from structure-reactivity relationships is still very empirical. For example, photolyses of polynuclear aromatics in water have quantum yields varying from 0.01 to 0.001 on going from naphthalene to benzo[a]pyrene; no structure-reactivity relationships were noted.[1,13] Quantum yields for many reactions may be expected to change on going from water to organic solvents, air or soil.[14] Caution should be used in applying data measured in one medium to another.

OXIDATION

Kinetic Processes and Measurements

Oxidations in the environment usually require the presence of O_2 but almost never involve direct reaction of O_2 with a chemical. Moreover, the rates of oxidation processes do not bear much relation to thermodynamic redox potentials. The essential rate steps in most oxidations involve reaction of a reactive species with the chemical; reactions include those with free radicals, ($RO_2 \cdot$, $RO\cdot$, $HO\cdot$) ozone and $^1O_2$ (singlet oxygen). To predict the importance of oxidation under environmental conditions, it is necessary to be able to identify the important oxidants and their concentrations in air, soil or water. Values for many oxidation rate constants are known reliably,[15-17] but the concentrations of oxidants and their identities in soil and water are still under investigation. In air $HO\cdot$ is the principal oxidant at $\sim 10^{-15} M$.[18]

Radical oxidation processes with $HO\cdot$ are very important in the atmosphere, where they control rates of oxidation of most chemicals. Ozone is important only in oxidation of some olefins in the atmosphere and possibly for oxidation of some sulfur or phosphorous compounds exposed on surfaces in smoggy areas. Mill et al.[19] have shown that $RO_2 \cdot$ can be important in sunlight photolyses of natural waters, and Zepp et al.[20] have shown that singlet oxygen is generated by photolysis of natural waters.

Laboratory tests that measure $k_{RO_2 \cdot}$, $k_{RO\cdot}$, $k_{HO\cdot}$, k_{O_3} and $k^1{}_{O_2}$ are needed to predict rates of oxidation in aquatic, atmospheric or soil ecosystems. When these values of k are combined with measured or estimated values of radical concentrations in selected environmental compartments, rates of oxidation (R_{OX}) and lifetimes for chemicals in oxidation may be estimated using some form of

$$R_{OX} = [C](k_{RO}[RO] + k_{RO_2}[RO_2] + k_{HO}[HO])$$ (20)

Test methods now being considered involve use of competitive kinetic techniques to evaluate relative rates of loss of two chemicals, one of which is a standard having a known reactivity toward a specific oxidant.[21] For two chemicals, 1 and 2, with reactivities k_1 and k_2, their relative rates of reaction are expressed by the simple relation which is independent of the oxidant concentration.

$$\frac{\ln(C/C_O)_1}{\ln(C/C_O)_2} = \frac{k_1}{k_2}$$ (21)

For these tests, we have proposed using an azocompound to generate $RO_2 \cdot$ in water, nitrous acid to generate $HO\cdot$ in air and a dye to generate $^1O_2$ in water.[21]

Predictive Methods for Oxidation

Several empirical structure-reactivity relationships for oxidation by $RO_2\cdot$, $HO\cdot$ and $^1O_2$ have been developed including the Polyani relation for $RO_2\cdot$,[22] Hammett sigma-rho relations for several oxy radicals[23-24] and singlet oxygen[25] and a large body of empirical data for all of these oxidants,[15-17,26,27] from which reasonably good estimates (a factor of 3-5) can be made of rate constants for new chemicals. Figure 5 shows the Polyani relationship for oxidation of hydrocarbons by sec-$RO_2\cdot$ and t-$RO_2\cdot$ radicals.[22]

Chemical structures most susceptible to oxidation by $RO_2\cdot$ and $^1O_2$ include phenols, aromatic amines, electron rich olefins and dienes, alkyl sulfides and eneamines with rate constants (M^{-1} s^{-1}) in the range 10^4-10^6. Good methods for evaluating rate constants for those reactions are still needed. Oxidations

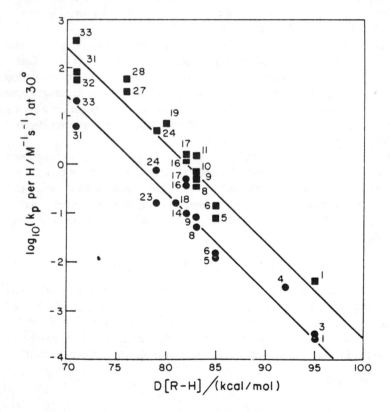

Figure 5. Log [k_p per active H/(M^{-1} s^{-1})] plotted against d[R-H]/(kcal/mol). The circles represent rate constants for tertiary peroxy radicals; the squares represent rate constants for secondary peroxy radicals [The numbers refer to the compounds listed in Table 5 of reference 22 (reproduced with permission of National Research Council of Canada)].

of saturated alkyl compounds including alkenes, haloalkanes, alcohols, esters and ketones are all too slow in water or soil to be important loss processes $(k_{OX} < 10M^{-1}\ s^{-1})$.[15-19] In air, almost all organic CH bonds and aromatic rings are readily oxidized by HO· radical $k_{OX} > 10^8\ M^{-1}\ s^{-1}$; only a few simple alkanes, haloalkanes and one or two carbon acids, alcohols and esters are relatively nonreactive toward HO· radical.[16] As a result, most chemicals in the air are oxidized via HO· radical processes.

HYDROLYSIS

Kinetic Measurements

Hydrolysis of organic compounds usually results in the introduction of a hyroxyl function (-OH) into a chemical, most commonly with the loss of a leaving group (-X).[28]

$$RX + H_2O \longrightarrow ROH + HX \qquad \text{(A-7)}$$

$$R\overset{O}{\overset{\|}{-}C}\text{-X} + H_2O \longrightarrow R\overset{O}{\overset{\|}{-}C}\text{-OH} + HX \qquad \text{(A-8)}$$

In water, the reaction is catalyzed mainly by hydronium and hydroxyl ions; but in moist soil, loosely complexed metal ions such as copper or calcium may also be important catalysts for certain types of chemical structures. Sorption of the chemical may also increase its reactivity toward H^+ or HO^-.

The general rate equation for hydrolysis in water is

$$R_h = k_h[C] = k_A[H^+][C] + k_B[OH^-][C] + k_N{}'[H_2O][C] \qquad \text{(22)}$$

where k_h is the measured first-order rate constant at a given pH. The last term is the neutral reaction with water (second-order rate constant $k_N{}'$), and in water it can be expressed as a pseudo-first-order rate constant k_N.

This equation can be modified to account for the incursion of bound or free metal-ion catalysis in soil or sediments by including one or more terms for the form

$$k_M K_A[M]_T / \left(K_A + [H^+]\right) \qquad \text{(23)}$$

where k_M is the metal-ion catalysis constant, $[M]_T$ is the total metal ion concentraton, and K_A is the equilibrium constant for dissociation of the hydrated ion complex. Since a metal may be complexed in soil in several ways, the descriptors needed for the complete rate equation could be quite complex.

Equation 22 shows that the total rate of hydrolysis in water is pH-dependent unless k_A and $k_B = 0$. The dependence of k_h on the pH of the solution is conveniently shown by the plot of log k_h as a function of pH (Figure 6).

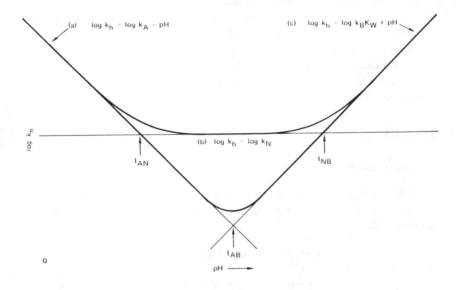

Figure 6. pH dependence of k_h for hydrolysis by acid; water- and base-promoted processes. (From reference 28; reproduced with permission of the authors.)

Mabey and Mill have recently reviewed kinetic data for hydrolysis of a variety of organic chemicals in aquatic systems and have reviewed the chemical characteristics of most freshwater systems.[28] These data have been used in turn to calculate persistence (half-life) of these same chemicals at $25°$ and at pH 7 in fresh water. Predictive test methods (screening and detailed) for hydrolysis to develop the essential kinetic data: k_h, k_A, k_N, k_B, and their temperature dependence (Arrhenius equation) have been prepared recently by Mill, Mabey and Hendry.[29]

A variety of hydrolysis reactions have been observed on soils and sediments.[30] In some cases, rates were markedly accelerated compared to bulk solution, but detailed understanding of mechanisms is limited and structure-reactivity relationships appear to be available for only a few compounds.

Hydrolyses in soils and sediments appear to be complicated by unusual pH relationships at the soil surface, by possible incursion of metal-ion catlyzed processes and by general acid- and base-catalyzed reactions possibly involving phenolic, amine or sulfide groups in soils.

Predictive Methods for Hydrolysis Kinetics

A significant literature exists covering semiempirical structure-reactivity relationships for most kinds of hydrolysis reactions.[31,32] The size of the data base available for estimations of hydrolysis kinetic constants is large enough to warrant some confidence in being able to predict within a factor of 2 or 3 a rate constant for a specific chemical wthin a family of closely related molecular structures.

BIODEGRADATION

Kinetic Processes and Measurements

Biological transformations comprise a group of enzymatically mediated reactions carried on by microorganisms found in soil and water including bacteria, fungae, protozoa and algae. The reactions include oxidation, reduction, hydrolysis and, in rare cases, rearrangements.[33] The biological processes are lumped together and treated as one kind of transformation process in fate assessment simply because there usually is no practical way to separate one process from another in the natural setting. Moreover, a specific chemical will usually undergo only one type of biotransformation process, although different chemicals will be subject to different processes, depending on chemical structure, the composition of the microorganism population and other enviormental factors governing population dynamics.

Although many classes of organic structures are susceptible to biological transformation, the persistence of an increasing number of synthetic chemicals in soil and water demonstrates clearly that microorganisms cannot unfailingly transform all molecular structures, at least in time frames useful to man.

Probably no other area of environmental transformations is as complex or poorly understood in detail as that of biological transformations carried out by mixed populations of organisms which continually change in composition. Most biological transformations studied in the laboratory use single-organism cultures with optimum nutrient sources. These model systems are essential for interpreting the processes found in natural waters or soils and are used with surprising success to describe the much more complex mixed cultures, but they should be recognized as extreme simplications.

Kinetic expressions governing biological transformations are usually based on the concepts of Monod;[34] the rate of loss of chemical C is

$$d[C]/dt = \frac{M}{y}[B] = \frac{\mu_m [C] [B]}{yK_c + [C]} \qquad (24)$$

where C is the concentration of chemical,

μ is the specific growth rate,

μ_m is the maximum growth rate,

Y is the cell yield,

B is the biomass per unit volume,

K_c is the concentration of chemical supporting a half-maximum growth rate $(0.5 \mu_m)$.

The rate constant, k_b, is conventionally defined as

$$k_b = \frac{\mu_m}{Y} \tag{25}$$

Implicit in these kinetic relations is the assumption that μ_m, K_c and Y are constants.

Although Monod kinetics are based on the use of a pure culture and the rate of disappearance of a growth-rate limiting single substrate, these kinetic expressions can be used to obtain useful rate constants with mixed-culture systems. It is possible to choose culture conditions that allow simplication of the Monod expression by making one of its variables constant, or one of its constants insignificant. The results of the simplification allow a straightforward estimation of the remaining constants.

Several procedures for meas'ring kinetics have been used with mixed-culture systems in which the chemical serves as the sole carbon source:

- Batch fermentations with low-level inocula of washed biodegrading cells
- Continuous chemostat fermentations
- Batch fermentations with large microbial populations and low substrate levels.

Each method has specific advantages; however, the simplest procedure is the last, in which the microbial population remains nearly constant and the chemical is present in very low concentration—conditions similar to those found in aquatic and soil environments.

Ordinarily the mixture of organisms used for kinetic studies is one that has been collected from a soil or water body of interest and has been acclimated to the chemical by exposure over weeks or months to low concentrations of the chemical in the presence of nutrients, including carbon and inorganic species, until the enzymes necessary to effect some transformation are expressed. If no activity is found for the culture after 6-8 weeks, another culture may be tried, but generally a chemical would be considered nontransformable by microorganisms, and k_b from Equation 26 would be considered zero.

Very recent studies by Paris and co-workers[3,5] show that surprisingly good agreement is found for biotransformation rate constants measured in kinetic studies using water samples from many different rivers and lakes. These results give greater confidence in applying these techniques in their present form to fate assessment.

Predictive Methodology

Well-informed microbiologists disagree among themselves as to whether broad structure-reactivity relationships can be developed for biodegradation of organic structures in the same way as chemists have done for abiotic chemical processes. Dagley has discussed the problem in a limited way with respect to selected pesticides, pointing out how structures common to otherwise diverse molecules—in this case carboxy esters—confer biodegradability on these compounds.[36]

However, Alexander has long argued that some structures are inherently recalcitrant toward biotransformation,[37] and it seems likely that insofar as oxidative biodegradation is concerned, highly halogenated compounds have thermochemically limited rates.

This area of microbiology needs careful attention from a variety of able workers in microbiology, enzyme kinetics and physical chemistry to develop useful correlations of structure and reactivity.

VOLATILIZATION

Theory and Measurement

Volatilization of chemicals from water to air is now recognized as an important transport process for a number of chemicals that have low solubility and low polarity; volatilization from surfaces is also a major transport process for many chemicals deliberately applied to fields. Despite very low vapor pressure, many chemicals can volatilize at surprisingly rapid rates owing to their very high activity coefficients in solution.

Mathematical expressions for the rate of volatilization from water, have been developed by Liss and Slater[38] and Mackay and Leinonen.[39] The rate constant for volatilization from water (k_{vw}) is given by the relation

$$k_{vw} = \frac{A}{V} \left(\frac{1}{K_L} + \frac{RT}{H_c K_G} \right)^{-1} \tag{27}$$

where A = surface area (cm^2)

V = liquid volume (cm^3)

H_c = Henry's law constant (torr M^{-1})

K_L = liquid film mass transfer coefficient (cm hr^{-1})

K_G = gas film mass transfer coefficient (cm hr^{-1})

R = gas constant (torr $^\circ K^{-1} M^{-1}$)

T = temperature ($^\circ K$)

Mackay and Wolkoff[40] showed that an estimate of H_c can be obtained from

$$H_c = P_{sat}/[C]_{sat} \tag{28}$$

where P_{sat} is the vapor pressure of pure chemical (or the hypothetical super-cooled liquid, if it is a solid) and $[C]_{sat}$ is the solubility of C in (mol liter^{-1}). Equation 27 simplifies if $H_c > 1000$

$$k_{vw} = \frac{A}{V}\left(\frac{H_C K_G}{RT}\right) \tag{29}$$

where mass transfer is liquid-phase limited; if $H_c \ll 1000$, then Equation 27 becomes

$$k_{vw} = \frac{A}{V}\frac{H_C K_G}{RT} \tag{30}$$

and the process becomes limited by gas-phase mass transfer.

For high volatility compounds, a simple relative measurement for volatility becomes possible because the rate constant for volatilization of the chemical is proportional to the rate constant for volatilization or reaeration of oxygen from the same solution over a range of turbulence

$$\frac{k_{vw}^C}{k_{vw}^{O_2}} = n \tag{31}$$

If the value of $k_{vw}^{O_2}$ in a real water body is known, then

$$k_{vw}^C \text{ (water body)} = n k_{vw}^{O_2} \text{ water body} \tag{32}$$

The ratio n can be measured conveniently in the laboratory by measuring simultaneously the concentration of O_2 redissolving in a deaerated stirred solution and the concentration of chemical volatilizing from the same solution, both as a function of time.[1]

Spencer has reviewed volatilization processes from soil surfaces.[41] The overall process is complicated by variable contributions from volatilization of the chemical from the water at the surface, evaporation of water itself and the wick effect that brings more water and dissolved chemical to the surface. Initially, volatilization of the chemical from the surface water will be rate controlling, and Equation 27 can be used to estimate the rate constant. This model will fail as a concentration gradient of chemical is established through the soil column, and at this time no simple laboratory measurement will reliably measure the process in a way than can be extrapolated to the field.

Predictive Methods

Smith *et al.*[1] have shown that for chemicals with $H_c > 1000$ torr M^{-1} a reasonably good estimate of the ratio $k_{vw}^C/k_{vw}^{O_2}$ can be made from the relation between the ratio of volatilization rate constants and the ratio of molecular diameters for O_2 and the chemical

$$\frac{k_{vw}^C}{k_{vw}^{O_2}} = \frac{D^{O_2}}{D^C} \tag{32}$$

For chemicals having $H_C < 1000$ torr M^{-1}, no satisfactory estimation procedure is available.

SORPTION TO SEDIMENT AND SOIL

Measurement Procedures

Sediments and soils are complex mixtures of aluminosilicate minerals (clays), metals oxides, water and humic materials. The proportions of these components will vary widely from one source to another, as will the particle size distribution.

Many organic chemicals, especially those that are nonpolar and insoluble in water, sorb strongly to sediments or soils. If the fraction of chemical sorbed to sediment or soil is large, the overall loss rate of chemical by other transformation processes will be slowed; in effect, sorption serves to buffer the concentration of chemical present in the aqueous phase (see Equation 12). In some cases, reversible sorption to sediments or soils may be followed by irreversible transformation of the chemical in the sorbed state such as reduction of carbon-halogen bonds.[44] Possibly other transformations may occur as well.

The equilibrium ratio of sorbed to nonsorbed chemical on a sediment may be expressed as an equilibrium constant (constant temperature)

$$K_s = [C]_s/[C]_w \tag{33}$$

The concentrations of C in sediment or soil are in $\mu g\ ml^{-1}$; for water 1 ml = 1 g, and K_s becomes dimensionless. Strongly sorbed chemicals such as benzo[a]pyrene or mirex have $K_s > 10^5$, and weakly sorbed chemicals such as nitroaromatics or quinoline have $K_s < 10^2$.

Values of K_s for a single chemical will vary with the composition of the sediment. To place sediments on a more nearly equal basis the value K_s can be expressed as

$$K_s = AK_{sc} \tag{34}$$

where A is the fraction of organic content expressed as mg C per mg sediment, thus K_{sc} is a sorption constant corrected for the organic content. It follows from Equation 34 that if a = 1, then K_s is equivalent to a partition coefficient such as the octanol-water coefficient (K_{ow}).

Smith et al.[1] have described procedures for measuring values of K_s or K_{sc} in which several sediment suspensions in water are prepared with saturated solutions of the chemical. The mixtures are equilibrated by shaking for 5-16 hr, centrifuged and analyzed for chemical both in the water and in the sediment, by extraction of both phases with a suitable solvent.

For screening purposes, one concentration of chemical and two loadings of sediment (low and high) are sufficient to obtain a reliable estimate of K_s. The accuracy of the method depends critically on having analytical methods capable of analyzing for levels of chemicals expected to be found in both phases and on careful planning to ensure that some of the chemical is found in both phases. To arrange to have about half the chemical in each phase requires some prior estimate of the value of K_s. This can be done using the relation between solubility and K_{oc} and then calculating the equilibrium amounts present in, say, 100-ml volume containing 0.050 g sediment. If K = 10^4 and the initial concentration of chemical of mw 300 is 5×10^{-5} M, then at equilibrium, 2.5×10^{-4} g will be in water and 1.25×10^{-3} g will be in sediment. Smith et al.[1] describe a nonlinear statistical analysis of the data obtained in this kind of experiment.

Predictive Methods

Recent studies and correlations by Karickoff et al.[45] and by Kanaga and Goring[46] bring out the close direct relationship between K_{sc} solubility, bioconcentration and the organic content of the sediment or soil. Other studies, notably by Chiou,[47] have shown a similar relationship between K_{ow} and solubility such that

$$\log K_{ow} = n \lg (\text{solubility}) + c \tag{35}$$

It follows that as a simple screening tool, estimation of K_{sc} from solubility would greatly simplify the investigators' task of deciding when to carry out more detailed measurements.

Smith and Bomberger[48] have taken the data of Karickoff et al.,[45] Kanaga and Goring[45] and Smith et al.,[1] rescaled the data to one coordinate set, and developed the following regression relation

$$\log K_{sc} = -0.782 \log[C] - 0.27 \tag{36}$$

$$[C] \text{ in moles liter}^{-1}$$

Figure 7 shows the combined data plotted as $\log K_{oc}$ (or K_{sc}) vs log (solubility) and the regression line. Using Equation 36 and the solubility of the

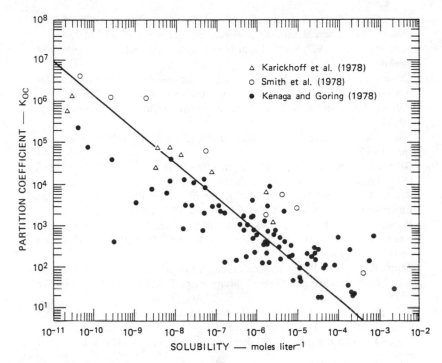

Figure 7. Soil or sediment partition coefficient of chemicals vs solubility in water (reference 48).

chemical in water (measured by a suitably careful method[1]), the investigator can estimate K_{sc} within a power of ten for most nonpolar chemicals—an accuracy sufficient, in most cases, for screening purposes.

REFERENCES

1. Smith, J. H., W. R. Mabey, N. Bohonas, B. R. Holt, S. S. Lee, T. W. Chou, D. C. Bomberg and T. Mill. "Environmental Pathways of Selected Chemicals in Freshwater Systems," Parts I and II, EPA Reports 600/7-77-113 and 600/7-7–074 (October 1977 and May 1978).
2. Baughman, G. L., and R. R. Lassiter. In: *Estimation of the Hazards of Chemical Substances to Aquatic Life,* J. Cairns, K. L. Dickson and A. W. Maki, Eds. (Philadelphia: ASTM, 1978) p. 35; Stern, A. M., and C. R. Walker, *ibid.,* p. 81.
3. Haque, R., and V. H. Freed. "Behavior of Pesticides in the Environment. Environmental Chemodynamics," *Residue Rev.* 52:89 (1974).
4. Hecht, T. A., J. H. Seinfeld and M. C. Dodge. "Further Development of Generalized Kinetic Mechanism for Photochemical Smog," *Environ. Sci. Technol.* 8:327 (1974).

5. Kopczynski, S. L., R. L. Kuntz and J. J. Bufalini. "Reactivities of Complex Hydrocarbon Mixtures," *Environ. Sci. Technol.* 8:327 (1974).
6. Schiermeier, F. A. "RAPS Field Measurements Are In," *Environ. Sci. Technol.* 12:644 (1978).
7. Calvert, J. G. "Test of the Theory of Ozone Generation in Los Angeles Atmosphere," *Environ. Sci. Technol.* 10:248 (1976).
8. Calvert, J. G., and J. N. Pitts. *Photochemistry* (New York: John Wiley, 1967).
9. Turro, N. J., *Molecular Photochemistry* (New York: W. A. Benjamin, Inc., 1967).
10. Balthrop, J. A., and J. D. Coyle, *Excited States in Organic Chemistry* (New York: John Wiley and Sons, Inc., 1975).
11. Zepp, R. G., and D. M. Cline, "Rates of Direct Photolysis in Aquatic Environments," *Environ. Sci. Technol.* 11:359 (1977).
12. Mabey, W. M., T. Mill and D. G. Hendry, "Test Protocols in Environmental Processes: Direct Photolysis in Water," EPA Report, Contract 68-03-2227 (January 1979).
13. Zepp, R. G., and P. F. Schlotzhauer, "Photoreactivity of Selected Aromatic Hydrocarbons in Water," Manuscript for Publication, November 1978.
14. Both photophysical and photochemical processes are affected by changes in medium; photolysis of acetone to CO at 25° has a quantum yield of 0.1 in the vapor phase and 0.001 in solution; see reference 8, p. 240 *et seq.*
15. Hendry, D. G., T. Mill, L. Piszkiewicz, J. A. Howard and H. K. Eigenmann, "A Critical Review of H-Atom Transfer in the Liquid Phase: Chlorine Atom, Alkyl, Trichloromethyl, Alkoxy and Alkylperoxy Radicals," *J. Phys. Chem. Ref. Data* 3:937 (1974).
16. Farhataziz and A. B. Ross, "Selected Specific Rates of Reactions of Transients from Water in Aqueous Solution, III," NSRDS-NBS 59 U.S. Government Printing Office (1977).
17. Foote, C. S. In: *Free Radicals in Biology*, Vol. II. W. A. Pryor, Ed. (New York: Academic Press, Inc., 1976), p. 85.
18. Singh, H. B., "Atmospheric Halocarbons: Evidence in Favor of Reduced Average Hydroxyl Radical Concentration in the Troposphere," *Geophys. Res. Letters* 4:101 (1977).
19. Mill, T., D. G. Hendry and H. Richardson In: *Aquatic Pollutants—Transformations and Biological Effects*, O. Hutzinger, L. H. Van Lelyveld and B. C. J. Zoeteman, Eds. (Oxford, England: Pergamon Press, 1978), p. 223.
20. Zepp, R. G., N. L. Wolfe, G. L. Baughman and R. C. Hollis. "Singlet Oxygen in Natural Water," *Nature* 278:421 (1978).
21. Mill, T., W. M. Mabey and D. G. Hendry. "Test Protocols for Environmental Processes: Oxidation in Water," EPA Report (Draft), EPA Contract 68-03-2227 (November 1978).
22. Korchek, S., J. H. B. Chenier, J. A. Howard and K. U. Ingold. "Absolute Rate Constants for Hydrocarbon Oxidation. XXI. Activation

Energies for Propagation and Correlation of Propagation Rate Constants with Carbon-Hydrogen Bond Strengths," *Can. J. Chem.* 50:2285 (1972).

23. Russell, G. A., and R. C. Williamson. "Nature of the Polar Effects in Reactions of Atoms and Radicals. II. Reactions of Chlorine Atoms and Peroxy Radicals," *J. Am. Chem. Soc.* 86:2357 (1964).

24. Walling, C., and B. B. Jacknow "Positive Halogen Compounds. II. Radical Chlorination of Substituted Hydrocarbons with t-Butyl Hypochlorite," *J. Am. Chem. Soc.* 82:6113 (1960).

25. Gollnick, K. In: *Singlet Oxygen,* B. Ranby and J. F. Rabek, Eds. (Chichester, England: John Wiley and Sons Ltd., 1978), p. 111.

26. Denisov, E. T., *Liquid Phase Reaction Rate Constants,* English Trans. by R. K. Johnson (New York: IFI/Planum, 1974).

27. Howard, J. A., "Absolute Rate Constants for Reactions of Oxyl Radicals," *Adv. Free Radical Chem.* 4:49 (1972).

28. Mabey, W. M., and T. Mill. "Critical Review of Hydrolysis of Organic Compounds in Water Under Environmental Conditions," *J. Phys. Chem. Ref. Data* 7:383 (1978).

29. Mabey, W. M., T. Mill and D. G. Hendry. "Test Protocols for Environmental Processes: Hydrolysis," EPA Report, EPA Contract 68-03-2227 (May 1978).

30. Hautala, R. R. Presented at Symposium on Nonbiological Transport and Transformation of Pollutants on Land and Water, National Bureau of Standards, Gaithersburg, Maryland, May 11-13, 1975.

31. Wells, P. R. "Linear Free Energy Relationships," *Chem. Rev.,* 63:171 (1963).

32. Taft, R. W. *Steric Effects in Organic Chemistry* (New York: John Wiley and Sons, 1956), Chapter 13.

33. Goring, C. A. I., D. A. Laskowski, J. W. Hamaker and R. W. Meikle. In: *Environmental Dynamics of Pesticides,* R. Haque and V. Freed, Eds. (New York: Pleun Publishing Corporation, 1974), p. 135.

34. Stumm-Zollinger, E., and R. H. Harris. In: *Kinetics of Biologically Mediated Oxidation of Organic Compounds in Aquatic Environments,* S. L. Faust and J. V. Hunter, Eds. (New York: Marcel Dekker, 1971), p. 555.

35. Paris, D. Private communication (1978).

36. Dagley, S. *Essays in Biochem.* 11:81 (1975).

37. Alexander, M., In: *Pesticides and Their Effects on Soil and Water* (Washington, D.C., American Chemical Society Special Publication No. 8, 1966), p. 78.

38. Liss, P. G., and P. G. Slater "Flux of Gases Across the Air-Sea Interface," *Nature* 247:181 (1974).

39. Mackay, D., and P. J. Leinonen. "Rate of Evaporation of Low Solubility Contaminants from Water Bodies to Atmosphere," *Environ. Sci. Technol.* 9:1178 (1975).

40. Mackay D., and A. Wolkoff. "Rate of Evaporation of Low-Solubility Contaminants from Water Bodies to Atmosphere," *Environ. Sci. Technol.* 7:611 (1973).

41. Spencer, W. F., W. Farmer and M. M. Cliath. "Pesticide Volatilization," *Residue Rev.* 49:1 (1973).
42. Bailey, G. W., J. L. White and T. Rothberg. "Adsorption of Organic Herbicides by Montmorillonites," *Proc. Soil Sci. Soc. Amer.* 32:222 (1968).
43. Hamaker, J. W., and J. M. Thompson. In: *Organic Chemicals in the Soil Environment, Vol. I,* C. A. Goring and J. W. Hamaker, Eds. (New York: Marcel Dekker, 1972), p. 49.
44. Williams, R. R., and T. F. Bidleman. "Toxaphene Degradation in Estuarine Sediments," *J. Agric. Food Chem.* 26:280 (1978).
45. Karickoff, S. W., D. S. Brown and T. A. Scott. "Sorption of Hydrophobic Pollutants on Natural Sediments," EPA Internal Report, Environmental Research Laboratory, Athens, GA (1978).
46. Kanaga, E. E., and C. A. I. Goring. "Relationship Between Water Solubility, Soil Sorption, Octanol-Water Partitioning and Bioconcentration of Chemicals in Biota," Preprint, Amer. Soc. Testing Materials, Third Aquatic Toxicology Symposium, New Orleans, LA October 17-18, 1978.
47. Chiou, G. T., B. H. Freed, D. W. Schmedding and K. L. Kohnert. "Partition Coefficients and Bioaccumulation of Selected Organic Chemicals," *Environ. Sci. Technol.* 11:475 (1977).
48. Smith, J. H., and D. Bomberger. Unpublished results (1979).

22

COMPUTER SIMULATION MODELS FOR
ASSESSMENT OF TOXIC SUBSTANCES

Alan Eschenroeder

Arthur D. Little, Inc.
Cambridge, Massachusetts 02140

Eileen Irvine, Alan Lloyd, Cindie Tashima and Khanh Tran

Environmental Research & Technology, Inc.
2030 Alameda Padre Serra
Santa Barbara, California 93105

INTRODUCTION

A large variety and growing number of chemical substances are placed into the earth's environment every year. Involuntarily, people are exposed to many of these substances. Some of these exposures involve risks to human health and the environment; however, many of the new substances have been incompletely characterized with respect to their biological effects, transport, transformations and ultimate sinks in the environment. It is only recently that these health hazards have become fully recognized by the major control agencies [*e.g.,* Environmental Protection Agency (EPA), Occupational Safety and Health Administration (OSHA) and National Institute for Occupational Safety and Health (NIOSH)]. This awareness is underscored by frequent announcements of the possible carcinogenic properties of already existing and utilized pesticides, food additives, preservatives, solvents, etc. (*e.g.,* nitrosamines, benzene and vinyl chloride).

In order to understand the problems associated with these areas and to eliminate the importance of various paths for the transport and ultimate fate of toxic chemicals, the sensitivity of this transport in the ecosystem to various parameters should be assessed. One part of the approach to this problem is the formulation of mathematical models for the prediction of the concentration levels in the physical and biological environment.

A primary objective of model development is the unification of cause and effect parameters by incorporating physical, chemical and biological data obtained from ongoing laboratory programs. The ultimate application of any such model is to provide scope and direction to the measurement programs, as well as to aid decision makers in their formulation of controls for the introduction of chemical substances into the environment.

The profile model must satisify many requirements. It must differentiate between the *point source* aspects of many industrial sources of these chemicals and the special *area source* problems typical of pesticide or herbicide applications. Special chemical properties that must be included in the model are those of persistence and potential bioaccumulation. A logical approach to formulation of the model should include air, water, soil and biota compartments with flows of substances described among these compartments. Within each compartment, chemical transformation and loss mechanisms can be represented. Consequently, the capability for multiple species equations should be provided within any compartment of the model scheme. Divisions indicative of trophic levels in food webs may also be employed to account for any bioaccumulation of a substance.

An environmental chemical model will help in organizing laboratory-derived data. The model generates a profile reflecting the fate and transport of the substance in the environment. Central to this theme is the concept of benchmark chemicals that may relate the properties of entire classes of substances to the behavior of a single surrogate compound. For example, single compounds might be selected to represent broad classes, such as organohalides, organophosphates and carbamates. Such an approach is useful for dealing with the large variety of chemicals which may have only small structural chemical variations within broad classifications. The ultimate application of a chemical environmental model may assist in decisions regarding manufacture, transport, handling, storage, exposure and use with respect to prevention of the escape or intentional exposure of substance to the environment. In order to maximize the cost-effectiveness of a given control program, it is necessary that strategies be based on considerations of the chemodynamics, biological effects and assimilative properties of the chemical. The model, therefore, must consider the pathways the chemicals may follow from their release to the receptor sites of concern in order to provide a broad perspective of risk assessment.

For maximum effectiveness, modeling should be related on three levels to the ongoing measurement programs. The first level is represented by fundamental laboratory programs involving physical, chemical and biological measurements of substances. Typical measurements would include vapor pressure, volatilization from soils, aqueous solubilities, dissociation and hydrolysis rates, octanol/water partition coefficients and degradation rates in sunlight. Laboratory-linked research should aid in the formulation of relationships between such fundamental properties and the coefficients needed for inputs into the compartment differential equation scheme which is basic to the model structure.

The second level of relationship between modeling and measurement should be that involving controlled laboratory ecosystem experiments. Because some constraint may be imposed on the operating conditions of these systems, they provide validation data bases to test the biological logic of a chemical model. Interaction between the model development and laboratory experimental plans is essential to make maximum advantage of this second level of scientific interaction.

The third, and perhaps the most important, interface between theoretical simulations and actual observations requires that the model be subjected to validation tests using observational data measured in field programs. Although the uncontrollable conditions presented by the natural environment must be accepted, there is some latitude of decision. Some elements which are amenable to decisions are the quantities measured, data acquisition frequency and choice of monitoring sites. Interaction between model exercises and field observation programs help to optimize the choices in various areas for any given chemical substance in interest. Conversely, it is essential to subject the model calculation to "real world" data obtained in the field.

In recent years, the development of a profile model for chemicals in the environment was initiated under the sponsorship of the National Science Foundation (NSF). The principal objective of the study is to develop a predictive mathematical simulation model, which outputs chemical concentration in components of the environment, using the input rate of the material, the physicochemical properties of the substance and the biological partitioning characteristics. Such a task is obviously a major one and represents an intermediate goal.

This chapter outlines the results of an exploratory study. In addition, it presents some of the areas requiring further work and areas for which data are most urgently required. The significance of this work has increased since it is being recognized that a growing number of chemicals can present threats in terms of possible mutagenic, carcinogenic and teratogenic hazards. In addition, with the introduction of the Toxic Substances Control Act (administered by the EPA), the identification of the experimental data which are required in assessing the fate of toxic chemicals has assumed increased importance.

The first part of the chapter includes a summary of a literature survey. This is followed by identification of chemical and biological aspects to be considered in assessing the movement of chemicals through the environment. Subsequent discussions cover the formulations of the model and results obtained from its demonstration for a case study. The chapter concludes with some suggestions for future work and refinements to be made in the model.

SURVEY OF EXISTING DATA AND FUTURE NEEDS FOR MODEL

Background

In the initial phase of development of chemical models, major emphasis has been placed on DDT, that is, 1,1,1-trichloro-2,2-bis (p-chlorophenyl)-ethane. While general application of this toxic and persistent chemical has been banned in the United States for a number of years, nevertheless, it is of interest in model development because of a large data base for concentrations and release rate. For example, DDT has been found in rivers, oceans, precipitation, air, soils, atmospheric dust and the entire world ecosystem.[1-4]

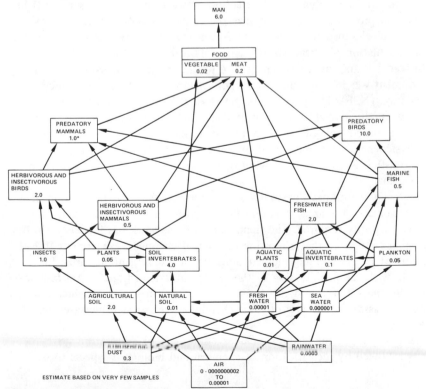

Figure 1. Typical levels of DDT in the environment.

Thus, Figure 1 shows typical levels of DDT in various parts of the environment, and Figure 2 given an overview of current knowledge associated with DDT. However, the mechanisms for flow of pesticides through the environment remain ill-defined. Major uncertainties include the transport rates and the pathways through the environment.

Atmospheric advection is believed to play a major role in the transport of DDT. This suggestion has been supported through the detection of DDT in dusts collected in various areas distant from sites of application. The

common trade name	DDT
formula:	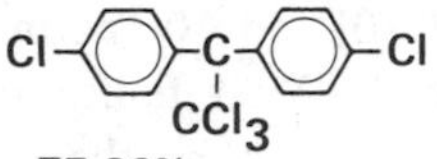 $Cl-⟨⟩-C-⟨⟩-Cl$ with CCl_3
purity:	75-80% main by-product o,p - DDT
production level:	not exactly known
use pattern:	use against more than 100 different insects on large variety of crops, in public health, as moth proofing agent quantities unknown
metabolites:	mainly DDE, DDD, DDA, DCB
occurrence residues:	up to 7ppm including analogues
outside area of use:	concentration up to more than 10 ppm in wildlife, fish, including analogues atmosphere: up to 200 $pp10^{12}$
LD_{50}:	DDT rat - oral 250 mg/kg bodyweight DDD " " 3400 " " DDE " " 1000 " " DDT monkey " >200 " "
chronic toxicity:	great number of data on DDT and analogues in animals and men - e.g. neoplastic disorders in mice (multigeneration) no abnormalities in men after 11 - 19 years occupational exposure to around 18 mg/man/day
wildlife effects:	accumulation in food chains, reduction egg-shell thickness, certain bird species

Figure 2. An overview of current DDT knowledge.

measurements were made by Risebrough *et al,*[7] Seba and Prospero,[8] and Prospero and Seba.[9] The study by Bidleman and Olney,[4] where hydrocarbons were measured in the area of the Sargasso Sea, in Bermuda and in an area of the North Atlantic, produced atmospheric and oceanic surface-layer) data. There is substantial interest, and several studies have determined the background atmospheric levels of DDT. Bidleman and Olney[4] have shown that the major concentration of DDT is present in the vapor phase, in contrast to the particulate phase. However, a recent study by Orgill *et al.*[10] has shown that soon after the application of DDT, particulate DDT concentrations were generally larger than the concentrations occurring in the vapor phase. The actual background levels of DDT remain uncertain, and values around 10^{-4} to 10^{-5} ng/m^3 have been reported by Seba and Prospero[8] for the Barbados-West Indies area. However, levels which are a factor of 100 to 1000 greater than these values were reported by Bidleman and Olney for a similar region. The latter workers suggested a possible explanation for their observed DDT levels exceeding by two orders of magnitude those of other workers. They suggested that the organochlorine compounds could escape detection by volatilizing from particles impacted on a filter. Hence, in their case any volatilized material would pass through the filter and be collected an a polyurethane foam used for this purpose. They suggested that earlier types of collectors used for DDT would have missed the bulk of these substances in the air, since they found that a large portion of DDT occurred in the vapor phase at background levels. This emphasized the importance of scrutinizing measurement techniques.

Data Needed as Input for Model Development

In order to generate a physicochemical model of transport, transformation and fate of chemicals through the environment, certain parameters are necessary as inputs. Some of these are specific to the particular case *e.g.,* application rate, whereas others are intrinsic properties of materials and can be used for model applications in many areas, *e.g.,* water solubility and photodegradation rate (if any) in the atmosphere. Some of these will be discussed separately for the case of DDT.

Application Rate

For the case of DDT, the value for the application rate will vary greatly depending on whether one is considering a global model, such as those developed by Randers and Meadows[1] or Cramer,[3] or whether one is focusing on a localized area, such as that studied by Orgill *et al.*[10] These latter workers examined the fate of DDT after it was applied to a Pacific Northwest forest to control tussock moth infestation.

The application rate is dependent on the area under consideration; hence, it does not lend itself to the objective of this study—namely, the assessment of the best values for certain parameters which can be widely employed in model applications.

Partition Coefficients

An important parameter commonly measured for pesticides is the partition coefficient. This is defined as the equilibrium concentration ratio of an organic chemical, in this case DDT, partitioned between an organic liquid, *e.g.*, *n*-octanol and water. This parameter is important in order to assess the relative amounts of pesticide going into aqueous solution compared with that going into fatty tissues.

The partition coefficients measured for DDT often differ by many orders of magnitude,[11,12] *e.g.*, from 9 x 10^3 to 1.6 x 10^6. Partitioning data in general, and the *n*-octanol/water partition coefficient in particular, have been useful to predict soil adsorption, biological uptake, lipophilic storage and biomagnification.[12-18]

Recently, Chiou *et al.*[19] reported a study whereby an empirical equation was formulated relating experimentally determined *n*-octanol/water partition coefficients to the aqueous solubilities of a wide variety of chemicals, including DDT. These authors state that their correlation allows an assessment of the partition coefficient from water solubility measurements with a predicted error of less than one order of magnitude. In view of the wide discrepancy in the previously reported values covering several orders of magnitude, this represents a significant improvement. They state that the accepted value for the aqueous solubility for DDT is between 1 and 5 ppb, although published values ranged from 0.2 to 1000 ppb. They also state that, using their equation, a predicted value for the partition coefficient in the million range would result. This value is in agreement with the value of 1.6 x 10^6 calculated by O'Brien.[11]

In addition, Chiou *et al.*[19] observed a correlation between the biomagnification factors in rainbow trout and the aqueous solubilities for some stable organic compounds, including DDT. This would appear to bear out results of previous studies by Metcalf and co-workers,[12,17,20] which showed a correlation of biomagnification in mosquito fish with aqueous solubilities of chemicals, including DDT. This type of relationship is useful in model development, since it provides some basis for assessing the magnitude of biomagnification from water solubilities, which is an experimentally measurable property. Thus, for rainbow trout, the relationship between biomagnifi-

cation and solubility is given by log(BF) = 3.41 - 0.508 logS, where BF
is the biomagnification factor in rainbow trout and S is the aqueous
solubility in μmoles per liter.

MODELING FORMULATIONS FOR
MOVEMENT OF CHEMICALS IN THE ENVIRONMENT

Data Needs for Model Formulation and
Relationship to Measurement Programs

A recognized difficulty with all modeling efforts at the outset is that
insufficient data exist to test the model completely. It is precisely this
problem that is best addressed by the application of models in the early
stages of the research. The model serves a logical structure for organizing exactly what data are required to provide a system description. With
this systematic guide, experiments and measurements can be planned to
fill in the gaps that are revealed by the attempt to model. Therefore,
despite deficiencies in the information base, it is never too soon to begin
structuring the designing models in order to link the prediction requirements
with the data-gathering efforts.

Model Structures For Biotic and
Abiotic Compartments

Frequently the flows and concentrations of a substance in the physical/
chemical environment are large compared with those in a biological system
of interest, and the two systems can be mathematically decoupled. That
is to say, the uptake or elimination of a particular chemical material by a
biological system may impose negligible perturbations of the levels of this
substance in air, soil or water. In the compartment model sense, even if
large bioaccumulation of a substance, the total mass involved may still
be small. For example, if the mass of all biota is considered to be 1, that
of the soil is approximately 100, the atmosphere, 10,000, and water,
1,000,000.[21] Despite bioaccumulation, the DDT mass estimated for marine
fish is 600 metric tons and plankton 3 metric tons, compared with a peak
application rate of 10^5 metric tons/yr. The estimated holding capacity of
the mixed layer (top 100 m) of the ocean is about 10 times the peak
annual application rate.

The decoupled systems of the physical/chemical environment and the
biological environment may be represented in simple form according to
the flow diagrams expressed in Figures 3 and 4. Figure 3 recognizes the
commonly used partitions in the physical environment differentiation
fresh water and sea water, for example, or differentiating the ocean

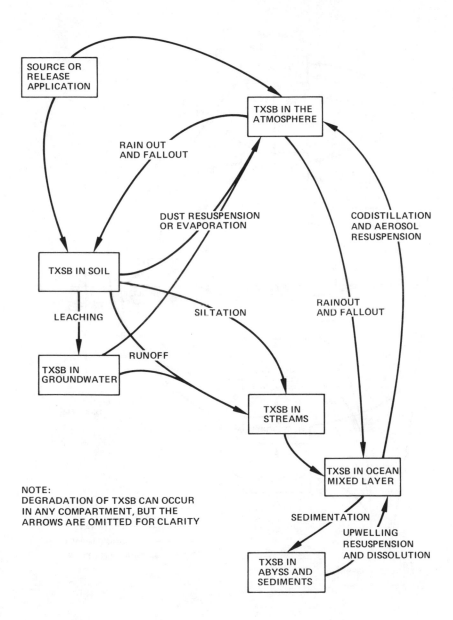

Figure 3. Compartments and flows in the physical/chemical environment.

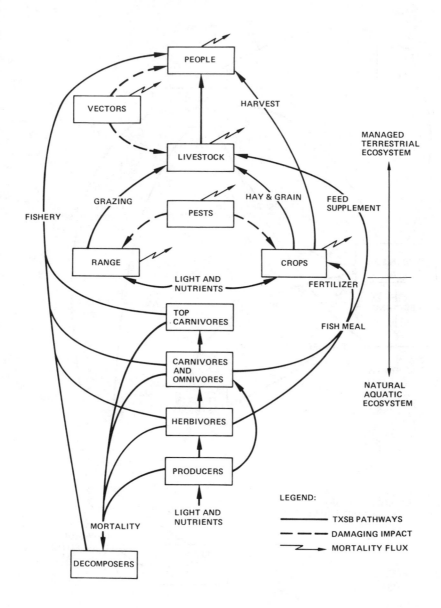

Figure 4. Compartments and flows in the biotic environment.

surface layer from the abyss and sediments. In a specific model application it may be possible to eliminate certain compartments or paths shown on Figure 3, depending on the properties of the substance.

Figure 4 differentiates between the natural ecosystem and the cultural ecosystem by the use of separate compartments. A particular example is carried through this diagram by the inclusion of pests and vectors as separate compartments. The flows of material through the various pathways indicated, therefore, may bring about either damages or benefits according to the functional specification of the model terms. For example, toxicity effects among the pests and vectors may be considered beneficial, whereas influences on other components can be damaging. A difference between Figure 3 and Figure 4 should be emphasized here. The flows in Figure 3 are solely those of the chemical substances, whereas the flow pathways indicated in Figure 4 have a dual role. They indicate both flows of energy expressed as feeding terms as well as flows of the chemical substances. Consequently, in the biological system two sets of equations must be carried in the food unless the toxic effects of the chemical rate are assumed and imposed on the population dynamics equations. The diagrams for model structures are based in part on previous suggestions by Harrison et al.,[23] Woodwell et al.,[2] Randers and Meadows, Cramer,[3] Gillett et al.[24] and Bolin.[25]

Logic, Inputs and Outputs

Governing Equations

In the physiochemical regime, species budget equations in the fluid phases are commonly expressed in the following terms:

$$\frac{\partial c_j}{\partial t} + \sum_{\ell=1}^{3} \frac{\partial(c_j v_{j\ell})}{\partial \xi_\ell} = w_j + S_j \qquad (j = 1,2,3, \ldots p) \qquad (1)$$

where
c_j = mass concentration of the jth species
t = time
ℓ = an index referring to each coordinate direction
$v_{j\ell}$ = velocity of jth species in ℓth direction
ξ_ℓ = distance in ℓth direction
w_j = net molar production rate of jth species per unit volume by chemical reaction
S_j = source strength for emitters of jth species at some location above the ground
p = number of species

The approach to solving these equations involves varying degrees of approximation. It may be seen that these are nonlinear partial differential equations with the nonlinearities confined to the source and sink terms because the velocity field is given.

In the model structure dealing with the biota, both energy flows and chemical substance flows must be considered. Equation 2 represents one form of a set of governing equations for biomass energy equivalents in each subcompartment of the biological flow diagram shown in Figure 4.

$$\frac{dx_i}{dt} = F_{oi} + \sum_{j=1}^{s} \tau_{ji} \frac{x_j x_i}{(x_j + a_{ji} x_i)} - \sum_{j=1}^{s} \tau_{ij} \frac{x_i x_j}{(x_i + a_{ij} x_j)} - \mu_i x_i$$

| Forcing flux | "i" feeding on other species | other species feeding on "i" | first order mortality and excretion |

$$-\kappa_{ii} x_i^2 \qquad -\lambda_{io} x_i \qquad -\rho_{io} x_i, \qquad (i = 1,2,3, \ldots, 5) \ (j \neq 1) \qquad (2)$$

| competitive mortality | export | respiration |

where
x_i = biomass energy equivalent of ith population
t = time
F_{oi} = forcing flux into ith population
τ_{ji} = feeding rate coefficient (i eats j)
τ_{ij} = feeding rate coefficient (j eats i)
μ_i = mortality rate coefficient
κ_{ii} = competition coefficient
λ_{io} = emigration or export rate coefficient
ρ_{io} = respiration rate coefficient
a_{ji} = saturation balance parameter
s = total number of populations
j = index of any species other than i (increasing values of index denote ascending position in the food chain)

The denominators in the feeding terms reflect saturation effects of an excess of either predator or prey. The competitive mortality terms give the familiar logistic growth behavior of leveling out a population which has increased greatly.

The governing equation for toxic substance concentration in the ith population is given by Equation 3 as follows:

$$\frac{d(y_i x_i)}{dt} = \sum_{j=1}^{s} \tau_{ji} \frac{x_i x_i y_j}{(x_j + a_{ji} x_i)} - \sum_{j=1}^{s} \tau_{ij} \frac{x_i x_i y_i}{(x_i + a_{ij} x_j)} - \mu_i' x_i y_i \qquad (3)$$

$$\kappa_{ii} x_i^2 y_i - \lambda_{io} x_i y_i + w_i$$

where y_i = concentration of toxic substance in ith population

$y_{j,}$ = concentration of toxic substance in jth population

μ_i' = loss coefficient for toxic substance from ith population

It will be noted in Equation 3 that most of the terms are identical to those in Equation 2 with the simple added feature of multiplying a concentration. For $w_i=0$, this equation is restricted to nonreactive substances. Clearly, under the influence of some life processes the substances will undergo changes in the food chain. Equation 3 can be generalized to cover that case by defining a hierarchy of y_i. This hierarchy might be designated as y_{ik}, where k is a species index ranging from 1 up to the number of chemical species involved. Thus, those cases would be represented by a coupled set of species mass balance equations for each trophic level.

Lethal toxicity effects might be represented by an expression like that in Equation 4 which imparts values to the mortality coefficient μ_i, that build

$$\mu_i = \mu_i^o \left[1 + \frac{\mu_i^{max} - \mu_i^o}{\mu_i^o} \exp(-y_{io}/y_i) \right] \qquad (4)$$

where μ_i^o = baseline mortality rate

μ_i^{max} = mortality rate at saturation of lethal effects

up from a baseline level, μ_i^o, to the level μ_i^{max} in accordance with an exponential law based on a characteristic concentration parameter, y_{io}, expressive of a lethal concentration 50% of the way between the extremes and mortality coefficient. This form is analogous to an Arrhenius chemical kinetic rate in which the rate of transformation is proportional to a fraction of normally distributed particle energies exceeding a threshold. The analog is that the individuals are normally distributed in their step-function response to dose. For instance of sublethal toxic effects, such as an impairment or enhancement of metabolism or mutagenesis, a damage function equation similar to Equation 4 can be constructed if the dose-response curves for the effect of interest are known. In some cases the damage function need not be coupled to the equations if it has no feedback into population dynamics.

Approaches to Solving the Governing Equations

Standard numerical methods of integration using finite difference formulas are widely available in package form and for linear equations are available in terms of user-oriented compilers permitting simple specifications of the system in terms of some intermediate language. Our approach to the solution employs an implicit numerical integration scheme of variable order developed by Gear[26] suitable for the solution of "stiff" equations. The term stiffness refers to widely differing time constants among the processes represented. Especially with reactive substances, this property may be a prominent obstacle to the solution of the equations because of wide variability in reactivity within various trophic levels. Another implicit method which is more forgiving of discontinuous changes is available for solving equations of this class. This method is called EPISODE and was developed by Hindmarsh and Byrne.[27] Both of these techniques are available as working codes and are fully documented so that the user can incorporate them as subroutines to be called without becoming concerned with the internal details of the programs themselves. They are designed to handle nonlinear equations.

Even with the availability of these powerful numerical integration techniques, it still may be desirable to incorporate approximations in the governing equations. For example, Equation 1, describing flows in the physical environment, can be solved for a system of geographical compartments, thereby reducing it to additional ordinary nonlinear differential equations. This coarse resolution representation involves rate coefficients for flow from one compartment to another that are obtained from Equation 1 for transport rates. Finer-scale approximations are also available in Equation 1, such as the neglect of transformations by dropping the source terms or the neglect of diffusion in certain directions by dropping components of the diffusive transport terms on the right-hand side.

Typical approximations in the equations for biomass and toxic substance concentration involve dropping spatial resolution entirely, lumping parameters by aggregation of species or levels in the food chain, neglecting competition, neglecting saturation effects in the feeding terms, and linearizing all terms. As each of these levels of approximation is introduced, certain input requirements vanish. With the simple geographical box representation, all flow variables except average velocity gradients become unnecessary. Increasing degrees of aggregation of biota within compartments accordingly decreases the requirement for numbers of rate coefficients in initial values for the trophic equations. This is commonly known as a "lumped parameter" approach in the field of simulation. Neglect of competition among species again permits the deletion of further coefficients, as does neglect of saturation effects in the feeding terms. Finally, if the instantaneous population effect of an acute dose of toxic substance is known *a priori,* it can be inserted in the population dynamics description without tracing the flows of the substance through the environment. This will be exemplified in the next section.

BIOTIC COMPARTMENT MODEL DEMONSTRATION

Prototype Ecosystem Description

As a point of departure in a demonstration, a prototype ecosystem must be chosen and a model specified for describing it. It is useful to choose a system which has been measured extensively for the variables of interest. An example approaching satisfaction of this requirement is that established by Odum[28] for the aquatic community in Silver Springs, Florida. Its equilibrium state has been observed to be stable for many years due to a large steady flow of water. External changes lead to periodic diurnal and annual excursions about the mean value of biological parameters. These are averaged out by Odum; coefficients described from these data will restrict the time resolution of the model calculation to the order of a year.

Turning to the specification of a model structure (as shown in Figure 5) a box model will be employed using coupled ordinary differential equations similar to those outlined in the previous section under Equation 2. The feeding terms have been simplified to omit the saturation effects in the denominators. Energy flows from each population account for changes in the biomass energy equivalent of that population. The state of the system, therefore, is described by a list of biomass energy equivalent of that population. The state of the system, therefore, is described by a list of biomass energy equivalents, one for each level of the trophic chain. The equation suggested by Patten[29] has been selected as a starting point for the model, and the nonlinear form of the rate equations has been chosen from the examples shown in that work.

Seven contributions described in the equations are:

1. flows of energy across the ecosystem boundary into a given population,
2. feeding of population "i" on other species,
3. mortality of population "i" due to being eaten,
4. spontaneous mortality of population "i" in the first order,
5. spontaneous mortality of population "i" which depends on the square of the ith biomass energy equivalents (competitive),
6. emigration of population "i" across the ecosystem boundary, and
7. net energy loss due to respiration.

According to Odum,[28] the plant respiratory biomass is represented mainly as an algea/sagittaria complex. The herbivore compartment is divided into microfauna such as snails and worms and some of the fish population

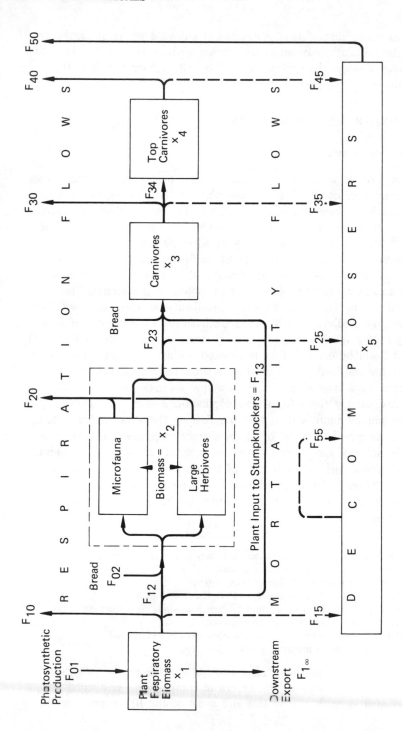

Figure 5. Energy flows in the prototype ecosystems.

(mainly mullets and half of the spotted sunfish or stumpknockers). The carnivores are composed of the other half of the spotted sunfish and catfish. Top carnivores consist mainly of bass and gar pike, whereas the decomposers are crayfish and bacteria. The arbitrary division of sunfish into two compartments is a model method of indicating a branched food chain by assuming equal partition in feeding. The role of competitive spontaneous mortality is represented parametrically; and by use of the equilibrium flows in the system as reported in the literature, τ_{ij} and the κ_{ii} coefficients assuming some fixed partitioning between first order mortality and competitive mortality were derived.

Modeling Acute Toxic Effects

Acute kill episodes can occur selectively to certain segments of the community depending on toxicity levels. A mathematical expression of this challenge to the ecosystem is simply the injection of a reduced value of the population in question among the initial values that are provided to the population dynamics equation. This further assumes that the killed fish are exported from the ecosystem. The recovery of the system is modeled following the episode. Figure 6 illustrates the recovery of top carnivores after 90% of them are killed in a single episode. The various curves show the influence of both chain branching and competition in the population dynamics. The most rapidly returning curve is that characterized by 50% competition. The meaning of "percent competition" is taken as that percentage of spontaneous mortality due to competitive mortality in the equilibrium system. The system with 10% competition appears to take nearly twice as long to recover as that with 50% competition. Finally, the Odum-Patten model with no competition and no branching appears to be very closely approximated by the 10% competition curve. The strong competitive effect occurs because the second-order mortality terms drop off faster than the first-order ones when the population is depressed.

Going deeper into the food chain, a 90% kill episode was imposed on the herbivore population in another simulation. Figure 7 illustrates the profound effect in this case of chain branching on ecosystem recovery. It should be noted that much of the difference between the carnivore and herbivore curves appears because of the reallocation of stumpknockers to account for their feeding options.

Competitive mortality as well as options in feeding affects the ecosystem recovery from the 90% herbivore kill. Figure 8 shows historical plots of the aftermath exhibiting a smaller die-off and a faster restoration for a highly competitive community. These rapid changes may not be represented accurately by the model because of the time-resolution prob-

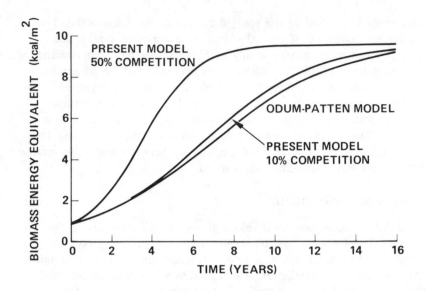

Figure 6. Recovery of top carnivores following a 90% kill.

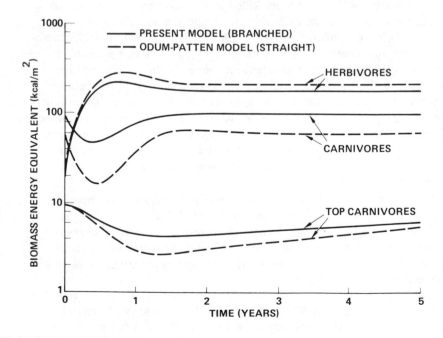

Figure 7. Response of the ecosystem following a 90% herbivore kill

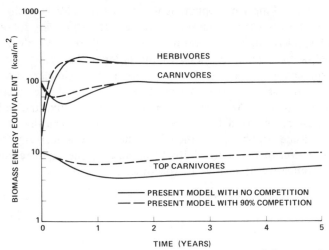

Figure 8. Effects of competitive mortality in the recovery of the ecosystem following a 90% herbivore kill.

lems referred to earlier. While herbivores and carnivores show only small percentage changes in the presence of higher competition, the top carnivores show relatively large change. Comparison of the 0 and 90% competition cases shows this clearly. For the high competition factor, about half as much die-off is observed compared with the case for low competition. Apparently the recovery time as well as the die-back is reduced by higher competition, whereas chain branching reduces the die-back without much change in recovery time.

Modeling Chronic Toxic Effects

For the study of chronic toxic effects, some preliminary calculations of biomagnification were carried out to test for plausibility.

An initial attempt of adjusting mortality/excretion loss terms (μ_i) in Equation 3 did not reflect the correct time response to get magnifications typical of those for chlorinated hydrocarbons. The input forcing flux was set at a value of 10^{-12} mass units of toxic material per kilocalorie of energy flow into the producer compartment. (This influx corresponds to an equivalent absorption of 1 ppt contamination of material with 1 kcal/g energy equivalent.) Downward adjustment of the mortality/excretion coefficients by 10^{-2} for the producers, 10^{-3} for the herbivores, 10^{-4} for the carnivores and 10^{-5} for the top carnivores gave equilibrium concentration values of 4 ppt, 30 ppt, 300 ppt, 1 ppm and 10 ppm of material in the five respective compartments. Since the relative values appeared reasonable in light of environmental data for DDT/DDE, a dynamic simulation was run with only a trace (1 ppt) of contaminant material initially in each compartment. The results were implausible because after 100 years following the onset of application, the top carnivores were still several orders of magnitude below the equilibrium level of the order of 10 ppm.

The second simulation of biomagnification took a much simpler form than the first one. It was noted that the producer level equilibrates rapidly to a concentration nearly in direct proportion to the input flux, and only that flux was adjusted. Observed DDT levels in plants were typically of the order of magnitude of 10^{-2} ppm; therefore, the flux was increased to place the concentration in that range initially because of the rapid response of the producer compartment relative to the others.

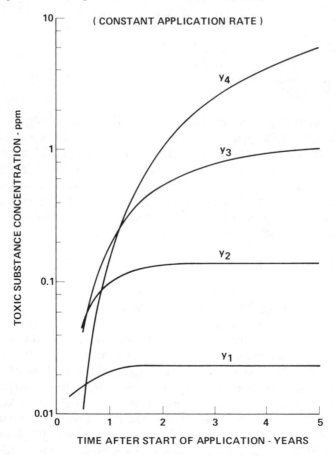

Figure 9. Propagation of a chemical substance through the food chain.

In the second version of the simulation, all of the other adjustments were removed and the mortality and excretion coefficients were restored to 10% of their original values (except for producers, which had the original value). The flow of contaminant into a level was related through predation coefficients, biomass energy equivalents and concentration (in mass of contaminant/unit biomass energy equivalent). The loss was similarly related to the prey terms and mortality/excretion flows with the adjustments mentioned above. As shown in Figure 9, the system response showed a build-up of contaminant in 4 yr to approximately 0.02 ppm in producers, 0.1 ppm in herbivores, 1 ppm in carnivores and 4 ppm in top carnivores. These degrees of biomagnification over this period of time after beginning of application are reasonable for chlorinated hydrocarbons in the environment. Figure 10 shows the recovery of the system for a 15-yr period following suspension of application. The inertia effect in the food chain causes an overshoot of concentration in the top carnivore population.

In order to model possible chronic toxicity effects, a dose/response curve was synthesized as was suggested in Equation 4. Figure 11 illustrates this function. It provides a feedback from the contaminant concentration to the population dynamics, each of which is represented by its own set of differential equations. The manner in which the function was applied provided an exponential transition from the normal mortality rate to some elevated rate characterizing the full force of the lethal effect. This choice is analogous to chemical rate processes where a normal distribution of velocities interacts with a step function occurring at a discrete energy threshold.

Three parameters are sufficient to describe this continuous sigmoid-shaped dose/response curve:

1. baseline mortality rate coefficient (normal),
2. "threshold" concentration of contaminant, and
3. saturation mortality rate coefficient (elevated).

Thus, this is equivalent to saying that among the individual members of a population compartment there is a distribution of dosages dependent on individual variations of age, exposure and metabolism. As this distribution moves to higher concentrations, even larger fractions of the population exceed the critical value where mortality rate jumps to an elevated rate.

One numerical experiment was run to simulate the chronic lethal toxic effect by first imposing an input of contaminant for five years and then removing the source of toxic material. The source flux adjustment described above was employed, and all of the rate coefficients were kept at normal levels except for the dose/response effect. This mortality rate behavior was assumed for the top carnivore compartment with a saturation mortality 10^3 times the normal baseline spontaneous mortality with a "threshold" of

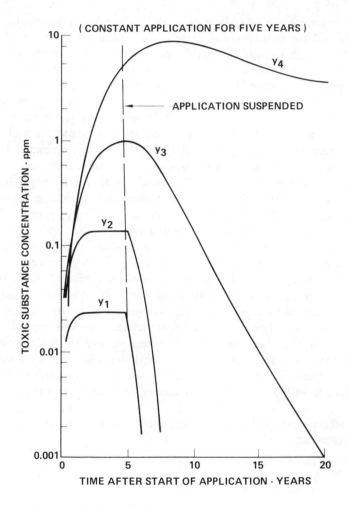

Figure 10. Recovery of a system following suspension of chemical application at 5 years.

1ppm. At this point, the biomass of top carnivores dies back nearly two orders of magnitude during the second and third years. The contaminant concentration in the top carnivore compartment became self-limited at about 1 ppm because the lower biomass decreased the integrated ingestion rate and the mortality loss rate was enhanced. These effects are illustrated in Figure 12.

Another numerical experiment was conducted by imposing the dose/ response effect described above on the herbivore population. Again the μ^{max}/μ° ratio was set at 10^3, but the level at which significant effect

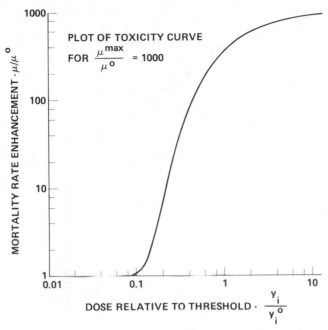

Figure 11. Assumed dose/response curve for study of toxic effects.

sets in, y_i^o, was assigned the value of 0.05 ppm. As expected, the damage to the ecosystem showed a much faster response than that in the case of the top carnivore mortality enhancement. Figure 13 shows the successive decreases in biomass energy equivalent among the populations above the producer level.

A further adjustment in the model involved returning the mortality/ excretion coefficients to their original values and then increasing the input flux of contaminant by a factor of ten greater than in the second simulation (Figure 14) in an attempt to reflect reasonable buildup of contaminant. The resultant concentration of contaminant after 5 yr is approximately 4.6 ppm in top carnivores, 1.7 ppm in carnivores, 0.42 in herbivores and 0.2 ppm in the producers.

Both the acute and chronic response simulations demonstrate the applications of modeling to the movement and effects of toxic materials in the environment. In some cases of ecosystem modeling literature, the mathematical difficulty of numerically integrating nonlinear, interacting systems has been overemphasized. These demonstration simulations had nonlinearities of the x^2, xy and exponential type in a set of equations of

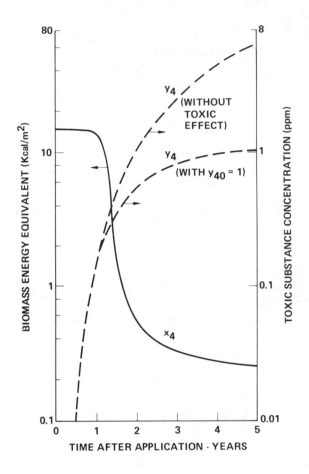

Figure 12. Simulation of lethal toxic effects on the ecosystem at the top carnivore level.

the forms $x_i = f_i(x_j, y_k, t)$ and $y_j = g_i(x_j, y_j, t)$. The simulations were performed in seconds on a CDC 6400 central processing unit for five years of simulated time. Multiplicity of boxes for geographical specifications of the system is the next step of complexity. Indeed, with the present **level** of computing efficiency established, these elements of realism are feasible

Specific Chronic Effects

Respiratory Damage Function

Chronic potential hazards to ecosystem processes can manifest themselves in an alteration of rates of other important processes besides mortality,

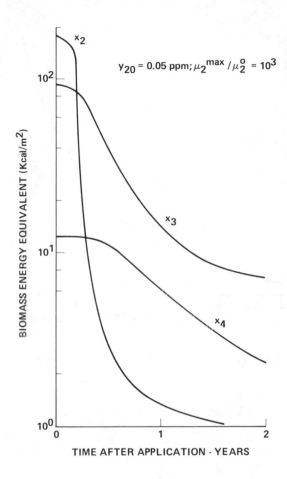

Figure 13. Simulation of lethal toxic effects on the ecosystem at the herbivore level.

such as respiration and feeding. The variation in these rates depends in part on the nature of the contaminant. Respiration can be inhibited by different chemical agents, such as arsenicals, which uncouple oxidative phosphorylation, or others, such as rotenone, which affect the electron transport system. A more general effect is an increase in respiratory rate due to the increased energy necessary to combat internal stress in normal functioning caused by existence of contaminants, and to carry on detoxification processes. A respiratory damage function (Equation 5) in the form of a dose-response curve was incorporated to reflect rate increase due to toxic concentration in the Silver Springs ecosystem.[30]

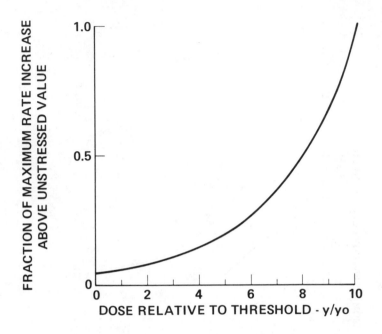

Figure 14. Effects of increasing contaminant input rate on the ecosystem.

$$R = R_{min} \left[1 + \frac{R_{max} - R_{min}}{R_{min}} \exp [c_1(y_i/y_o - c_2)] \right] \tag{5}$$

where R_{min} = unstressed respiratory rate

R_{max} = maximum respiratory rate

c_1, c_2 = rate constants dependent on trophic level which affect rate and extent of curve.

This curve (Figure 14) reflects a buildup of respiration from an unstressed rate, R_{min}, to a maximum rate, R_{max}, as a function of increased toxic substance concentration, y_i, relative to a threshold concentration, y_o, and providing feedback into the biomass/energy equations due to toxic stress.

This respiratory damage function was applied to the top carnivores equation in the Silver Springs simulation (Figure 15), with the maximum rate set for factor of 10 greater than the unstressed rate and the respiratory threshold set at 1 ppm (c_1 = 0.3, c_2 = 10.35). This assumption produced a rapid increase in toxins, leading to an extremely high increase in toxins after four years and undetermined after that point as the exponential error bound of the machine was exceeded.

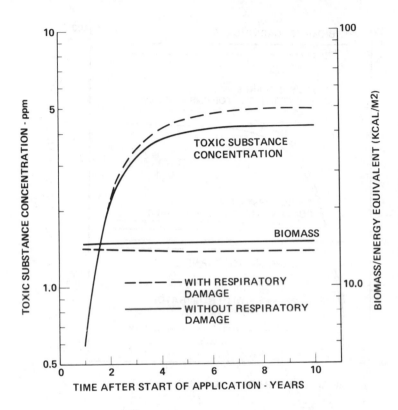

Figure 15. Effects of respiratory damage stress at the top carnivore level.

The maximum rate was then set to twice the normal, which probably more realistically reflects the small variance and large impact of this parameter on the system. While data on temperature and oxygen are readily available, very little data are available as to the sensitivity of the respiratory rate to toxic concentrations.

The damage function (Figure 14) produced an increase in the toxic substance concentration in top carnivores (Figure 15) caused by the decrease in biomass due to respiration and a slightly reduced ingestion rate, but without any corresponding toxic loss. Because the biomass reduction was minor, neither biomass nor toxic concentrations in other trophic levels showed effect.

The damage function was then applied to carnivores (Figure 16), also with a maximum respiration representing twice the normal rate. Again the toxic material concentration in this group was increased but to a larger extent than in the previous case, and an impact was evident on

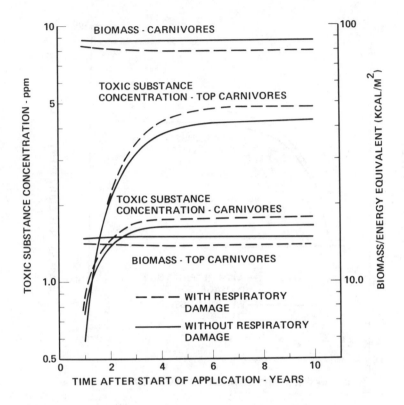

Figure 16. Effects of respiratory stress at the carnivore level.

the top carnivores which also had an increased toxic substance concentration. The decrease in biomass (20%) without toxic material loss in the carnivores increased the toxic substance concentration and resulted in an increase in toxic substance concentration at the next higher level.

Feeding Damage Function

Another potential chronic hazard to ecosystem processes may be an alteration in rates of feeding. Feeding rates have been shown to vary with toxic concentration as mobility, predative abilities and other vital behavioral patterns, as well as physiological malfunctioning, may be affected by toxic presence. To reflect toxic stress effects on normal feeding rates, a possible damage function, Equation 6, was included in the model [30]

$$r'_{ij} = F_{ij} \ [1 - \exp \ [-c_1(c_2 - y_i/y_u)]]$$

where F_{ij} = normal feeding rate and F'_{ij} = altered feeding rate. In this case, the function (Figure 17) was represented as a decrease in normal feeding rate F_{ij} with increase in toxic substance concentration y_i relative to a threshold concentration y_o, reflecting the toxic stress on the trophic category.

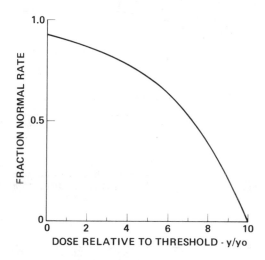

Figure 17. Effects of toxic stress on feeding rates at the carnivore level.

The feeding damage function was applied to carnivores to determine the change in resultant toxicity due to this effect. The threshold representing onset of effect was set at 1 ppm (c_1 = 0.25, c_2 = 10, where c_1 and c_2 are rate variant constants dependent on trophic level which affect the rate and extent of the curve). The incorporation of this function produced an increase in toxic substance concentration in carnivores (Figure 18) due to the reduction of carnivore biomass, which overcompensated for the reduction of carnivore toxins. This biomass reduction served to reduce the biomass of top carnivores, which although they fed on a group with reduced toxicity which depressed their ingestion of toxins, also acted to increase their toxic concentration.

Lethal Toxic Effects—Food Chain Propagation

With lethal toxic mortality effect (Figure 11) on the carnivore level, the carnivore biomass drops, and the herbivore biomass increases (due

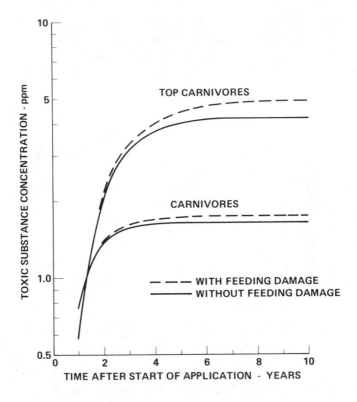

Figure 18. Time distribution of toxic substance concentration under conditions of feeding rate stress at the carnivore level.

to reduced predation by carnivores) is compensated by increased predatory pressures due to increases in herbivores and so ultimately decreases. Increased mortality of the carnivores, therefore, causes a decreased biomass in other levels, except herbivore. The toxic substance concentration (Figure 19) however, shows initially a slight reduction in plants, an increase in herbivores, a huge decrease in carnivores and a reduction (50%) in top carnivores. Thus, the toxic substance concentration change parallels the biomass change, with the toxic effect mediated by the altered ingestion caused by altered biomass values.

A similar pattern is seen with an increased mortality on herbivores, where the toxic substance concentration in plants increases, the herbivore and carnivore toxic substance concentration decreases greatly (the biomass of both is greatly reduced), and the toxic substance concentration in top carnivores decreases, but to a much lesser extent. It appears that the lethal mortality effect is propagated by an increased toxic substance

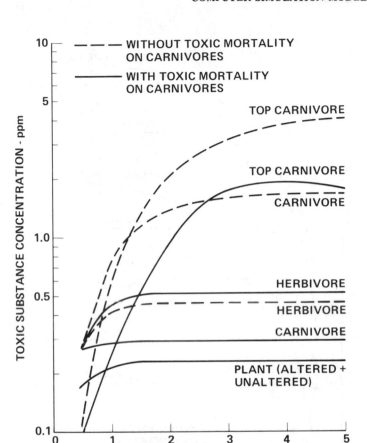

Figure 19. Effects of lethal toxic mortality at the carnivore level on the toxic substance concentrations in the ecosystem.

concentration in the category below the one perturbed; there is essentially no effect two levels down, a substantially decreased toxic substance concentration in the perturbed level and the level above, and a much slighter but still apparent effect on the next higher level.

Lethal toxic mortality in plants (Figure 20) results in a decreased toxic substance concentration in all categories, due to the parallel biomass reduction that derives from the reduced ingestion. An interesting effect is that the effect on the other levels (which drop to 15% or 20% of equilibrium value) is so much more pronounced than the mortality of the plants (which drop to 40% normal) that the toxic material level in the herbivores becomes lower than that in plants. This may be due to a low lethal toxic threshold and reflects the relative importance of such values.

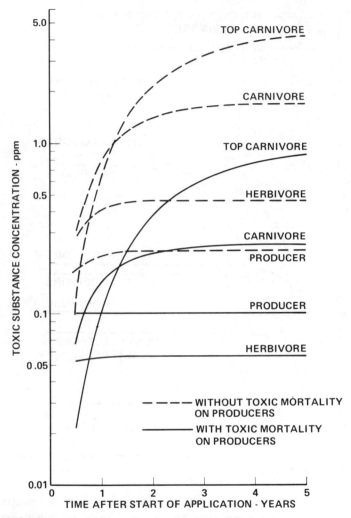

Figure 20. Effects of lethal toxic mortality at the plant level on toxic substance concentrations in the ecosystem.

Simultaneous Application of Toxic Mortality Function on Four Trophic Categories

A lethal toxic mortality function (Figure 11) was applied simultaneously to plants, herbivores, carnivores and top carnivores to determine the interaction of the various effects. The characteristic concentration parameter y_{io} (see Equation 4), used at each level was that which had been used previously in individual simulations (7×10^{-8}, 3×10^{-7}, 1×10^{-6} and 1×10^{-6} for plants through top carnivores, respectively). Although the relative value of the parameters in the case of herbivores and carnivores appeared high compared to the other categories, a lower parameter caused a great die-off. Figure 21 shows the new toxic concentration values computed in the simulation, compared with the values achieved for each category when a mortality function was applied to each category separately. The resultant effect in an increase in the plant toxic concentration (by 79%) and a reduction in the concentration in the other categories (by 47, 50 and 88% in herbivores, carnivores and top carnivores, respectively.) The largest effect was felt by the producers and top carnivores, although in opposite directions. This interactive effect depends on the values used

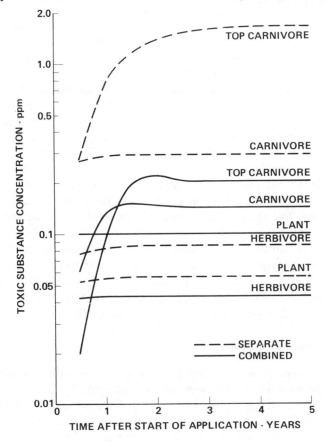

Figure 21. Effects of simultaneous toxic mortality at all levels of the ecosystem.

as characteristic concentration parameters for the various categories and points to the need for tested figures.

The increase in plant toxic substance concentration seems to be due to the combined effects of increases caused by the toxic mortality in herbivores, which overwhelms the reduction due to toxic mortality among plants. The increasing reduction seen up the food chain is understandable because each category perturbed causes a toxic decrease in all higher categories, with only a slight counteractive increase due to the decreased predation caused by the toxic die-off of the next higher category.

Combined Effects of Damage Functions—
Mortality, Feeding and Respiration

The combined effects of altered rates of mortality, feeding and respiration due to toxic stress were investigated by application of all three damage functions simultaneously to the carnivore trophic level. The altered rates caused an interaction which resulted in a decrease in toxic substance concentration beyond that caused by the application of only lethal toxic mortality (Figure 22). When the respiratory and feeding damage functions are applied in combination with a lethal toxic mortality function, they enhance the depression of toxic substance concentration resulting when the mortality function is applied alone. This can be understood by the fact that these two functions do not cause as large an effect as the lethal toxic mortality function, and the increase in toxic substance concentration which they produce would serve only to enhance the toxic mortality loss (biomass and toxic) which would overcompensate for the slight toxic increase.

Hence, the effect on toxic substance concentration caused by this sublethal damage to respiration and feeding rates is reversed when combined with the more dramatic effect caused by lethal toxic mortality. Thus, it can be seen that the exact nature of such processes must be known relative to each other to predict what the resultant interactive effect may be.

Simulation of Second Ecosystem

As a further investigation of system simulations a model for a second system was designed following the principles of the first.

The system simulated was Cone Spring,[31] a cold spring in Iowa with relatively constant physical and chemical conditions. It differs from Silver Springs in that its food web builds on the base of detritus rather than primary producers. The general flow of energy is: primary producers, detritus, decomposers, detrital feeders and carnivores. Another important difference in this system is that microscopic forms usually lumped with

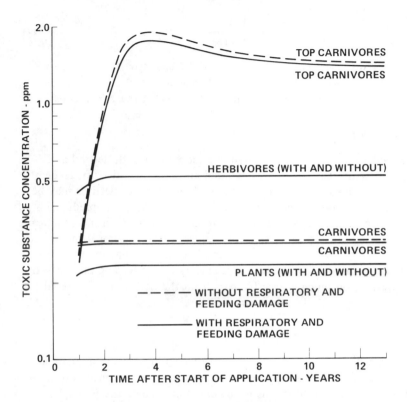

Figure 22. The combined effects of mortality, feeding and respiration toxic stresses applied to the carnivore level.

decomposers have been identified as primary consumers. Also, it represents a branched food chain as detritus feeders consume both detritus and bacteria. In addition, export plays quite an important role in this system.

This system was represented by a five-compartment model with a direct forcing flux into both the plant and detrital compartments. The feeding terms and division between linear and nonlinear mortality were constructed as in the previous simulation. The main structural difference lies in the fact that mortality terms from the various equations result in flows to the detrital rather than to the bacterial compartment. Also, the input to the bacterial compartment is from the detrital compartment. The structure of the detrital feeders and carnivore equations is basically the same as that of the carnivores and top cornivores in Silver Springs. Process rates were determined as in the previous simulation, where the rate constants were determined by dividing each energy flux by the equilibrium energy equivalent (or their products) appropriate to the levels involved. For example,

$$\tau_{12} = \left| \frac{F_{12}}{x_1 x_2} \right| \qquad (7)$$

where τ_{12} = the feeding rate of trophic level 2 upon level 1
F_{12} = the energy flux from level 1 to level 2
x_1, x_2 = the biomass-energy equivalents of levels 1 and 2, respectively

Biomass Results

The model predicted a smaller standing biomass in both detrital and bacterial compartments than measured and an increase in detritus feeders and carnivores, possibly because the energy input term to detritus may be underestimated as postulated earlier in this chapter. Also, the predicted standing crop of carnivores and detritus feeders is larger than measured. This may be due to too low respiratory values for these compartments, also a possibility cited in the chapter. Another explanation for these differences in standing crop may be a discrepancy in the feeding rates which would transfer too much energy/biomass from the detrital and bacterial categories to the detrital feeders and carnivores.

Toxic Concentrations

To simulate the flow of toxins in the Cone Spring ecosystem, equations were designed to reflect the change in toxic substance concentration in each trophic level. The equations were constructed along the principles used in the Silver Springs simulation, that is, by constructing an equation for the rate $d(xy)/dt$ for each trophic level and solving for dy/dt. The simulation did show a toxic buildup along the food chain (Figure 23), with the concentration in the carnivores approximately 1.6 ppm after 5 yr, in detrital feeders and bacteria 0.4 and 0.36 ppm, respectively, in detritus 0.09, and in plants 0.0012, which appear to be reasonable values.

This simulation was achieved with a forcing flux of 1.118×10^{-4} and initial concentrations of 1.0×10^{-10}.

However, in this run the toxic steady-state values were not reached after 5 yr (except in plants which come almost automatically into equilibrium) but were still increasing in the detrital level and also the trophic levels higher up the food chain which relied on the detritus after 20 yr.

Some Sensitivity Aspects of the Model

Mortality

Mortality in the model was represented by a nonlinear and a linear term, partitioned to have equal effect at equilibrium. Mortality rates were varied

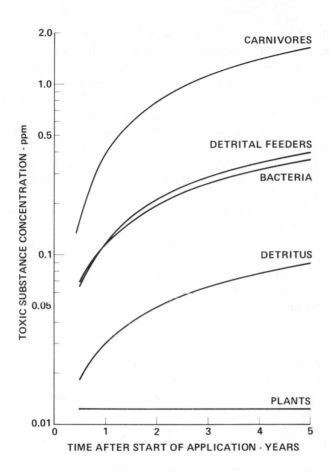

Figure 23. Toxic substance concentrations in the Cone Spring ecosystem.

to see relative influence on the model. An increase in plant linear mortality by a factor of 10 reduced both biomass and toxic substance concentrations in all trophic levels and shortened the response time for equilibrium to be reached. (The biomass was 67% of previous value in plants, herbivores and carnivores, and the toxic concentration was 50%, 70%, 70% and 70% of previous values up the chain). The reduction became more extreme when the nonlinear competitive mortality was also increased by a factor of 10, but only by a small percentage loss in each category. Increase of mortality rate coefficients by a factor of 10 in herbivores, however, resulted in a slight increase in plant and decomposer biomass and concentration, and reduction of both in the other trophic levels. The same pattern was seen in increased carnivore and top carnivore mortality as well, with decreased toxic substance concentration in the trophic levels above the one with increased mortality and possibly increased toxic substance concentration in the one below. Lethal toxic mortality on top carnivores did not show effects on lower levels.

Comparative Effects of Linear and Competitive Mortality. Generally, (if no other parameters were perturbed) a tenfold increase in linear mortality caused about a 1.5 times greater die-back of standing crop than the increase in competitive mortality. The combined effects of increasing both competitive and linear mortality caused only a small percentage loss in excess of linear mortality alone. This occurred because with a decreasing biomass, the competitive nonlinear effect decreases with biomass faster than the linear representation.

Comparison of Increased Mortality Rate Constants in Plants and Herbivores. While increased mortality of plants decreases the standing crop and toxic concentration in plants, increased mortality on herbivores increases it. Mortality, in this system simulation, is relatively more important in herbivores, as a factor of 10 increase results in a 52% change in plant biomass, while the same magnitude change in herbivores results in an 86% change. Thus the relative importance of proper rate constants may be seen.

Respiratory Effects

Whereas mortality has a tendency to decrease toxic substance concentration in a given compartment, increased respiration tends to increase it. For example, the respiration rate of the carnivores was doubled and resulted in a 2.3 increase in toxic substance concentration.

Feeding Effects

The effects of varying several of the feeding parameters were investigated as to subsequent effects on biomass and toxic substance concentration. When the feeding rate of herbivores was reduced, it appeared that a 50% reduction led to about a 10% increase in plant biomass, while a 90% reduction led to a 14% increase, seeming to indicate that plant biomass was not very sensitive to feeding rate, but was kept in check by it.

However, such feeding reductions caused a large drop in herbivore biomass, to 45% and 19% normal in the respective cases. This biomass reduction was propagated up the chain, affecting both carnivore and top carnivores, although to a lesser extent. In the case of top carnivores, this effect was combined with a lethal toxic mortality on the carnivore level which enhanced the biomass decrease in the 50% feeding case. However, the enhancement was not apparent in the 10% case. This reduction in feeding rate at lower levels can be quite an important effect through the food chain, since a 50%

decrease at the herbivore level can reduce carnivore and top carnivore biomass by at least 30%.

A 50% reduction in the rate of carnivores feeding herbivores will have a quantitative impact on the carnivores and the next lower trophic level that is simular to the effect noted with a decreased herbivore feeding rate, causing a 62% reduction in carnivore biomass and a 12% increase in herbivore (vs 55% and 10% in herbivores and plants). This also seems to indicate that the herbivore population is kept in check by the carnivore feeding rate, although not being extremely sensitive to it. The deduction in rate of carnivores feeding on herbivores has an alternating effect down the chain, (with reduced carnivore biomass, increased herbivore) leading to decreased plant biomass. The reduction in biomass is propagated up the chain, however, and seems enhanced in top cairnvores by the existence of lethal toxic mortality. The addition of a decreased branched feeding rate (50%) of carnivores eating plants tends to increase biomass in both plant and herbivore compartments.

The toxic effects of feeding rate reduction are perhaps more difficult to generalize. Toxic concentrations resulting from perturbation of feeding rate vary according to the interaction of several factors. A reduction in the feeding rate term may decrease some input toxins, while the resultant biomass loss may act to decrease some toxic loss term (such as competition and predation by the next level), tending to increase concentrations. Since the biomass value is represented as a factor in the denominator of several terms in the equaiton, whether these terms represent a gain or a loss depends on the relative value of feeding and growth rate changes vs biomass changes.

The magnitude of the herbivore feeding rate greatly influences the toxic substance concentration in the various components of the food chain. When the rate is reduced to 75% normal, there is an increase of toxic substance concentration in all compartments (by 8, 5, 7 and 14% in plant, herbivores, carnivores, and top carnivores, respectively). However, a 50% decrease causes a concentration decrease of 2% in the herbivores but increases in plants, carnivores and top carnivores (15, 5 and 3%, respectively). With 25% and 10% normal rates, there are increased concentrations in plants and decreased concentrations (less reduction up chain and less with 25 vs 10%) in the other trophic levels (see Table I).

Chain branching also seems to have an effect on patterns of toxic buildup. If the rate of carnivores eating herbivores is reduced by 50%, then there is a slight loss of toxic concentration in all categories, in no case greater than 6%; but if the rate of carnivores eating plants is reduced by 50%, there is an increase in toxic concentration in all categories (except decomposers, no change in plants). And if both rates are reduced simultaneously, there is an even greater buildup in herbivores, carnivores and top carnivores (11%, 20% and 11%, respectively)

Table I. Concentration of Toxic Substance in Perturbed Case Expressed as a Percentage of Concentrations in Baseline Case

	Percent Baseline Feeding Rate Constant for Herbivores Eating Plants			
	75%	50%	25%	10%
Plants	108	115	120	123
Herbivores	105	98	63	27
Carnivores	107	105	78	50
Top Carnivores	114	103	89	76

Chain branching also seems to have an effect on patterns of toxic buildup. If the rate of carnivores eating herbivores is reduced by 50%, then there is a slight loss of toxic concentration in all categories, in no case greater than 6%; but if the rate of carnivores eating plants is reduced by 50%, there is an increase in toxic concentration in all categories (except decomposers, no change in plants). And if both rates are reduced simultaneously, there is an even greater buildup in herbivores, carnivores and top carnivores (11%, 20% and 11%, respectively).

POSSIBLE REFINEMENTS TO MODEL

To improve realism in the description of biological phenomena, certain refinements could be included in the model. However, further model complexity requires additional data, much of which is not presently available. Investigation of such features for inclusion has necessitated a survey of both possible forms of representation and the availability of data for operation and validation.

Possible refinements to the model include the incorporation of:

- trophic level subcategories based on physiological or functional distinctions, such as weight, age, developmental state, sex, feeding type or labor category;
- seasonality and random environmental fluctuations in forcing functions Foi;
- the consideration of different toxic compounds separately rather than as a lumped group;
- representation of antagonistic and synergistic effects of toxins;
- the introduction of direct absorption to trophic levels other than the first; and
- representation of toxic transformation products and their flow (as DDD and DDE).

Trophic Level Subcategories Based on
Physiological or Functional Distinctions

The difficulty of compartment models is that much of the heterogeneity within the compartment must be reduced to an average structure or function such as average rates. However, such rates will vary with physiological states, such as age, weight, sex, developmental state, feeding type or labor category, and their value will determine the pattern of energy flow. The subdivision of a compartment into different physiological states would allow for a more precise set of rate functions for each subcompartment and greater potential realism. Also, the impact of toxic effects could be more precisely pinpointed.

Studies of the uptake of cadmium by snails[32] showed that immature snails were three times more sensitive to cadmium than mature snails. (Chronic exposure to concentrations even lower than 0.41 ppm cadmium would produce mortality in juvenile *P. gyrina* while for mature snails the concentration required was 1.0 ppm.) Some surveys of DDT contamination seem to indicate a strong correlation between DDT residues and age, and many workers have found sexual distinctions in DDT accumulation, with males accumulating larger concentrations than females.[33] The effects of creating subcompartments would reflect toxic effects on a system for which a specific subcategory would be a more susceptible target because of its distinctive set of rates or sensitivity.

It seems that the most realistic and effective subcategory division lines would be along a weight and/or age relationship, since the importance of these functions has been mentioned by a number of investigators in regard to both biological function rates and toxic sensitivity studies. Such a refinement could be incorporated in the model by replacing the equation for dx_i/dt with a set of equations for dx_{ik}/dt where the k's represent different subcategories, such as weight classes, or geographic locations as designated by a distribution function. Each equation governing dx_{ik}/dt would have its own distinctive set of rate constants. Counterpart equations governing dy_{ik}/dt would be constructed.

This additional structure would incorporate more biological realism into the model but would also require more data. At the present time, there are some data on rates for different weight or age classes, especially in fisheries publications; however, the data tend to exist for only one compartment, say herbivores. There seem to be little or no information on entire ecosystems which provide such specific rate data. This is an area where the need for data to expand the model is quite important.

Assumptions could be made for a particular ecosystem as to how the rates would vary from major comparments to subcompartments according to other studies on the particular group of organisms in question. However, assumptions would also have to be made as to the particular age/weight distribution function to be used as well, if the measurements were not available.

Seasonality as a Possible Contribution to Cyclic Equilibrium

Many biological functions, such as reproduction, which affect the standing biomass of a trophic level, are functions of seasonal variation. This variation contributes to an equilibrium which is not static due to constant population levels but varies with cyclic population oscillations. Although population dynamics are difficult to incorporate in our model, it is theoretically possible to incorporate a periodic time dependence for a whole trophic level in the form of a deterministic seasonality. Such a variation of each trophic level would give stable cycle equilibria. May[34] has suggested a possible representation such as

$$\frac{dN(t)}{dt} = r(t)N(t)$$

where $N(t)$ is population size or biomass-energy equivalent of a trophic level, and $r(t)$ is a periodic function, with period T having value 0 when averaged over a complete life cycle. The $r(t)$ function is subject to the condition that

$$\int_o^T r(t)dt = 0$$

Such a function applied to the complete trophic level, however, would indicate the cumulative effect of all component populations, which may not average out to an apparent seasonal variation. Seasonal data on standing crop in ecosystem studies are necessary to determine an appropriate seasonality variance function.

Perhaps a more realistic approach would be to bias the different process rates, such as respiration and feeding, or the input and output terms, such as photosynthetic production and export, with a seasonality function effect. In Tilly's study of Cone Spring,[31] annual variation of organic export and detritus standing crop were measured, the seasonal fluctuation of which had a large effect on the energy flow of the system and should be represented in a simulation.

Random Environmental Fluctuation

Perturbations to an ecosystem, such as a scarcity of food which causes reduction of a population, have often been attributed to fluctuations in environmental factors. As this may represent a stress to a system which may enhance or alter toxic stress, it is important that possible effects of environmental fluctuations be investigated. Such random fluctuations may act to destroy the repetitive determinancy of a model, giving instead stochastic-type fluctuations. Such an effect may be included by allowing one or more parameters to have an element of random fluctuation. This could be achieved by changing a rate parameter R

$$R = R_o + R(T)$$

where R(T) is randomly fluctuating white noise.[34] Such an environmental variation might be applied to the mortality, respiration or feeding rates, or to primary productivity or system export.

Consideration of Separate Toxic Categories Rather Than a Lumped Group

Different contaminants, with their differing physicochemical properties, will have different patterns and rates of movement and effect. Due to the large number of contaminants, the most feasible method of categorizing would be to group the chemicals according to their similarity to certain benchmark compounds. Each group applicable to the species simulation might then be represented by a subcategory equation for the rate dx_{ik}/dt, where each k represents a different contaminant group. Similar relationships for dy_{ik}/dt would be derived and applied.

Reflection of Synergistic and Antagonistic Effects of Toxins

The synergistic and antagonistic effects of toxins may be reflected by an increase in apparent concentration. The interactive effects can be represented by a rate of mortality that depends on the concentrations of more than one substance. The functional form of the expression should be derived from observations in order to maintain realism.

Introduction of Direct Toxic Absorption to Higher Trophic Levels

Studies have revealed that bioaccumulation occurs not only through the food chain, but also by direct absorption through gills, lungs and cuticle. This absorption is especially apparent in aquatic ecosystems and has been shown to contribute to bioaccumulation in fish. The respiratory intake of trace metals in man has been seen to play perhaps a more important role than ingested intake.[35] Thus, it is important to include a direct input gain term to all trophic categories. The magnitude of these terms will be dependent on the specific situation investigated. They will enter as forcing functions denoted by F_{ij}''

Representation of Toxic Transformation Products
and Their Flow as DDT, DDD and DDE

As the transformation of contaminants may occur in the biota or a-biota, and have persistent or toxic effects quite different from those of the parent material, it would be important for a model to reflect both the rate of transformations and the flow of major transformation products. This could perhaps be done by a set of subcategory-coupled equations governing various species for the rates dx_{ik}/dt with a loss from one equation feeding into a gain in the next through source and sink terms that have the same net effects as the forcing functions.

Data Difficulties

In order to refine the model, data to reflect the additional information are needed. While information exists for some phenomena, as for physiological differences in rates, it is often only measured for a particular phenomenon or a particular species. The difficulty of obtaining data on a whole system which includes such measurements is immense. Also, data on stressed systems, both qualitative and quantitative are sorely lacking. It is hoped that further measurement programs with lab microcosm and field experiments will provide needed information. The modeling exercises provide a direct means for highlighting specific data requirements.

ACKNOWLEDGMENTS

This material is based upon research supported by the National Science Foundation under Grant No. ENV76-80329. Any opinions, findings and conclusions or recommendations expressed in this publication are those of the authors and do not necessarily reflect the views of the National Science Foundation.

We wish to thank a number of people for their help on various aspects of the program described in this report. Specifically, we wish to thank Dr. Marvin Stephenson and Carter Schuth of the National Science Foundation and Dr. Rizwanul Haque of the Environmental Protection Agency for general guidance in carrying out this program; Mr. Lambert Joel of the National Bureau of Standards for helpful discussions on data sources for model development; Mr. Ken Marshall of the Environmental Studies Department of the University of California, Santa Barbara, for help in carrying out a literature search on DDT; Mr. Harry Bowie for assistance in drawing some of the diagrams; and Ms. Kathy Wilkowski and Mrs. Catherine Ferranti for painstakingly typing the manuscript.

REFERENCES

1. Randers, J., and D. Meadows. "System Simulation to Test Environmental Policy: A Sample Study of DDT Movement in the Environment," *Toward Global Equilibrium - Collected Papers*, D. Meadows, Ed. (Cambridge: Wright-Allen Press, 1971).
2. Woodwell, G., P. Craig and J. Johnson. "DDT in the Biosphere: Where Does it Go?" *Science* 174:1101 (1971).
3. Cramer, J. "Model of the Circulation of DDT on Earth " *Atmos. Environ.* 7:241 (1972).
4. Bidleman, T.F., and C.E. Olney. "Chlorinated Hydrocarbons in the Sargasso Sea Atmosphere at Surface Water," *Science* 183:516 (1974).
5. Edwards, C. A. *Persistent Pesticides in the Environment*, 2nd ed. (Cleveland: CRC Press, 1973).
6. Klein, W. "Environmental Pollition by Pesticides," *Adv. in Environ. Sci. Technol.* 6:65 (1976).
7. Risebrough, R.W., R. J. Huggett, J. J. Griffin and E. D. Goldberg. *Science* 159:1233 (1968).
8. Seba, D. B., and J. M. Prospero. *Atmos. Environ.* 5:1043, (1971).
9. Prospero, J. M., and D. B. Seba. *Atmos. Environ.* 6:363 (1972).
10. Orgill, M. M., G. A. Schmel and M. R. Peterson. "Some Initial Measurements of Airborne DDT on Pacific Northwest Forests," *Atmos. Environ.* 10:827 (1976).
11. O'Brien, R. D. *Environmental Dynamics of Pesticides*, R. Haque and V. M. Freed, Eds (New York: Plenum Publishing Corporation, 1975).
12. Lu, P. Y., and R. L. Metcalf, *Environ. Health Perspect.* 10:269 (1975).
13. Briggs, G. G. "A Simple Relationship Between Soil Adsorption of Organic Chemicals and Their Octanol/Water Partition Coefficients," Proceedings of the 7th British Insecticide and Fungicide Conference, 1973.
14. Kenaga, E. E. *Res. Rev.* 44:73 (1972).
15. Davies, J. E., A. Barquet. V. J. Freed, R. Haque, C. Morgade, R. E. Sonneborn and C. Vaclevek. *Arch Environ. Health* 30:608 (1975).
16. Metcalf, R. L., I. P. Kapoor, P. Y. Lu, C. K. Schuth and P. Sherman. *Environ. Health Perspect.* 4:35 (1973).
17. Metcalf, R. L., J. B. Sanborn, P. Y. Lu and D. Nye. *Arch. Environ. Contam. Toxicol.* 3:151 (1975).
18. Neely, W. B., D. R. Branson and G. E. Blau. *Environ. Sci. Techol.* 8:(13):1113 (1974).
19. Chiou, C. T., V. H. Freed, D. W. Schmeddling and R. L. Kohnart. "Partition Coefficient and Bioaccumulation of Selected Organic Chemicals," *Environ. Sci. Technol.* 2:475 (1977).
20. Metcalf, R. J. "Organochlorine Insecticides, Survey and Prospects," *Adv. Environ. Sci. Technol.* 6:223 (1976).
21. Haque, R., and V. H. Freed. "Behavior Pesticides in the Environment· Environmental Chemodynamics," *Residue Reviews*, F. A. Gunther, Ed. (New York: Springer-Verlag, 1974).

22. SCEP. *Man's Impact on the Global Environment: Report of the Study of Critical Environmental Problems* (Cambridge: MIT Press, 1970).

23. Harrison, J., O. Loucks, J. Mitchell, D. Parkhurst, C. Tracy, D. Watts and V. Yannacone. "Systems Studies of DDT Transport," *Science* 170: 503 (1970).

24. Gillett, J., J. Hill, IV, A. Jarvinen and W. Schoor. "A Conceptual Model for the Movement of Pesticides through the Environment," U. S. Environmental Protection Agency, Ecological Research Series, EPA-660/3-74-024 (1974).

25. Bolin, B. "Transfer Processes and Time Scales in Biochemical Cycles," *Ecol. Bull.* 22:17 (1976).

26. Gear, C. "Algorithm 407: DIFSUB for Solution of Ordinary Differential Equations [D2]," *Commun. ACM,* 14:185 (1971).

27. Hindmarsh, A., and G. Byrne. "EPISODE: An Experimental Package for the Integration of Systems of Ordinary Differential Equations," Lawrence Livermore Laboratory Computer Documentation UCID-30112, (1975).

28. Odum, H. T. "Trophic Structure and Productivity of Silver Springs, Florida," *Ecol. Monog.* 27:55 (1957).

29. Patten, B. C. In: *Systems Analysis and Simulation in Ecology, Vol. I,* B. C. Patten, Ed. (New York: Academic Press, Inc., 1971).

30. Kelly, R. A. In: *Conceptual Model of the Delaware Estuary, System Analysis and Simulation in Ecology, Vol. IV,* B. C. Patten, Ed. (New York: Academic Press, Inc., 1976).

31. Tilly, L. J. "The Structure and Dynamics of Cone Spring," *Ecol. Monog.* 38:169 (1968).

32. Wier, C. F., and W. M. Walter. "Toxicity of Cadmium in the Freshwater Snail, *Physa gyrina (Say),*" *J. Env. Qual., Vol. IV* (1976).

33. Hayes, W. J., Jr., G. E. Quinby, K. C. Walker, J. W. Elliot and W. M. Upholt. "Storage of DDT and DDE in People with Different Degrees of Exposure to DDT," *Arch. Ind. Health.* 18:398 (1958).

34. May, R. M. *Stability and Complexity in Model Ecosystems* (Princeton, NJ: Princeton University Press, 1973).

35. Bryce-Smith, D. "Lead Pollution—A Growing Hazard to Public Health," *Chem. Brit.* pp. 54-56 (1971).

23

PHARMACOKINETICS OF
ENVIRONMENTAL CONTAMINANTS

Peter M. Bungay and Robert L. Dedrick

Biomedical Engineering and Instrumentation
Branch
National Institutes of Health
Bethesda, Maryland 20014

H. B. Matthews

Laboratory of Pharmacokinetics
National Institute of Environmental
Health Sciences
Research Triangle Park, North
Carolina 27709

INTRODUCTION

Any effective organization of data on chemical toxicity in biological systems should include schemes that permit the comparison, extrapolation and prediction of toxic effects in different animal species and in humans. The demonstrated utility of physiologically based mathematical models for quantitatively rationalizing interspecies differences in pharmacokinetic behavior suggests that such models could enhance our capability for chemical hazard assessment.

To clarify the context for the models, it is helpful to keep in mind the distinction between pharmacokinetics and pharmacodynamics. Pharmacokinetics concerns the time course of the acquisition, distribution, chemical conversion and elimination of a substance and its coreactants and metabolic products. The variation over time in the therapeutic and toxic effects of the substance, *i.e.*, the pharmacodynamics, are plausibly linked to the pharmacokinetics, although the association would have to be established. A common supposition is that the pharmacodynamics are related to local concentration

369

levels of the chemical or its metabolites within or on the body, and not the amount *per se* to which the whole body is exposed. A corollary supposition is that the temporal variations in the therapeutic and toxic effects are associated with the reversibility or irreversibility of interactions between the chemical or metabolites and other substances in the local milieu. In accordance with these suppositions, it would be desirable in pursuing an understanding of pharmacodynamics to possess descriptions of the local pharmacokinetics. This chapter reviews one approach to obtaining the descriptions in the form of mathematical models which can simulate profiles in time of the local concentrations throughout the body. These mathematical tools incorporate, as much as possible, known anatomical and physiological information, and hence are referred to as physiological pharmacokinetic models.

The adjective "physiological" also distinguishes this type of model from those of classical pharmacokinetics. The latter are abstract curve fits whose end products are values for rate constants. Since these parameters do not possess a model-independent identity, they are not susceptible to independent measurement, unlike the parameters used in physiologically based models.

PHYSIOLOGICAL MODELING CONCEPTS

The models are based on concepts employed in chemical reaction engineering. In formulating such models, a mass conservation equation is written for each important, anatomically recognizable subdivision, or "compartment", of the body. "Importance" relates to whether the compartment is a target organ (for a drug) or a site of toxicity, or whether it significantly influences the kinetics. A set of conservation equations is obtained for each of the chemical constituents of interest. The equations are interrelated by terms representing mass exchange between compartments or interconversion between constituents by chemical reaction.

Figure 1 shows a typical organ in schematic form suggesting how lumped parameter mass balances are derived. The organ has been arbitrarily subdivided for illustrative purposes into blood and tissue compartments. Each compartment is assumed to be well mixed. The concentration of the chemical in the blood within the compartment is thus taken to be the concentration in the outgoing blood, C_{B_o}. The rate at which the chemical is supplied to the organ convectively by the blood is the product of the blood flow rate, Q, and the incoming arterial concentration, C_{B_i}. The rate at which the chemical is carried away by the blood is similarly QC_{B_o}.

In general, the chemical also exchanges between the blood compartment and the tissue compartment. The exchange can be effected by various processes occurring in the wall of capillary blood vessels or in tissue cell membranes. The mechanisms may be complex and are frequently unknown. For

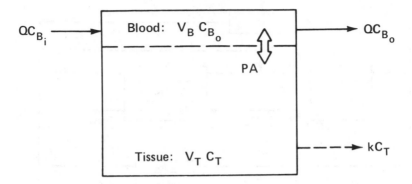

Figure 1. Schematic drawing of a typical organ subdivided into homogeneous compartments to permit generalized physiologically based pharmacokinetic analysis. Q=blood flowrate; PA=permeability-area product; k=metabolism or excretion rate constant; V_B=volume of organ blood compartment; V_T=volume of tissue compartment; C_B=concentration in incoming blood; C_{B_o}=concentration in outgoing blood.

modeling purposes it is usually satisfactory to describe the overall effect of such processes in simple mathematical terms. In Figure 1 the exchange between blood and tissue is assumed to be symmetrical, as indicated by the double-bar arrow. The symbol PA denotes the product of a permeability, P, and an exchange surface area, A. The rate of exchange is assumed to be linear in the difference in free concentration across the interface (with PA as the proportionality constant), although nonlinear rate expressions have been employed where appropriate.[1] Because of binding to blood or tissue constituents, the total concentration in a compartment is typically not equal to the free concentration. Total concentrations are usually easier to determine in experiments than free concentrations; hence, the models are usually formulated in terms of the former together with parameters, like equilibrium intercompartment distribution ratios, to take binding into account.

The conversion of the chemical agent through reaction and the processes for secretion or excretion from the tissue (or the body) are often not well understood. As in the case of compartmental exchange processes, simple operational kinetic expressions are frequently adequate for representing rates of conversion and excretion. In Figure 1, tissue elimination is taken, for illustration, to be a process which is linear in the tissue concentration, with k being the first-order rate constant.

Figure 2 shows the flow diagram for a whole-body model that would be suitable for many lipophilic environmental agents. Fat, muscle and skin compartments are included since these are important storage organs for lipid-

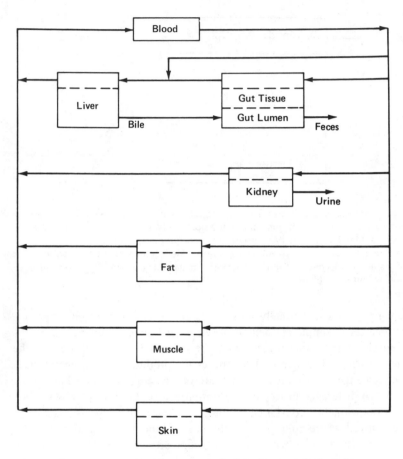

Figure 2. Schematic diagram for a typical whole-body physiological pharmacokinetic model.

soluble compounds. The liver, gut and kidney compartments appear because of their role in metabolism and elimination.

Any given model is a simplified representation of a very complex biological system which, insofar as practicable, incorporates directly verifiable physical and biochemical information: blood flow rates, organ sizes, constants representing chemical binding and other parameters generally independent of the chemical's route of entry into the body. By comparison of model simulations with experimental data, the magnitude of unknown kinetic parameters can be determined and hypotheses of unknown process mechanisms can be tested.

The distribution processes are usually qualitatively similar from one animal to another, at least for mammals. Much of the quantitative difference in pharmacokinetic behavior observed for the same agent in various animal species is accounted for by differences in purely physical, independently measurable parameters, such as flow rates, volumes and exchange areas. Such parameters have been found to scale with body weight to a power ranging approximately from 2/3 to 1; hence, rough pharmacokinetic estimates can be made before detailed experiments are performed. Once the distribution, metabolism and excretion processes for a number of animal species are satisfactorily described, the models provide a rational basis for interspecies comparisons and scale up predictions of the pharmacokinetics in man.

Few physiological pharmacokinetic models have been developed for environmental contaminants. Aspects of models for two agents—a hexachlorobiphenyl isomer and Kepone—will be discussed. Although both compounds are highly chlorinated hydrocarbons, they exhibit quite different distribution behavior.

HEXACHLOROBIPHENYL

The polychlorinated biphenyl (PCB) compounds are among the most widely dispersed and persistent environmental contaminants. The persistence to a large extent is dependent upon their reactivity in biological systems, which varies considerably among the compounds of this class. Pharmacokinetic models have been developed[2,3] for four of the PCBs in the rat. For illustrative purposes, only the model for 2,2',4,4',5,5'-hexachlorobiphenyl (6-CB) will be discussed here.

The 6-CB model predicts the pharmacokinetics of the parent compound and its principal metabolite in the rat—a glucuronide derivative. Since the metabolite is more polar than the parent, its distribution is quite different, as can be seen by comparison of the equilibrium tissue-to-blood distribution ratios for the two chemical species given in Table I. The very large distribution coefficient for the parent in adipose tissue is typical of many highly lipophilic substances. Fat is the storage site for most of the 6-CB in the body.

Table I. 6-CB Tissue-to-Blood Distribution Ratios[2,3]

Compartment	Parent	Metabolite
Adipose	400	2
Liver	12	4
Muscle	4	0.3
Skin	90	2

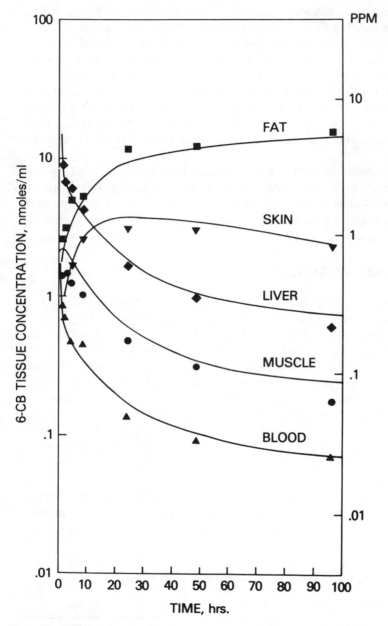

Figure 3. Variation in rat tissue concentrations of 6-CB during four days following intravenous administration of a single dose of 0.6 mg/kg body weight. Solid lines are generated by computer solution of pharmacokinetic model equations. Points represent experimental data.[2,3]

Since blood flow per unit volume of fat is very low compared with the blood perfusion rates of many other body compartments, the time scale of the distribution transients is very long. This can be seen in Figure 3, which shows the time course of 6-CB concentration in the major tissues following the intravenous administration of a single dose. The 6-CB is quickly taken up by the highly perfused organs, like the liver, and is then redistributed to the fat. Because of its low blood perfusion, the fat takes up the 6-CB so slowly that the fat concentration is still rising four days afterward.

The model incorporates a description of the kinetics of metabolism (which occurs in the liver) and the kinetics of elimination (which is predominantly by way of the feces. The respective metabolism and elimination coefficients were determined by matching the model simulations with experimental data. Comparison of the rate coefficients leads to the conclusion that the rate of elimination of 6-CB is controlled by the rate of metabolism.

Data have recently been collected on PCB disposition in the dog and the monkey in addition to previous data from the rat. Attempts will be made to generalize the models thus far developed to determine if they will permit prediction of the pharmacokinetics in man.

CHLORDECONE (KEPONE)

Like the hexachlorobiphenyl, Kepone (chemical formula $C_{10}Cl_{10}O$) is a highly chlorinated hydrocarbon; however, the disposition of Kepone in rats is markedly different from 6-CB. The key distribution coefficients are strikingly different as can be seen by comparison of the values reported in the literature[4,5] for Kepone (Table II) with those given for parent 6-CB in Table I. The coefficient for distribution into fat is quite low for Kepone in contrast to that for 6-CB and many other highly halogenated hydrocarbons. In addition, the liver is the most important storage organ for Kepone.

Table II. Kepone Tissue-to-Blood Distribution Ratios[a]

Compartment	Rat[4,5]	Man[6]
Adipose	8-31	4-12(6.7)[a]
Liver	28-126	5-31(15.0)[a]
Muscle	4-8	2-4.5(2.9)[a]
Bile	1-5	1-4(2.5)[a]

[a]Average values for man in parentheses.

Values reported by Cohn et al.[6] for Kepone distribution ratios in man have been included in Table II. For the tissues in which the distribution is likely to be dominated by thermodynamic partitioning, the ratios are similar for rat and man. Similarity in physical properties improves the prospects for using the pharmacokinetic models to extrapolate results from one species to another.

Kepone is eliminated from the body primarily in the feces. Kepone and its metabolites are excreted by the liver into the bile at a sufficient rate to account for its appearance in the feces. However, our data and that of others[5,6] indicate that there exist for Kepone, and probably other lipophilic agents, additional pathways into the gut lumen besides in the bile. The nonbiliary transport appears to aid in the increased rate of elimination of Kepone upon oral administration of adsorbents. The nonabsorbable ion exchange resin, cholestyramine, was fed to rats previously given a dose of Kepone.[5] The granular resin was also administered to Kepone-poisoning victims in a controlled clinical trial.[6] As an adsorbent, cholestyramine is not selective for Kepone. However, both in the animal experiments and in the clinical trial the rate of elimination from treated subjects was approximately double the rate measured in the absence of treatment.

In order to understand better the elimination process Kepone transport has been studied in the rat. An indispensable part of the study has been the development of a pharmacokinetic model in which the gastrointestinal tract is treated in greater detail than indicated by the schematic in Figure 2. By a combination of experiments and computer simulation, it was found that the exchange of Kepone between the blood and gut contents is consistent with a diffusion-limited, linear process. This process is characterized by permeability-area coefficients for exchange across the blood-gut tissue and gut tissue-gut lumen interfaces. The coefficients vary along the intestinal tract. The estimates obtained for these coefficients are sufficiently high to suggest[7] that the potential rate of elimination of Kepone may be well above the rates obtained in previous studies of adsorbent therapy.[5,6] Further studies need to be undertaken with other lipid soluble toxicants and more selective adsorbents to test the limits and generality of this therapeutic approach.

SUMMARY

Physiologically based pharmacokinetic models complement experimental investigation by (1) giving precise meaning to distribution and disposition pharmacokinetic parameters, (2) assisting in the design of experiments to determine unknown parameter values, and (3) providing a mechanism for comparing the pharmacokinetic behavior in different animal species and scaling up the results of animal studies to man. The models provide a mathematical structure into which can be embedded mathematical descriptions of the dynamics of toxicological action when such descriptions become available. Combined pharmacokinetic and pharmacodynamic models could prove to be a useful component of chemical hazard assessment methodologies.

ACKNOWLEDGMENT

The authors wish to express their appreciation to Mrs. Teresa Troy for assistance in the preparation of the manuscript.

REFERENCES

1. Dedrick, R. L., D. S. Zaharko and R. J. Lutz. "Transport and Binding of Methotrexate *in Vivo*," *J. Pharmaceut. Sci.* 62:882-890 (1973).
2. Lutz, R. J., R. L. Dedrick, H. B. Matthews, T. E. Eling and M. W. Anderson. "A Preliminary Pharmacokinetic Model for Several Chlorinated Biphenyls in the Rat," *Drug Metabol. Disp.* 5:386-396 (1977).
3. Anderson, M. W., T. E. Eling, R. J. Lutz, R. L. Dedrick and H. B. Matthews. "The Construction of a Pharmacokinetic Model for the Disposition of Polychlorinated Biphenyls in the Rat," *Clin. Pharmacol. Therapeutics* 22:765-773 (1977).
4. Egle, J. L., S. B. Fernandez, P. S. Guzelian and J. F. Borzelleca. "Distribution and Excretion of Chlordecone (Kepone) in the Rat," *Drug. Metabol. Disp.* 6:91-95 (1978).
5. Boylan, J. J., J. L. Egle and P. S. Guzelian. "Cholestyramine: Use as a New Therapeutic Approach for Chlordecone (Kepone) Poisoning," *Science* 199:893-895 (1978).
6. Cohn, W. J., J. J. Boylan, R. V. Blanke, M. W. Fariss, J. R. Howell and P. S. Guzelian. "Treatment of Chlordecone (Kepone) Toxicity with Cholestyramine. Results of a Controlled Clinical Trial," *New England J. Med.* 293:243-248 (1978).
7. Bungay, P. M., R. L. Dedrick and H. B. Matthews. "Pharmacokinetics of Halogenated Hydrocarbons," *Proc. N.Y. Acad. Sci.* 320:257-270 (1979).

24

BIOCONCENTRATION AND ELIMINATION OF SELECTED WATER POLLUTANTS BY BLUEGILL SUNFISH (*Lepomis macrochirus*)*

Michael E. Barrows, Sam R. Petrocelli and Kenneth J. Macek

E G & G
Bionomics Aquatic Toxicology Laboratory
Wareham, Massachusetts 02571

John J. Carroll

U. S. Environmental Protection Agency
Criteria and Standards Division
Criteria Branch
Washington, DC 20460

INTRODUCTION

The effect of chemicals in the aquatic environment has been a subject of growing concern to the populace as well as to industry, and to the layman as well as to the scientist. In an effort to understand the complexities involved with the interaction between chemicals and the physical and biological environment, a variety of testing methods and procedures have been developed to aid in the search for answers. One method which has proven useful is the investigation of the bioconcentration of chemicals in fishes. The ability of fish to bioconcentrate chemical residues in their tissues to levels above the concentration of the chemical in their aqueous environment has been clearly demonstrated.[1-5] Although the study of the phenomenon, of itself, does not provide an assessment of the potential hazard to

---

*Mention of trade names or commercial products does not constitute endorsement or recommendation.

the environment associated with the use of such chemicals, information on uptake from water and retention in tissues by fish can be a useful tool in assessing the relative propensity of a chemical to enter and persist in aquatic food chains, a potential source of human exposure. More recently, a comprehensive study was conducted to address the controversy as to whether bioaccumulation and/or biomagnification of residues through the food chain or bioconcentration of chemical residues from water through the gills of fishes is the major source of residues occurring in top piscine carnivores within aquatic environments.[6] The data presented clearly indicate that the potential for residue accumulation by fish through food chains is relatively insignificant (<10%) for most compounds when compared to the tissue residues resulting from the bioconcentration process, that is, by direct uptake from water. They further suggested that only those chemicals which are relatively persistent in fish tissues appear to have any potential for significant transfer through the food chains.

This chapter describes the results of investigations of the bioconcentration potential and persistence of 33 chemicals representing 14 chemical classes in a representative aquatic species, the bluegill sunfish (*Lepomis macrochirus*). Tests were conducted at the Aquatic Toxicology Laboratory of E G & G Bionomics, Wareham, Massachusetts. The chemicals tested are members of the group of 65 Consent Decree Chemicals for which the U.S. EPA is currently recommending water quality criteria.

MATERIALS AND METHODS

Chemicals

Of the 33 chemicals tested, 27 were carbon 14-labeled organic compounds which represented 13 chemicals or chemical classes, *i. e.,* acenapthene, acrolein, acrylonitrile, carbon tetrachloride, chloroform, chlorinated benzenes (6), chlorinated ethanes (5), chlorinated ethylenes (2), *bis*(2-chloroethyl) ether, isophorone, phenols (2), n-nitrosodiphenylamine and phthalates (4). The position of the radiolabeled carbon atom and molecular weight for each compound are presented in Table I. Each compound was received in individual, sealed vials from New England Nuclear, Boston, Massachusetts, and held under refrigerated conditions until used.

The six remaining compounds tested were metal salts, *i.e.,* antimony trioxide, arsenic trioxide, beryllium chloride, selenous acid, silver nitrate and thallous sulfate, received from the Ventron-Alpha Research Chemical Company, Danvers, Massachusetts. The characteristics of each compound, *i.e.,* molecular weight and the percentage of metal in each compound, are presented in Table II.

Table I. Characteristics of Organic Compounds Evaluated for Bioconcentration Potential

Compound	Position of Carbon-14 Label	Molecular Weight
Acenaphthene	$1,2\text{-}^{14}C$	154.21
Acrolein	$3\text{-}^{14}C$	56.06
Acrylonitrile	$1\text{-}^{14}C$	53.06
1,2-dichlorobenzene		147.01
1,3-dichlorobenzene		147.01
1,4-dichlorobenzene	$\text{ring-}^{14}C(U)$	147.01
1,2,4-trichlorobenzene		181.46
1,2,3,5-tetrachlorobenzene		215.90
Pentachlorobenzene		250.35
Carbon tetrachloride	^{14}C	153.84
Chloroform	^{14}C	119.39
1,2-dichloroethane	$1,2\text{-}^{14}C$	98.96
1,1,1-trichloroethane	$1\text{-}^{14}C$	133.42
1,1,2,2-tetrachloroethane		167.86
1,1,1,2,2-pentachloroethane	$1,2\text{-}^{14}C$	202.31
Hexachloroethane		236.74
Bis(2-chloroethyl)ether	$^{14}C(U)$	143.02
1,1,2-trichloroethylene		131.40
Tetrachloroethylene	$1,2\text{-}^{14}C$	165.85
Isophorone	$3\text{-}^{14}C$	138.23
N-nitrosodiphenylamine		74.08
2-chlorophenol		128.56
2,4-dimethylphenol		122.16
Dimethyl phthalate	$\text{ring-}^{14}C(U)$	194.19
Diethyl phthalate		222.26
Butylbenzyl phthalate		312.39
Di-2-ethylhexyl phthalate		390.54

Table II. Characteristics of Metals Evaluated for Bioconcentration Potential

Compound	Metal Salt	% Metal	Molecular Weight
Antimony	Antimony trioxide (Sb_2O_3)	83.5	291.50
Arsenic	Arsenic trioxide (As_2O_3)	75.7	197.82
Beryllium	Beryllium chloride ($BeCl_2$)	11.3	79.93
Selenium	Selenous acid (H_2SeO_3)	61.2	128.98
Silver	Silver nitrate ($AgNO_3$)	63.5	169.87
Thallium	Thallous sulfate (Tl_2SO_4)	81	504.85

Test Species

Four populations of bluegill sunfish *(Lepomis macrochirus)* were used during the entire program of studies. Three of these populations were obtained from a commercial fish farmer in Connecticut and had mean (± standard deviation, SD) wet weights ranging from 0.37 (± 0.18) g, to 0.94 (± 0.34) g and mean (± SD) standard lengths ranging from 25 (± 3) mm, to 32 (± 4) mm, respectively. One population of bluegill was obtained from a commercial fish farmer in Nebraska and had a mean (± SD) wet weight of 0.95 (± 0.36) g and a mean (± SD) standard length of 35 (± 4) mm. Thirty fish representative of each test population were weighed and measured for the calculation of means and standard deviations for each group.

All groups of fish were maintained in the holding facilities at Bionomics for a minimum of 30 days prior to use in any experiment. During that acclimation period, the cumulative mortality for each population was less than 3% and fish appeared to be in excellent physical condition prior to use.

Test Systems

Studies were conducted utilizing modifications of a continual-flow proportional dilution apparatus described by Mount and Brungs,[7] which provided for the automatic intermittent introduction of either a carbon 14-labeled organic compound or metal salt and diluent wellwater (pH of 7.1, total hardness of 35 mg/liter as calcium carbonate, dissolved oxygen >60% of saturation) into each test aquarium.

During these studies, two types of exposure systems were utilized. The predominant system (used in 87% of all tests) was a "closed" system comprised of 14 glass aquaria, each measuring 40 x 20 x 25 cm (length x width x height) and designed to hold each test solution at a depth of 19 cm (15 liters). Due to the nature of the compounds tested in this system, *i.e.,* volatiles, suspected carcinogens, mutagens and teratogens, test systems were maintained in an environmentally controlled hazardous substances room (HSR) which was designed to evacuate the air at a rate of of 10 room volumes per hour. Each of the test aquaria was in itself a closed system equipped with a sealed lid and a water trap at both influent and effluent streams. A chemical introduction system, consisting of an injector mechanism and a 50-ml gas-tight glass syringe, was positioned over each aquarium (except the control) in such a manner that with each cycling of the modified diluter, a calibrated volume of chemical stock solution was thoroughly mixed with 500 ml of diluent wellwater prior to the introduction into each test aquarium. The control aquaria were not equipped with chemical introduction systems and received only wellwater with each cycling of the diluter. The system was timed and calibrated to deliver 500 ml of Bionomics' wellwater (previously described) to each aquarium at a mean (± SD) cycle rate of 196 (± 20) times a day, which is equivalent to a turnover rate of 6-7 aquarium volumes per day, in order to maintain the dissolved oxygen concentration above 60% of saturation.

The experimentation with the metals, which represented a low workplace hazard was conducted in an "open" system consisting of 8 glass aquaria, each measuring 61 x 30 x 30 cm and designed to hold each test solution at a depth of 15 cm (30 liters). This system was maintained in our regular laboratory facility. The chemical introduction system for each aquarium (except controls) consisted of a 4-liter Mariotte stock bottle and a mechanical metering device which was positioned in such a way that with each cycling of the modified diluter, calibrated volume of stock solution was thoroughly mixed with 1 liter of diluent wellwater prior to the introduction into each test aquarium. This system was timed and calibrated to deliver 1 liter of Bionomics' wellwater to each aquarium at a mean (± SD) cycle rate of 158 (± 8) times a day, which is equivalent to a turnover rate of 5-6 aquarium volumes per day.

The levels of exposure were selected on the basis of acute toxicity data generated at Bionomics and were intended to be sublethal during the continuous exposure period. Sufficient carbon 14-labeled material [200-500 microcuries (μCi)] was synthesized for each compound so as to provide an adequate range of minimum detectable limits for fish and water samples which ultimately served as an aid in increasing method sensitivity. The stock solution for each ^{14}C compound was prepared by transferring the contents of the sealed vial or multiples thereof, depending on the nominal test concentration, to a laboratory flask and diluting with 200 ml of pesticide-grade acetone (Burdick and Jackson). The final volume of each stock solution was dependent on the amount of test material synthesized and the calibrated delivery volume of the chemical introduction system. Each metal salt was tested on a 100% active ingredient basis after correction for the proportion of the metal present in the salt, and a stock solution was prepared by dissolving an appropriate weight of the salt in 4 liters of glass-distilled deionized water. The concentrations of each metal stock solution were dependent on the nominal test concentration and the calibrated delivery volume of the mechanical metering apparatus.

Test Initiation

At the beginning of each study, 100 bluegill were placed into an aquarium of the appropriate test system and continuously exposed in a dynamic (flowing water) system to a sublethal concentration of either a carbon 14-labeled organic compound or a metal salt in aqueous solution. Representative water and fish samples were collected periodically during each test until apparent equilibrium between concentrations in fish tissue (whole body) and exposure water was observed or, alternatively, for a maximum exposure period of 28 days.

Following the observation of an apparent equilibrium condition or at the end of the 28-day exposure period, the remaining fish were transferred into an aquarium through which pollutant-free water flowed at a rate equivalent to that during exposure. The purpose of this depuration period was to determine the rate at which chemical residues were eliminated from fish tissues as an indication of the biological persistence and potential for trophic level transfer of each chemical. In order to evaluate the persistence of chemical residues in fish tissues, chemical analyses were performed on fish sampled during this elimination phase to determine the half-life of chemical residues in the tissues. Half-life was defined as the period of time required for one-half of the mean chemical residues measured in fish tissues at equilibrium or at the end of the exposure period to be eliminated following termination of exposure.

During each study, fish were fed a dry pelleted ration *ad libitum* 3 times a week and at least 24 hours prior to sampling. Fecal material was siphoned from the aquaria when deemed necessary. Mild aeration was provided to aquaria in the open system to maintain the dissolved oxygen (DO) concentration above 60% of saturation (5.6 mg/liter). Due to the restrictions applied to 14 C-compounds in the closed system and the high turnover rate of aerated wellwater, no aeration was used. An emergency air system was provided to each test unit and would operate only if there were a diluter malfunction. The water temperature in each aquarium was measured daily throughout the study period. The DO concentration and pH were measured periodically during the study using a YSI DO meter and combination oxygen-temperature probe and a Model #175 Instrumentation Laboratory pH meter and Ingold combination electrode.

Sample Schedule and Techniques

The specific activity of each carbon 14-labeled stock solution was measured radiometrically prior to test initiation. During each test exposure, representative water and fish samples, including samples from control aquaria, were collected on days 0, 1, 2, 4, 7, 10, 14, 21 and 28 (if exposure continued for the entire 28 days) and analyzed for ^{14}C-residues or metal concentrations as appropriate. Fish remaining in each aquarium at the end of the exposure period were transferred to pollutant-free wellwater for 7 days of depuration. During that period, fish were sampled on days 1, 2, 4 and 7 to estimate the half-life of chemical residues in tissues.

Triplicate 30-μl aliquots of each radio-labeled stock solution were analyzed prior to test initiation. Triplicate 5-ml water samples were taken from each radio-labeled compound aquarium and triplicate 100-ml water samples, preserved with 1% nitric acid, were taken from each metal test aquarium on all sample days during the exposure period. During each sampling interval (exposure and depuration), 5 fish were removed from each test aquarium, blotted dry, and analyzed either radiometrically or by atomic absorption on a whole-fish basis. Due to the nature of the ^{14}C-compounds tested, each fish was wrapped in preweighed Parafilm® at each sampling to minimize the volatilization of each chemical from the fish tissue during the combustion process.[8]

Radiometric Methods

Sample Preparation

In order to measure the ^{14}C-activity of stock solutions and water samples, the appropriate volumes of sample (30 μ or 5 ml) were transferred to glass scintillation vials containing 15 ml of Monophase® (Packard Instrument Company). All samples were placed in a Model 2002 Packard Tri-Carb Liquid Scintillation Spectrometer, and the activity of each sample was recorded as the number of counts per minute (cpm).

In order to quantitate the extent of ^{14}C-residue accumulation, whole fish (0.14-1.25 g, wet weight) wrapped in Parafilm were weighed, wrapped in a Kimwipe® and placed in the ignition basket for combustion in a Packard Model 306 Oxidizer. The resulting $^{14}CO_2$ was trapped as a carbonate in a mixture of Carbosorb® and scintillation fluid in a glass scintillation vial. Each vial was then placed in the liquid scintillation spectrometer and the cpm recorded.

Method Sensitivity and Precision

Recovery rates of the oxidizer were determined prior to analyzing each set of samples by combusting and counting the activity of a standard reference material (^{14}C-methyl methacrylate). Recovery rates were determined to be 99-100%, and experimental data were not adjusted for percentage recovery. Counting efficiencies of all samples were determined according to the channel ratio method as described by Kobayashi and Maudsley.[9] The counting efficiency for each water or fish sample was determined by comparing the sample channel ratio to that from a series of quenched standards prepared at E G & G Bionomics.

Background levels of radiation for water and fish were determined by analyzing control samples prior to the initiation of each study and were measured to be 20 cpm for water and fish tissue. All test samples were counted for a minimum of 100 min or until 5000 counts were attained. Using these criteria and the calculations described in *Standard Methods for the Examination of Water and Wastewater*,[10] it was determined at the 95% confidence level that a minimum net cpm for all samples of 20 cpm had an associated counting error of 7.9% This percentage was the maximum acceptable counting error and was associated with the minimum detectable limit. Counting error for each sample depended on the net cpm of that sample; the counting error decreased as the sample activity increased.

Atomic Absorption Spectrophotometric Methods

Water samples were collected in high-density polyethylene bottles which were cleansed and prepared for use according to U.S. EPA.[11] Concentrated Ultrex® nitric acid was added to each water sample (1% by volume) and the samples stored at room temperature prior to analysis.

Tissue samples were weighed, placed in 1:1 nitric acid-washed glass scintillation vials, and stored at 4°C prior to analysis. At the time of analysis, the tissue was transferred to micro-Kjeldahl flasks and acid digested using a mix of 2:1 (v/v) 30% hydrogen peroxide and nitric acid, except that for arsenic and thallium samples, perchloric acid was used rather than the nitric acid. The concentrated digest was filtered quantitatively into a volumetric flask and diluted with deionized-distilled water to 25.0 ml. The diluted digest was transferred to an acid washed glass scintillation vial prior to analysis.

Water samples and diluted tissue digests for beryllium, silver and thallium were analyzed by atomic absorption spectroscopy according to U.S. EPA.[11] Antimony, arsenic and selenium water samples and tissue digests were analyzed by automated hydride generation[12] and atomic absorption spectroscopy. The method was modified by using glass beads, which furnished a larger surface area in the stripping column. Also, heating the column was not found to be necessary. A 2% (w/w) sodium borohydride (pH 10 ± 0.5) solution was used because of the increased quantity of oxidizing reagents present in the sample solution.

All analyses were performed with a Perkin-Elmer Model 305A atomic absorption spectrophotometer equipped with a factory-installed deuterium-arc background corrector. Instrument response to light absorption by the metal of interest was measured with a Perkin-Elmer Model 56 recorder (0.5 mV full scale).

RESULTS

The mean (± standard deviation) water temperature measured in the test systems during the three-month testing period was 16 (±1) °C. The

DO concentration ranged from 5.9 to 8.6 mg/liter (57% to 89% saturation) and the pH ranged from 6.3 to 7.9.

The results of investigations of the bioconcentration of carbon 14-labeled organic compounds from water by bluegill sunfish indicated that for the majority of compounds tested the ^{14}C-residue in fish tissues, on a whole-body basis, reached equilibrium with respect to the aqueous ^{14}C-residue concentration within the initial 3-10 days of exposure. It was further observed that these steady-state equilibria were maintained throughout the remainder of each exposure period. In other words, by the end of the first week of exposure, the net rate of accumulation of ^{14}C-residues from the water into the fish tissue was equivalent to the net rate of elimination of these residues, and therefore a steady-state or equilibrium condition prevailed. Based on the results of these analyses, a steady-state bioconcentration factor (BCF) was calculated as the quotient of the mean chemical concentration measured in fish tissues during equilibrium divided by the mean measured chemical concentration in water during the entire exposure period (Table III).

When no steady-state condition (equilibrium) was observed during the limits of the exposure period, then a maximum BCF was calculated to indicate the apparent maximum accumulation potential of that compound during this study. Of the compounds tested, ^{14}C-acrylonitrile, ^{14}C-tetra-chlorobenzene, ^{14}C-N-nitrosodiphenylamine and the four metals (i.e., arsenic, beryllium, selenium and thallium) were the only ones which did not appear to reach steady state in fish tissue during the four-week exposure period (Table III). The maximum bioconcentration factor was calculated as the quotient of the maximum mean chemical concentration observed in fish tissue divided by the mean measured chemical concentration in water throughout the exposure period.

A summary of the BCF for 31 out of the 33 compounds tested is presented in Tables III and IV. Two compounds, antimony and silver, apparently did not bioconcentrate in bluegill tissue. No residues of these two metals were detected in tissues of exposed fish at concentrations greater than those concentrations measured in the tissues of the control fish. Results of analyses (radiometric or atomic absorption spectrophotometric) indicated that 61% or 19 out of the 31 compounds tested had BCFs of less than 100X. Some of the chemical classes which exhibited a low potential for bioconcentration were the metals with BCFs ranging from 4 to 34X, the dichlorobenzenes (60-89X), chlorinated ethanes (2-139X) and chlorinated ethylenes (17-49X). Of the 12 remaining compounds tested, only 2 exhibited a BCF greater than 1000X: tetrachlorobenzene (1800X) and pentachlorobenzene (3400X).

Results of the analyses of tissue samples collected during the depuration period indicate that the half-lives of most of the chemicals tested in fish

Table III. Accumulation and Persistence of Chemical Residues in Bluegill Sunfish Continuously Exposed to 31 "Priority Pollutants"

Compound	Duration of Exposure (days)	Mean Water Concentration (µg/l)	Bioconcen- tration[a] Factor (X)	Half-Life[b] in Tissues (days)
Acenaphthene	28	8.94 ± 2.13	387	<1
Acrolein	28	13.1 ± 2.64	344	>7
Acrylonitrile	28	9.94 ± 1.16	48[c]	>4 <7
1,2-dichlorobenzene	14	7.89 ± 1.20	89	<1
1,3-dichlorobenzene	14	107 ± 10.9	66	<1
1,4-dichlorobenzene	14	10.1 ± 0.75	60	<1
1,2,4-trichlorobenzene[d]	28	2.87 ± 1.08	182	>1 <3
1,2,3,5-tetrachlorobenzene	28	7.72 ± 0.59	1800[c]	>2 <4
Pentachlorobenzene	28	5.15 ± 2.20	3400	>7
Carbon tetrachloride	21	52.3 ± 12.7	30	<1
Chloroform	14	110 ± 6.6	6	<1
1,2-dichloroethane	14	95.6 ± 11.1	2	>1 <2
1,1,1-trichloroethane	28	73.4 ± 14.3	9	<1
1,1,2,2-tetrachloroethane	14	9.62 ± 1.09	8	<1
Pentachloroethane	14	7.93 ± 0.49	67	<1
Hexachloroethane	28	6.17 ± 1.95	139	<1
Bis(2-chloroethyl)ether	14	9.91 ± 0.43	11	>4 <7
1,1,2-trichloroethylene	14	8.23 ± 0.42	17	<1
Tetrachloroethylene	21	3.43 ± 1.53	49	<1
Isophorone	14	92.4 ± 10.5	7	1
N-nitrosodiphenylamine	14	9.21 ± 0.98	217[c]	<1
2-chlorophenol	28	9.18 ± 2.02	214	<1
2,4-dimethylphenol	28	10.2 ± 0.76	150	1
Dimethylphthalate	21	8.74 ± 1.90	57	>1 <2
Diethylphthalate	21	9.42 ± 2.89	117	>1 <2
Butylbenzylphthalate	21	9.73 ± 1.75	663	>1 <2
Di-2-ethylhexyl phthalate[d]	42	5.82 ± 0.90	114	3
Arsenic (as As_2O_3)	28	130 ± 20	4[c]	1
Beryllium (as $BeCl_2$)	28	270 ± 30	19[c]	<1
Selenium (as H_2SeO_3)	28	120 ± 10	20[c]	>1 <7
Thallium (as Tl_2SO_4)	28	80 ± 10	34[c]	>4

[a]Bioconcentration factor is the quotient of the mean measured residues of the compound in fish tissues (whole body) during the equilibrium period divided by the mean measured concentration of the compound in exposure water.

[b]Half-life of compound in tissues if the time in days required for the mean measured residue concentration in tissues to be reduced to half that which was measured during the equilibrium period in the uptake phase.

[c]Maximum bioconcentration factor.

[d]Compounds tested prior to EPA task order.

Table IV. Distribution of Bioconcentration Factors Calculated for 31 Chemicals in Studies in which Bluegill Sunfish were Continuously Exposed to Carbon 14-Labeled Organic Compounds or Metal Salts in Aqueous Solution

Range of Bioconcentration Factors (X)	Distribution of Bioconcentration Factors	
	Number	Percent of Total
< 50	14	45
> 50 < 100	5	16
> 100 < 500	9	29
> 500 < 1000	1	3
> 1000	2	6

tissue were extremely short. Approximately 58% of the compounds tested, including the metals, had a half-life of less than 1 day. That is, within 24 hours following transfer of previously exposed fish to clean water, a minimum of 50% of the chemical residues detected in the tissues of members of these populations had been eliminated. Of the remaining 13 compounds tested, 7 (22%) had a half-life of >1<4 days, 3 (10%) had a half-life of >4<7 days, and only 3 compounds appeared to persist in fish tissues longer than the 7-day limit of the depuration period.

DISCUSSION

In general, we observed three basic patterns of accumulation of chemical residues in the whole body of bluegill continuously exposed to a constant level of chemical in flowing water. In the first instance, the net rate of accumulation initially exceeded the net rate of elimination for a period of time, but eventually the rates became approximately equal and an equilibrium situation was established. It is assumed that unless some external change occurs to shift the equilibrium, the tissue residue concentration will remain constant throughout continued indefinite exposure. Of the 31 chemical compounds investigated during this study, the majority (74%) exhibited this pattern of accumulation.

The second basic pattern of accumulation was one in which initially the net rate of accumulation exceeded the net rate of elimination as in the first case. However, in this instance, the equilibrium period was very short-lived, and the net rate of elimination exceeded the net rate of accumulation, whereupon tissue residue declined despite continuous aqueous exposure.

The third pattern observed was charcterized by a situation in which the net rate of accumulation exceeded the net rate of elimination throughout the period of exposure, and the tissue concentration continued to rise at what appeared to be a linear rate.

Results of studies with the 31 different compounds revealed that the estimated bioconcentration factors were relatively low as compared with those of other compounds which have been tested by similar techniques. Macek[1] reported BCFs for compounds such as DDT, dieldrin, heptachlor, hexachlorobenzene and Aroclor 1254, ranging from 9000 to 37,000X. As is evident from the data presented in Tables III and IV, we have observed a relatively wide range of BCFs (*i.e.,* 2x-3400X) for the 31 "priority pollutants" tested, yet few of them exhibited a potential to bioconcentrate to the same order of magnitude as for many chemicals (including pesticides) which have proven to be significant environmental pollutants.

The data regarding the rate of elimination of chemical residues from the whole body of bluegill after transfer to pollutant-free flowing water clearly indicate that the biological half-lives of 90% of the chemicals tested in this study were less than 7 days. When compared with those compounds reviewed by Macek *et al.*[1] (*e.g.,* DDT, Aroclor 1254, etc.) which exhibited biological half-lives ranging from 30 to 160 days, it is evident that for the priority pollutants tested biological persistence is relatively short and the potential for trophic-level transfer is low provided the source of contamination is not continuous or the duration of exposure prolonged.

In this chapter, a number of general observations relating to the time to establish equilibrium, the BCF obtained and the estimated half-life have been discussed to evaluate the bioconcentration potential of each compound. An attempt has also been made to evaluate any possible correlation between chemical structure and bioconcentration potential.

Specific members of chemical classes exhibited bioconcentration potential which appeared to be directly related to the degree of chlorine substitution. That is, as the number of substituted chlorine atoms on the molecule increased, the bioconcentration potential, as indicated by the BCF, similarly increased. The degree to which the substitution of chlorine atoms on the molecule affected the bioconcentration potential apparently depended on the nature of the molecule. For example, the three dichlorinated benzenes tested had similar BCFs. BCFs for 1,2-, 1,3- and 1,4-dichlorobenzene were 89X, 66X and 60X, respectively. The BCF for 1,2,4-trichlorobenzene was 182X; for 1,2,3,5-tetrachlorobenzene was 1800X and for pentachlorobenzene was 3400X. Therefore, it appeared that adding an additional chlorine atom to dichlorobenzene, thus producing a trichlorobenzene, resulted in a two- to threefold increase in the bioconcentration potential; addition of another chlorine atom to trichlorobenzene resulted in a tenfold increase in bioconcentration potential, which the substitution of still another chlorine atom, producing a pentachlorobenzene, results in an additional doubling of the bioconcentration potential from that of the tetrachlorobenzene.

The bioconcentration potential of the chlorinated ethanes appeared to be unaffected by the degree of substitution on the molecule up to and including tetrachloroethane. However, the addition of the fifth chlorine atom to the molecule resulted in a BCF for pentachloroethane of 7-33 times greater than those of the less substituted chloroethanes. Furthermore, addition of another chlorine atom results in hexachloroethane, with a doubling of the BCF as compared with pentachloroethane.

In comparing the overall bioconcentration potential of the major categories of compounds tested (*i.e.,* organic aromatics, organic aliphatics and metals), a basic correlation can be made. In general, the data indicate that the organic aromatic compounds were bioconcentrated to a greater extent (27-3400X) than were the organic aliphatic compounds (2-387X) and the metals were bioconcentrated the least (4-34X).

Results of the studies of uptake and elimination of the 31 "priority pollutants" by bluegill sunfish indicated that the majority of compounds tested exhibit a relatively low bioconcentration potential as compared with some of the other chemicals for which this evaluation has been made. Therefore, if bioconcentration factors are relatively low ($\leq$1000X), time to equilibrium is relatively short and the biological half-lives are relatively short ($\leq$7 days), it would be anticipated that these compounds would not be concentrated or retained to any great degree by aquatic organisms.

ACKNOWLEDGMENTS

We would like to express our gratitude and sincere appreciation to Kenneth Buxton and Gary Wilson for the analytical support necessary during a comprehensive study such as this and especially Scott Ladd and John Grygalonis, the very competent biologists who helped perform the investigations of the kinetics of these chemicals in a representative freshwater species of fish.

These studies were performed under U.S. Environmental Protection Agency Contract #68-01-4646.

REFERENCES

1. Macek, K.J., M.E. Barrows, R. F. Krasny and B.H. Sleight, III. "Bioconcentration of 14C-Pesticides by Bluegill Sunfish During Continuous Aqueous Exposure." Symposium on Structure Activity Correlation in Studies of Toxicity and Bioaccumulation with Aquatic Organisms, Canada (1975), pp. 119-141.
2. Macek, K.J., and S. Korn. "Significance of the Food Chain in DDT Accumulation in Fish," *J. Fish. Res. Bd. Can.* 27(8). 1496-1498 (1970).
3. Hansen, D.J., P.R. Parrish, J.I. Lowe, A.S. Wilson, Jr. and P.D. Wil-

son. "Chronic Toxicity, Uptake, and Retention of Aroclor 1254 in Two Estuarine Fishes," *Bull. Environ. Contam. Toxicol.* 6(2): 113-119 (1971).

4. Parrish, P.R., G.H. Cook and J.M. Patrick. "Hexachlorobenzene: Effects on Several Estuarine Animals," Proc. of 28th Annual Conference of the Southeast Assoc. Game & Fish Commissioners (1975), pp. 179-187.

5. Reinert, R.E., L.J. Stone and W.A. Willford. "Effect of Temperature on Accumulation of Methyl Mercuric Chloride and p,p' DDT by Rainbow Trout *(Salmo gairdneri),*" *J. Fish. Res. Bd. Can.* 31(10): 1649-1652 (1974).

6. Macek, K.J., S.R. Petrocelli and B.H. Sleight, III. "Considerations in Assessing the Potential for, and Significance of Biomagnification of Chemical Residues in Aquatic Food Chains," Proc. of the 2nd Annual Symp. on Aq. Tox. ASTM SP. Tech. Pub. (1977).

7. Mount, D.I., and W.A. Brungs. "A Simplified Dosing Apparatus for Fish Toxicological Studies," *Water Res.* 1(1): 21-29 (1967).

8. Krasny, R.F., and W.G. Wilson. "A Technique for the Combustion and Liquid Scintillation Counting of Volatile Carbon 14-labeled Organic Compounds in Fish Tissue," *Bull. Environ. Contam. Toxicol.* (Submitted). (1978).

9. Kobayaski, Y., and D.V. Maudsley. *Biological Applications of Liquid Scintillation Counting* (New York: Academic Press, Inc., 1974).

10. *Standard Methods for the Examination of Water and Wastewater,* 14th ed. (New York: American Public Health Association, 1975), pp. 648-653.

11. "Methods for the Chemical Analysis of Water and Wastewater," Environmental Monitoring and Support of Laboratory, Environmental Research Center, Cincinnati, OH, U.S. EPA Report-625/6-74-003a (19 6), pp. 78-135.

12. Pierce, F.D., T.C. Lamoreaux, H.R. Brown and R.S. Fraser. "An Automated Technique for the Sub-Microgram Determination of Selenium and Arsenic and Surface Waters by Atomic Absorption Spectorscopy," *Appl. Spectros.* 30(1): 38-42 (1976).

25

ECOLOGICAL AND HEALTH EFFECTS OF THE PHOTOLYSIS PRODUCTS OF CHLORINATED HYDROCARBON PESTICIDES

M. A. Q. Khan, Muhammad Feroz,
Andrew A. Podowski and
Lawrence T. Martin

Department of Biological Sciences
University of Illinois at Chicago Circle
Chicago, Illinois 60680

INTRODUCTION

Chlorinated hydrocarbons; aliphatics, such as chloromethanes, hexachlorocyclohexane and hexachlorocyclopentadiene and other cyclodienes; and aromatics, such as DDT, methoxychlor, PCBs, pentachlorophenol and 2,4-D, are common contaminants of air, water, soil, food, plants and animals in the United States and Canada.[1-4] High-residue levels of some of these in certain parts of this country (for instance, Kepone in the James River[5]; PBBs in Michigan; and PCBs, DDT and cyclodienes in Lake Michigan, the Mississippi River and Hawaii[6-13]) have caused concern about their ecological impact and hazards to human health. Consequently, several laboratories are presently engaged in work to develop a better understanding of their behavior in biotic and abiotic environments.

The residues are subject to a variety of physicochemical and biochemical forces in environment, and light is an important factor among them. It may photochemically alter environmental contaminants into molecular species innocuous to biological systems, sometimes, however, these transformations result in compounds more toxic than the parent molecules. The familiar examples are photoisomerization reactions of aldrin, dieldrin, chlordane and heptachlor (Figure 1).[14-17] The problem is compounded by the involvement of biological systems which are capable of producing metabolites from photolysis products of organochlorines, having even more enhanced biological activity. This chapter summarizes the results of investigations carried out in this laboratory to study the behavior of cyclodiene insecticides under different types of light, and the toxicity and biotransformations of their photolysis products.

PHOTOLYSIS PRODUCTS OF ALDRIN AND DIELDRIN

Photolysis of aldrin and dieldrin under sunlight and ultraviolet light leading to the formation of major photolysis products in the form of photoaldrin and photodieldrin (Figure 1) has been well documented.[14,18-24] These terminal environmental products[25] are several times more toxic than their respective parent cyclodienes to insects,[19] aquatic invertebrates, fish,[26] and mammals.[27,28] Studies of the metabolism of photodieldrin in mammals[29-34] and insects[35,36] have shown that, although this compound is less stable than dieldrin, it is metabolized to more toxic photodieldrin ketone[33,35] along with a number of other metabolites (Figure 2, Table I). One of these metabolites, photoaldrin-*trans*-diol, is a suspected carcinogen.[34]

Bluegill fish and *Daphnia pulex* eliminate the absorbed photodieldrin at a faster rate than other cyclodienes[37,38] and this may be related to its greater water solubility[39] as well as breakdown to polar metabolites which have been shown in bluegill.[38] However, the radioactive products excreted in water by fish contain significant amounts of photodieldrin ketone (35% of the radioactivity in water) along with free and conjugated photoaldrin-*trans*-diol. Because of low levels of radioactivity, no metabolites of photodieldrin could be detected in algae and daphnids.[15,40]

A study of the effects of environmental factors and inorganic surfaces showed that photodieldrin volatilization is quite common and this depends on the type of the substrate. On the other hand, photodieldrin appears to be quite persistent on plant surfaces and in algae.

The residues of photodieldrin have been reported in soils,[41] plants[42-44] and food,[45] and since it is very stable in the environment, contamination of living organisms with this toxicant may lead to the formation of more toxic and carcinogenic metabolites.

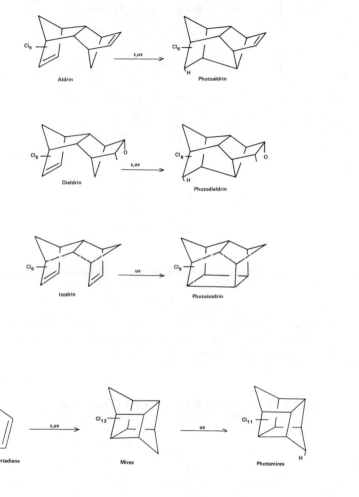

Figure 1. Structural formulae of some chlorinated hydrocarbon pesticides and their photolysis products. s = sunlight, UV = ultraviolet light. (See Figure 4 for structure of heptachlor and photoheptachlor.)

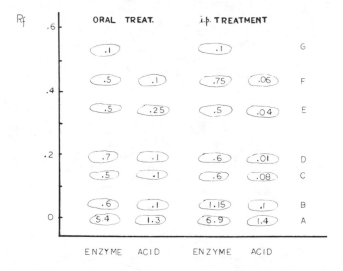

Figure 2. Thin-layer chromatographic presentation of ^{14}C-photodieldrin and its metabolites in the aqueous phase of the urine (containing 48-54% of the administered radioactivity in 9 days) of male rabbits treated with a single oral or intraperitoneal dose of photodieldrin. The chromatogram represents the metabolites and their relative amounts (in parentheses, expressed as percent of the administered dose) after HCl and *beta*-glucuronidase hydrolysis. Spot A is photoaldrin-*trans*-diol, F is photodieldrin ketone, G is photodieldrin. The structures were confirmed by TLC, GLC, infrared spectroscopy, and GC/MS.[31,32]

Table I. Excretion of Radioactivity in the Urine of Male Rabbits
Treated with a Single Dose of ^{14}C-photodieldrin[31]

| | % of the Administered Dose | | | |
| | Intraperitoneal | | Oral | |
Days After Treatment	Organic Phase	Aqueous Phase	Organic Phase	Aqueous Phase
1	0.040	3.79	—	—
2	0.102	10.53	0.024	0.77
4	0.250	21.21	0.082	18.26
6	0.110	6.92	0.050	19.35
7	0.060	4.55	0.031	2.64
8	0.011	3.27	0.014	4.29
9	0.048	4.03	0.013	3.00
TOTAL	0.621	54.30	0.214	48.31

PHOTOLYSIS OF HEPTACHLOR AND
ITS SIGNIFICANCE

Exposure of films of heptachlor mixed with equimolar benzophenone on a glass surface to a 30-watt source of long-wave ultraviolet light caused complete conversion to photoheptachlor in 216 hr; this could be completed in only 80 hr with a 96-watt source (Figure 3, Table II). Replacing benzophenone with chlorophyll slowed down photoisomerization but trapped four other minor products.[16] Benson *et al.*[46] have reported about 2% photoisomerization in sunlight in 500 hr. Exposure of residues of heptachlor on intact bean leaves to a 96-watt source of long-wave UV light for 1.5 hr showed about 10% photoisomerization in addition to the formation of four other products[16] (Figure 3). Presence of photoheptachlor in the air of a grass pasture, 3 to 5 hr after spraying with heptachlor on a sunny day, has been reported.[47] Photolysis of heptachlor to photoheptachlor and heptachlor epoxide has been reported in solution in acetone[16,46] and methanol-water.[48] These data clearly indicate that sunlight and very low intensity long-wave UV light can convert heptachlor on solid surfaces and plant leaves and in air and water to photoheptachlor. Photoheptachlor, therefore, appears to be an environmental contaminant.

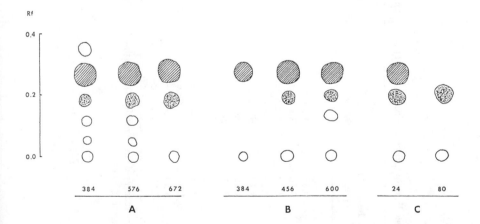

Figure 3. Thin-layer chromatogram of photolysis products, resulting from exposure of heptachlor + benzophenone in acetone (A) and heptachlor in acetone (B) to a 30-watt long wave ultraviolet light source. Rf values are: heptachlor 0.30, photoheptachlor 0.20, heptachlor epoxide 0.11; other products are not identified (0.25mm silica gel, F-254 plates developed 3X with *n*-octane). C shows photolysis of heptachlor + benzophenone films exposed to a 96 watt source. The numerals represent the number of hours the irradiation was carried out.[16]

Table II. Photoisomerization of ^{14}C-heptachlor and ^{14}C-cis-chlordane under Different Conditions[16]

Experiment	Insecticide (mg)	Solvent	UV Intensity (W)	Exposure Time (hr)	Amount Recovered		Total Recovery[a] (%)
					Photoisomer (mg)	Parent Compound (mg)	
^{14}C-heptachlor							
1	0.80	Acetone	30	240	0.20	0.40	82.00
2	0.40	Benzene	30	240	0.025	0.22	75.95
3	2.30	Benzene	96	336	0.17	1.60	78.89
4	29.10	None[b]	72	194	6.38	9.44	60.10
5	9.40	None[b]	96	120	4.51	2.32	91.48
6	1.50	None[b]	96	144	0.56	–	–
7	0.21	None[b]	96	216	0.11	Trace	61.20
^{14}C-cis-chlordane							
8	0.24	Acetone	30	180	0.058	0.165	93.09
9	1.00	Acetone	30	180	0.24	0.68	92.92
10	3.50	Acetone	30	504	2.50	–	–
11	2.40	Acetone	96	1,344	0.28	1.37	96.2

[a]Based on the radioactivity recovered (in all products) from thin layer plates, expressed as a % of the starting material.
[b]Irradiated as a film, others in 1 ml of the solvent.

Toxicity bioassays using aquatic invertebrates, fish, houseflies, rats and mice show that photoheptachlor is twice as toxic as heptachlor to invertebrates except *Gammarus*, to which it is 233 times more toxic (Table III). It is extremely toxic to goldfish (264 times), bluegill (47 times) and rats (19 times) (Table IV). These differences in toxicity may indicate the possibility of direct toxic effects of this terminal residue on aquatic food-chains as well as indirect effects in which aquatic invertebrates can accumulate concentrations of photoheptachlor deleterious to fish feeding on these organisms.

Table III. 24-hr LC_{50} Values of Cyclodienes and their Photoisomers for Aquatic Invertebrates Using a Static System[16]

| | 24-hr LC 50:ppb[a] | | | | | |
Insecticide	D. pulex	D. magna	Gammarus	Asellus	Aedes	Housefly
Cis-chlordane	58	4000	>10,000	–	–	4
Trans-chlordane	269	680	>10,000	–	–	11
Photo-*cis*-chlordane	551	1750	3500	–	–	–
Oxychlordane	930	2000	600	–	–	–
Heptachlor	330	1700	1400	100	11	1
Photoheptachlor	270	880	6	60	5	.28

[a]Data for *Aedes aegypti* and housefly from References 26,49. Values for housefly are in ppm (topical application).

Table IV. Toxicity of Cyclodienes and their Photoisomers to Fish, Rat and Mouse

| | 24-hr LC_{50} (ppb) or LD_{50} (mg/kg) | | | |
Insecticide	Bluegill	Goldfish	Rat	Mouse[a]
Cis-chlordane	17.5	27.0	83	30
Trans-chlordane	140	440	–	130
Photo-*cis*-chlordane	12	13.5	41	20
Oxychlordane	9	15	–	–
Heptachlor	64	185	71(162[c])	–
Photoheptachlor	1.35	0.70	3.8	–
Heptachlor Epoxide	5.30[b]	–	60[c]	18[c]

[a]Data from Reference 50.
[b]Personal communication. Fish Pesticide Research Lab., Columbia, MO.
[c]From Reference 49.

Because photoheptachlor is an extremely toxic compound, its fate in rats, rabbits and houseflies was investigated. Half-life of an intraperitoneally injected dose (0.8 mg/kg male rabbit, 0.93 mg/kg rats) was 70 days in male rabbits[51] (Table V), 14 days in male rats, and more than 28 days in female rats[52] (Table VI). Thus, photoheptachlor seems to be retained in the body of rats longer than heptachlor.[53]

Table V. Excretion of Radioactivity by Male Rabbit and Rat
Following i.p. Injection of ^{14}C-photoheptachlor[51,52]

	% of Dose Excreted			
	Rabbit[a]		Rat[a]	
Week	Urinary	Fecal	Urinary	Fecal
1	5.80	0.0	3.63	26.80
2	7.13	0.0	2.43	17.34
3	6.28	0.0	1.27	8.49
4	6.38	0.0	0.24	4.54

[a]Single dose 0.8 mg/kg for rabbits and 0.93 mg/kg for rats.

Table VI. Excretion of Radioactivity by Male and Female Rats Treated
with a Single Dose (0.93 mg/kg) of ^{14}C-photoheptachlor[52]

	Males[a]			Females[b]		
	Amount Excreted[c] Daily		Total (Cumulative:	Amount Excreted[c] Daily		Total (Cumulative:
Day	Feces (μg)	Urine (μg)	Dose) (%)	Feces (μg)	Urine (μg)	Dose) (%)
1	6.6	5.9	0.8	1.8	1.7	0.2
2	50.9	9.7	5.0	12.4	3.4	1.3
4	77.8	7.5	17.3	13.7	3.9	3.9
7	47.8	7.6	30.4	13.0	3.9	7.4
14	28.5	2.2	50.2	10.8	1.3	13.9
21	14.4	1.9	60.0	9.8	2.1	19.8
28	10.3	1.2	65.1	10.6	3.2	24.9

[a]Four males (1,579 g total body wt).
[b]Six females (total wt = 1,573 g).
[c]Equivalents of ^{14}C-photoheptachlor.

In rabbits, the radioactivity is excreted exclusively in urine in the form
of four metabolites, one of which has been identified (Figure 4) as a mono-
dechlorinated hydroxy product present in free as well as conjugated form.[51]
Rats excrete at least 9 metabolites in urine and 11 in feces, and males are
more active than females in metabolism and excretion.[52] Of the 66% of
the dose excreted in urine and feces in one month by males, 4 and 34%,
respectively, appears to be the unchanged photoheptachlor. In females the
excreted radioactivity (28% of dose) as photoheptachlor is 14% in feces
and 3% in urine (Table VI). The remaining excreted radioactivity is in the
form of lipophilic and hydrophilic products. Chromatographic behavior of
one of these metabolites suggests that it may be photoheptachlor ketone,
a possibility in photoheptachlor metabolism postulated by Rosen et al.[54]
Tissue residues of photoheptachlor in male rats are highest in visceral fat
(1.1 ppm), followed by duodenum (0.5 ppm) and kidneys (0.32 ppm) and
adrenals (0.32 ppm). In female rats, the highest concentration in fat (5.4 ppm)

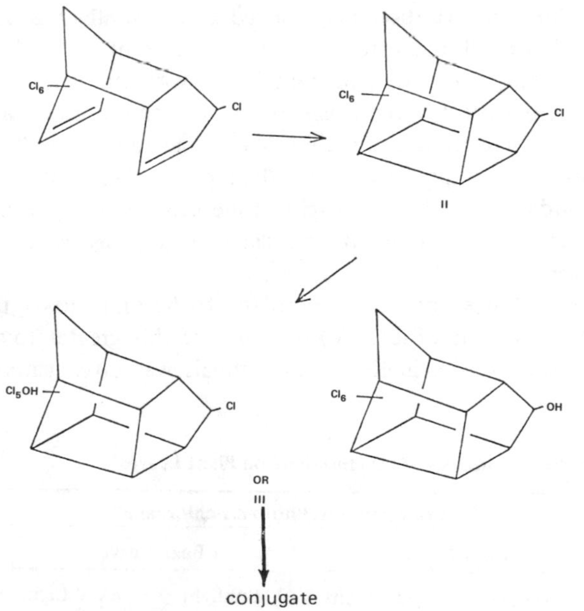

Figure 4. Conversion of heptachlor (I) to photoheptachlor (II) and one of the metabolic routes of II in male rabbits resulting in the formation of III (two alternate
structures) and its conjugate.[51]

is followed by skin (1.23 ppm), gonads (0.79 ppm), adrenals (0.58 ppm), duodenum (0.33 ppm) and muscles (0.33 ppm). In male rabbits the tissue concentrations in decreasing order are: fat (1.29 ppm), kidney (0.32 ppm), liver (0.10 ppm), duodenal content (0.08 ppm), brain (0.04 ppm) and testes (0.02 ppm).[51,52]

Because of the accumulation of photoheptachlor in fat, a long-term physiological and/or biochemical effect(s) may be possible, especially in females whose gonads show significant concentrations.

PHOTOLYSIS OF *cis*-CHLORDANE AND ITS SIGNIFICANCE

Films of *cis*-chlordane on glass surface and intact plant leaves are converted to photo-*cis*-chlordane in the absence of a sensitizer.[16,46] In the case of the former, 70% and 30% photoisomerization can occur in the presence or absence of benzophenone in 106 hr with a 96-watt source of long-wave UV light.[16] This photoisomerization may be reduced if sunlight is used[46] and increased if chlorophyll and rotenone are both present.[16] Irradiation of residues of *cis*-chlordane on bean or corn leaves with sunlight or a 30-watt source of long-wave UV light can cause 75% photoisomerization in 1 hr (Table VII). Irradiation of *cis*-chlordane in acetone[16] or dioxane-water[48] shows formation of products other than photo-*cis*-chlordane (Figure 5). These studies indicate that photo-*cis*-chlordane can be an environmental contaminant resulting from residues of chlordane on surfaces (soils, plants, etc.) and in solutions.

The bioassay tests showed photo-*cis*-chlordane to be more toxic than *cis*-chlordane to fish and rats (Table IV). Because of this greater toxicity, the fate of photo-*cis*-chlordane in fish and mammals was investigated.

Table VII. Photolysis of *cis*-chlordane on Plant Leaves[16]

Exposure Time (hr)	% Conversion to Photo-*cis*-chlordane[a]			
	Corn Leaves		Bean Leaves	
	Sunlight	UV Light	Sunlight	UV Light
1	15(78)	8	24(67)	11
2	18(70)	9	11(54)	7
3	20(62)	7	10(42)	5

[a]Values in parentheses obtained with rotenone. The quantitation of leaf washings was done by gas chromatography.

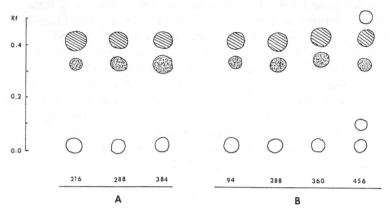

Figure 5. Thin-layer chromatogram of photolysis products resulting from the exposure of *cis*-chlordane without (B) and with benzophenone (A), to a 96-watt source of long wave ultraviolet. Rf values (0.25 mm silica gel G, F-254 plates, three runs in *n*-heptane) were: *cis*-chlordane 0.39, photo-*cis*-chlordane 0.29. Numerals at the bottom indicate duration (hr) of photolysis.

Exposure of goldfish and bluegill to 5 ppb of photo-*cis*-chlordane showed that photo-*cis*-chlordane, as compared with *cis*-chlordane, was absorbed at a faster rate[55,57] (Figure 6). However, the transfer of the preexposed fish

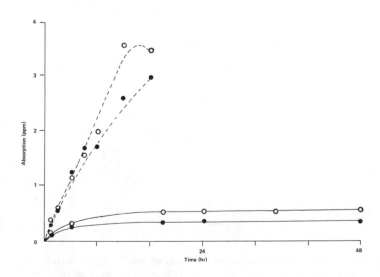

Figure 6. Absorption of [14]C-*cis*-chlordane (solid circles) and ^{14}C-photo-*cis*-chlordane (hollow circles) by goldfish (broken lines)[55] and bluegill (solid lines)[57] exposed to 5 ppb of the insecticides.

to clean water showed that photo-*cis*-chlordane was excreted at a faster rate than *cis*-chlordane by goldfish[55] (Figure 7) and bluegill.[57] Bluegill metabolized the absorbed photo-*cis*-chlordane to a number of products[57] (Figure 8). These and other fish metabolize and excrete *cis*-chlordane at a much slower rate.[57,58,59] Male rats receiving a single oral or intraperitoneal dose of photo-*cis*-chlordane also rapidly metabolize and excrete this compound. About 66% and 51% of the administered dose (oral and intraperitoneal, respectively) was excreted in 8 days, mostly in the form of polar metabolites[60] (Table VIII). Thus, although the acute toxicity of photo-*cis*-chlordane is greater than that of *cis*-chlordane, this compound is more efficiently metabolized and disposed of by fish and mammals.

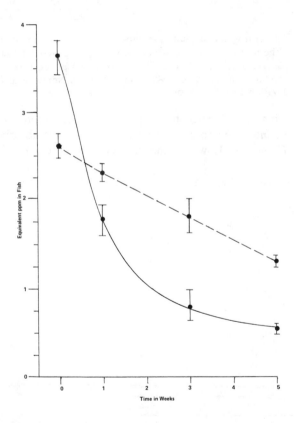

Figure 7. Elimination of *cis*-chlordane (broken line) and photo-*cis*-chlordane (solid line) by treated goldfish (5 ppb in 4 liters for 12 hr) on transfer to clean water.[55]

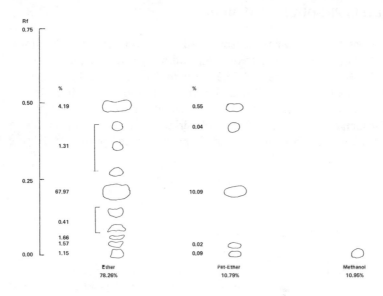

Figure 8. Tracing of a TLC plate (0.25 mm silica gel G, F-254, duplicate development with heptane) of organic extracts of bluegills treated with photo-*cis*-chlordane.[57]

Table VIII. Excretion of Radioactivity by Male Rats Treated with a Single Dose (3.12 mg/rat) of ^{14}C-photo-*cis*-chlordane[60]

	Cumulative Excretion (% of dose[a])					
	Oral			Intraperitoneal		
Day	feces	urine	Total	feces	urine	Total
1	54.86	2.60	54.45	1.72	3.72	5.44
2	58.92	3.77	62.29	6.16	8.32	14.48
4	60.80	4.47	65.27	18.72	14.77	33.49
7	61.49	4.90	66.39	28.48	19.79	48.27
8	61.64	4.98	66.62	39.35	20.64	50.99
14	62.06	5.29	67.35	35.55	23.59	59.14
21	62.23	5.45	67.68	37.81	24.96	62.87
Unextractable (21 days)	17.63	-	17.63	22.10	-	22.10
Cage Washes (21 days)	-	-	0.69	-	-	3.21
TOTAL	-	-	86.75	-	-	93.73

[a]Average of two rats.

PHOTOLYSIS PRODUCT OF ISODRIN

Isodrin has been shown to be converted to photoisodrin by ultraviolet light[61],[64] Photoisodrin is less toxic than isodrin to insects, aquatic invertebrates, fish and mammals.[65] The lower toxicity to houseflies appears to be due to its rapid metabolism and excretion by houseflies. In male mice about 90% of the organosoluble radioactivity in excreta was in the form of 4 metabolites along with photoisodrin; the remaining aqueous metabolites appeared to be hydroxylated and conjugated products[61](Figure 9).

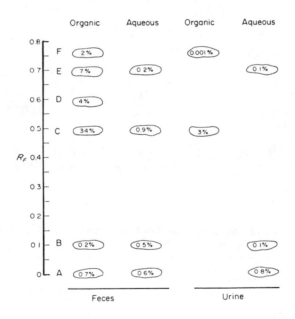

Figure 9. Thin-layer chromatographic presentation of ^{14}C-photoisodrin and its metabolites in feces (organic, aqueous) and urine (organic, aqueous) of mice. Spots A-E are unidentified metabolites. Spot F represents photoisodrin. Various fractions were applied on TLC plate and developed with chloroform: methanol (9:1) and autoradiographed.[61] Relative amounts (as % of dose) are indicated in the spots.

CONCLUSIONS

The photoisomers of aldrin, dieldrin, heptachlor and chlordane appear to be possible environmental contaminants produced under natural conditions. All these photoisomers are generally more toxic than their parent compounds. Photodieldrin and photoheptachlor are of special significance

Table IX. Half-Lives of Cyclodienes and their Photoisomers in Mammals and Fish

Animal	Dosage[a]	Half-Life (days)	Reference
Photoheptachlor			
Rat M	0.93 mg/kg, i.p.	single dose	14	52
Rat F	0.93 mg/kg, i.p.	single dose	>30	52
Rabbit M	0.8 mg/kg, i.p.	single dose	70	51
Photoisodrin			
Mouse M	7 μg/20g, p.o.	single dose	1	61
Photo-cis-chlordane			
Rat M	7.9 mg/kg, i.p.	single dose	7	60
Goldfish	3.594 ppm	12-hr exposure	7	55
Bluegill	0.556 ppm	96-hr exposure	70	57
Photodieldrin			
Rabbit M	20 mg/kg, i.p.	single dose	8	31, 32
Rabbit M	20 mg/kg, p.o.	single dose	11	31, 32
Rat M	0.05 mg/kg, i.p.	5 day/wk for 12 wk (9wk)	7	29, 30
Rat M	0.05 mg/kg, p.o.	5 day/wk for 12 wk (9wk)	14 (7)	29, 30
Rat M	0.016 mg/kg, p.o.	for 15 wk	13	29, 30
Rat F	0.024 mg/kg, p.o.	for 15 wk	63	29, 30
Rat F	0.05 mg/kg, p.o.	5 days/wk for 12 wk (9wk)	150 (150)	29, 30
Monkey M	2 mg/kg, i.v.	single dose	30	34
Monkey F	2 mg/kg, i.v.	single dose	55	34
cis-chlordane			
Goldfish	2.657 ppm	12-hr exposure	31	55
Rat F	0.2 mg/kg, p.o.	single dose	<1	62
Heptachlor			
Goldfish	0.864 ppm	38 μg injected/44 g fish	29	63
Rat M	0.053 mg/kg, p.o.	single dose	2	53

[a]The values for fish show the concentration in fish at the time of transfer to clean water. Where indicated, M = males, F = females.

Table X. Bioaccumulation of Cyclodienes by Goldfish, Bluegill and *Xenopus*

Biomass	Exposure Concentration (ppb)	Exposure Time for Maximum Absorption (hr)	Species	Bioaccumulation Ratio[a]
cis-chlordane				
80 g/8 liter	5.0	96	*Xenopus*	108
50 g/6 liter	5.0	24	bluegill	322
6 g/7.8 liter	5.0	16	goldfish	990
300 g/16 liter	5.0	72	cichlid	65
Photo-*cis*-chlordane				
6 g/7.8 liter	5.0	16	goldfish	1180
50 g/7 liter	5.0	24	bluegill	1143
Photodieldrin				
65 g/7 liter	20.0	48	bluegill	1457
Hexachlorocyclopentadiene				
1 g/3 liter	5.0	48	goldfish	323 (371)
1 g/3 liter	5.0	96	goldfish	123 (1297)

[a]The ratio was calculated from the final concentration of the cyclodiene in water at the time of fish analysis. The values in parentheses are for an intermittent system.

because they are much more toxic to certain animals, *e.g.*, photoheptachlor is 264 times more toxic than hpetachlor to goldfish. These compounds accumulate in body fat, gonads, liver, etc., and have long half-lives (Table IX). Their concentrations in gonads in females may have some effects on the offspring. Even photo-*cis*-chlordane, which is more biodegradable than *cis*-chlordane, shows higher bioaccumulation values (Table X). Therefore, their long-range effects on food chain organisms may be more severe. Photodieldrin (and possibly photoheptachlor) is metabolized to ketone by dehydrochlorination, as well as to hydroxylated products. Photodieldrin ketone is more toxic than photodieldrin[36,56] and photoaldrin-*trans*-diol is a suspected carcinogen.[34] Toxicological significance of similar photoheptachlor products remains unknown. Photoheptachlor is several times more effective than other cyclodienes in altering liver structure and function. At low dosages it increases liver weight, reduces barbiturate sleeping time and increases cytochrome P-450 content and activities of the hepatic microsomal drug-metabolizing enzymes[66] (Table XI). Heptachlor epoxide, also an environmental[7,16] and biological[7,53] product, is carcinogenic to mice.[67] Photo-*cis*-chlordane appears to have less pronounced effects than those reported for *cis*-chlordane and other cyclodienes[68,69] (Table XII).

In addition to the compounds mentioned above, the photolysis of two other environmental contaminants deserve serious consideration. The first one is pentachlorophenol whose resudies, which are quite common in the United States,[3,70] can be photolyzed to octachlorodibenzo-*p*-dioxin (on a wood substrate)[17] which is one of the very toxic environmental contaminants.[71] The second one is hexachlorocyclopentadiene, whose residues are rather common in Kentucky and Tennessee, which can be photolytically converted to a Mirex-type compound (unpublished data, this laboratory).

Table XI. Effect of Pretreatment with Photoheptachlor on Male Rat

Test	Number of[a] Animals	Control	Treated	Increase (%)
Sleeping Time (hr)	4	(5.33)	(3.89)	- (27)
Liver wt/Body wt (%)	6	(4.24)	(4.48)	(5.8)
	6	3.33	4.04	21.3
mg Microsomal Protein/gm Liver	3	9.6	9.60	-
n-mole P-450/mg protein	3	0.64	0.66	6.89
pNA:n-mole/mg/min	3	1.88	3.38	79.31
AP:n-mole/mg/min	3	3.29	3.86	17.29

[a]Rats were treated with 0.1 mg/kg photoheptachlor in corn oil. Sleeping time determined after 20 days following treatment. Values in parentheses are for i.p. injection every other day; other values for oral dosage once per week. pNA=*para*-nitroanisole and AP=aminopyrene demethylations.

Table XII. Effect of Photo-*cis*-chlordane on Hepatic Microsomal Drug-Metabolizing
Enzymes in Male Rats

Treatment	Sleep[a] (hr)	L/B[b] (%)	mg Protein/ g Liver	n-moles/mg/min[c]		
				p-NA	AP	Aniline
Corn Oil[d]	9.52	3.36	12.24	0.680	0.721	0.424
Photo-*cis*-chlordane[d]						
0.5 mg/kg	7.36	3.26	11.38	0.944	0.660	0.525
2.0 mg/kg	6.52	3.21	8.36	0.967	0.756	0.804

[a]Determined in 3rd week (250 mg/kg Hexobarbital).
[b]Liver wt/body wt.
[c]Microsomal mixed-function oxidase activity: pNA = *para*-nitroanisole, AP = aminopyrene demethylations, aniline hydroxylation.
[d]Daily i.p. injection 5 days/wk for 10 weeks.

Also, the oxygen analog of Mirex, Kepone, which has contaminated one of the most important estuaries (James River) in this country,[5] has seriously threatened the fishery and shellfish industry.

ACKNOWLEDGMENTS

This research was supported by a USPHS grant (ES-01479) from the National Institute of Environmental Health Sciences.

REFERENCES

1. Wheatley, G. A. "Pesticides in the Atmosphere," In: *Environmental Pollution by Pesticides*, C. A. Edwards, Ed. (New York: Plenum Publishing Corporation, 1973), pp. 365-408.
2. Duggan, R. E., and M. B. Duggan. "Pesticide Residues in Food." In: *Environmental Pollution by Pesticides*, C. A. Edwards, Ed. (New York: Plenum Publishing Corporation, 1973), pp. 334-364.
3. Bevenue, A., and H. Beckman. "Pentachlorophenol: A Discussion of its Properties and its Occurrence as a Residue in Human and Animal Tissues," *Residue Rev.* 19:83-134 (1967).
4. Edwards, C. A. "Nature and Origins of Pollution of Aquatic Systems by Pesticides. In: *Pesticides in Aquatic Environments*, M. A. Q. Khan, Ed. (New York: Plenum Publishing Corporation, 1977), pp. 11-38.
5. Hugget, R. In: "Fate and Transport of Toxic Chemicals in Environments," U.S.EPA workshop, Norfolk, VA, December 1978.
6. Zitko, V. "The Fate of Highly Brominated Aromatic Hydrocarbons in Fish." In: *Pesticide and Xenobiotic Metabolism in Aquatic Organ-*

isms, M. A. Q. Khan, J. J. Lech and J. J. Menn, Eds. (Washington, DC: American Chemical Society, 1979), pp. 177-182.

7. Metcalf, R. L., and J. Sanborn, "Pesticides and Environmental Quality in Illinois," *Bull. Ill. Nat. Hist.* Survey 31, Article 9 (1975), pp. 381-436.

8. Peterman, P. H., J. J. Delfino, D. J. Dube, T. A. Gibson and F. J. Prixnar. "Chloro-Organic Compounds in the Lower Fox-River, Wisconsin," *Proc. Int. Symp. on Analysis of Hydrocarbons and Halogenated Hydrocarbons in the Aquatic Environment*, B. K. Afghan, Ed. (New York: Plenum Publishing Corporation, 1979).

9. Hesse, J. L. "Polychlorinated Biphenyl Usage and Sources of Loss to the Environment in Michigan," In: National Conference on Polychlorinated Biphenyls, USEPA Report 560/6-75 (1976).

10. Haile, C. L. "Chlorinated Hydrocarbons in Lake Ontario and Lake Michigan Ecosystems," Ph.D. Thesis, Water Chemistry Program, University of Wisconsin, Madison, WI (1977).

11. Finley, M. T. Ph.D. Thesis, Mississippi State University (1970).

12. Bevenue, A., J. W. Hylin, Y. Kawano and T. W. Kelley. "Organochlorine Residues in Water, Sediment, Algae, and Fish in Hawaii, 1970-1971," *Pestic. Monit. J.* 6:56-64 (1972).

13. Tanita, R., J. M. Johnson, M. Chun and J. Maciolek. "Organochlorine Pesticides in the Hawaii Kai Marina, 1970-74," *Pestic. Monit. J.* 10: 24-30 (1976).

14. Rosen, J. D., D. J. Sutherland and G. R. Lipton. "The Photochemical Isomerization of Dieldrin and Endrin and Effects on Toxicity. *Bull. Environ. Contam. Toxicol.* 1:133-140 (1966).

15. Khan, M. A. Q., H. M. Khan and D. J. Sutherland. "Ecological and Health Effects of the Photolysis of Insecticides," In: *Survival in Toxic Environments*, M. A. Q. Khan and J. P. Bederka, Jr., Eds. (New York: Academic Press, Inc., 1974), pp. 333-355.

16. Podowski, A. A., B. C. Banerjee, M. Feroz, M. A. Dudek, R. L. Willey and M. A. Q. Khan. "Photolysis of Heptachlor and *cis*-Chlordane and Toxicity of their Photoisomers to Animals," *Arch. Environ. Contam. Toxicol.* 8:509-518 (1979).

17. Lamparski, L. L., and R. H. Stehl. "Photolytic Effects on Chlorinated Dibenzo-*p*-dioxin Concentration in Pentachlorophenol-Treated Wood," paper presented at the National Meeting of American Chemical Society, Honolulu, Hawaii, 1979.

18. Rosen, J. D. *Chem. Comm.* (1967), p. 189.

19. Rosen, J. D., and D. J. Sutherland. *Bull. Environ. Contam. Toxicol.* 1:127-134 (1967).

20. Harrison, R. B., D. C. Holmes, J. Roburn, and J. O'G. Tatton. *J. Sci. Food Agric.* 18:10 (1967).

21. Henderson, G. L., and D. G. Crosby. "The Photodecomposition of Dieldrin Residues in Water," *Bull. Environ. Contam. Toxicol.* 3:131-134 (1968).

22. Benson, W. R. *J. Agric. Food Chem.* 19:66-70 (1971).
23. Lykken, L. "Role of Photosensitizers in Alteration of Pesticide Residues in Sunlight." In: *Environmental Toxicology of Pesticides*, F. Matsumura, G. M. Boush and T. Misato, Eds. (New York: Academic Press, Inc., 1972), pp. 449-470.
24. Ivie, G. W., and J. E. Casida. "Photosensitizers for the Accelerated Degradation of Chlorinated Cyclodienes and Other Chemicals Exposed to Sunlight on Bean Leaves," *J. Agric. Food Chem.* 19:410-416 (1971).
25. Egan, H. *J. Assoc. Offic. Analyt. Chem.* 52:299 (1969).
26. Khan, M. A. Q., N. Maitra, D. J. Sutherland, J. D. Rosen and R. H. Stanton. "Toxicity-Metabolism Relationship of the Photoisomers of Cyclodiene Insecticides," *Arch. Environ. Contam. Toxicol.* 1:159-170 (1973).
27. Brown, V. K. H., J. Robinson and A. Richardson. "Preliminary Studies of the Acute and Subacute Toxicities of a Photoisomerization Product of HEOD," *Food Cosmet. Toxicol.* 5:771-780 (1967).
28. "Evaluation of Some Pesticide Residues in Food," Food and Agricultural Organization of the United Nations, FAO/PL, 1967/M/11/1/ Food Additives (1968), pp. 68-80.
29. Dailey, R. E., A. K. Klein, E. Brouwer, J. D. Link and R. C. Braunberg. "Effect of Testosterone on Metabolism of ^{14}C-Photodieldrin in Normal, Castrated, and Oophorectomized Rats," *J. Agric. Food Chem.* 20:371-375 (1972).
30. Dailey, R. E., M. S. Walton, V. Beck, C. L. Leavens and A. K. Klein. "Excretion, Distribution, and Tissue Storage of a ^{14}C-Labeled Photoconversion Product of ^{14}C-Dieldrin," *J. Agric. Food Chem.* 18:443-450 (1970).
31. Reddy, G., and M. A. Q. Khan. "Urinary Metabolites of ^{14}C-Photodieldrin in Male Rabbits," *J. Agric. Food Chem.* 26:292-294 (1978).
32. Reddy, G., and M. A. Q. Khan. "Metabolism, Excretion, and Tissue Distribution of ^{14}C-Photodieldrin in Male Rabbits, Following Single Oral and Intraperitoneal Administration," *J. Agric. Food Chem.* 23: 861-866 (1975).
33. Klein, A. K., R. E. Dailey, M. S. Walton, V. Beck and J. D. Link. *J. Agric. Food Chem.* 18:705-709 (1970).
34. Nohynek, G. J., W. F. Mueller, F. Coulston and F. Korte. "Metabolism, Excretion, and Tissue Distribution of ^{14}C-Photodieldrin in Non-Human Primates Following Oral Administration and Intravenous Injection," *Ecotoxicol. Environ. Safety* 3:1-9 (1979).
35. Khan, M. A. Q., J. D. Rosen and D. J. Sutherland. "Insect Metabolism of Photoaldrin and Photodieldrin," *Science* 164:318-319 (1969).
36. Khan, M. A. Q., D. J. Sutherland, J. D. Rosen and W. F. Carey. "Effect of Sesamex on the Toxicity and Metabolism of Cyclodiene

Insecticides and Their Photoisomers in the Housefly," *J. Econ. Entomol.* 63:470-475 (1970).

37. Khan, H. M., S. Neudorf and M. A. Q. Khan. "Absorption and Elimination of Photodieldrin by *Daphnia* and Goldfish," *Bull. Environ. Contam. Toxicol.* 13:582-587 (1975).

38. Sudershan, P., and M. A. Q. Khan. "Metabolic and Elimination Products of ^{14}C-Photodieldrin from Bluegill Fish," *Pestic. Biochem. Physiol.* (in press).

39. Khan, H. M. M.S. Thesis, University of Illinois at Chicago Circle, Chicago, IL (1974).

40. Reddy, G., and M. A. Q. Khan. "Fate of Photodieldrin under Various Environmental Conditions," *Bull. Environ. Contam. Toxicol.* 13: 64-72 (1975).

41. Suzuki, M., Y. Yamamoto, and T. Watanaba. "Photodieldrin Residues in Field Soils," *Bull. Environ. Contam. Toxicol.* 12:275 (1974).

42. Lichtenstein, E. P., K. R. Schulz and T. W. Fuhreman. "Effects of a Cover Crop Versus Soil Cultivation on the Fate and Vertical Distribution of Insecticide Residues in Soil 7 to 11 Years After Soil Treatment," *Pestic. Monit. J.* 5:218 (1971).

43. Klein, W., J. Kohli, I. Weisgerber and F. Korte. "Fate of Aldrin-^{14}C in Potatoes and Soil under Outdoor Conditions," *J. Agric. Food Chem.* 21:152 (1973).

44. Kohli, J., S. Zarif, I. Weisgerber, W. Klein and F. Korte. "Fate of ^{14}C-Aldrin in Sugarbeets and Soil under Outdoor Conditions," *J. Agric. Food Chem.* 21:855 (1973).

45. Robinson, J., A. Richardson, B. Bush and K. E. Elgar. "Photoisomerization Product of Dieldrin," *Bull. Environ. Contam. Toxicol.* 1:127 (1966).

46. Benson, W. R., P. Lombardo, I. J. Egry, R. D. Ross, Jr., R. P. Baron, D. W. Mastbrook and E. A. Hansen. "Chlordane Photoalteration Products: Their Preparation and Identification," *J. Agric. Food Chem.* 18:857-862 (1971).

47. Taylor, A. W., D. E. Glotfelty, B. C. Turner, R. E. Silver, H. P. Freeman and A. Weiss. "Volatilization of Dieldrin and Heptachlor Residues from Field Vegetation," paper presented at the 173rd national meeting of the American Chemical Society, New Orleans, LA (1977).

48. Vollner, L., H. Parlar, W. Klein and F. Korte. "Beitrage zur Okologischen Chemie—XXXI. Photoreaktionen der Komponenten des Technischen Chlordanes," *Tetrahedron Lett.* 27:501-509 (1969).

49. Brooks, G. T. *Chlorinated Insecticides: Biological and Environmental Aspects, Vol. II* (Cleveland, OH: CRC Press, 1974), pp. 197.

50. Ivie, G. W., J. R. Knox, S. Khalifa, I. Yamamoto and J. E. Casida. "Novel Photoproducts of Heptachlor Epoxide, *trans*-Chlordane, and *trans*-Nonachlor," *Bull. Environ. Contam. Toxicol.* 7:376 (1972).

51. Feroz, M., and M. A. Q. Khan. "Fate of ^{14}C-Photoheptachlor in Rabbits," *J. Agric. Food Chem.* 27:108-113 (1979).

52. Feroz, M., and M. A. Q. Khan. "Biotransformations of Photoheptachlor in the Rat. I. Excretion, Storage, and Isolation of Metabolites," *Pestic. Biochem. Physiol.* (in press).

53. Tashiro, S., and F. Matsumura. "Metabolism of *trans*-Nonachlor and Related Chlordane Compounds in Rat and Man," *Arch. Environ. Contam. Toxicol.* 7:113-127 (1978).

54. Rosen, J. D., D. J. Sutherland and M. A. Q. Khan. "Properties of Photoisomers of Heptachlor and Isodrin," *J. Agric. Food Chem.* 17:404-405 (1969).

55. Ducat, D. A., and M. A. Q. Khan. "Absorption and Elimination of [14]C-*cis*-Chlordane and ^{14}C-Photo-*cis*-Chlordane by Goldfish (*Carassius auratus*)," *Arch. Environ. Contam. Toxicol.* 8:409-417 (1979).

56. Klein, W., R. Kaul, Z. Parlar, M. Zimmer and F. Korte. *Tetrahedron Lett.* 3197 (1969).

57. Sudershan, P., and M. A. Q. Khan. "Metabolism of *cis*-chlordane and Photo-*cis*-chlordane in Bluegill," *J. Agric. Food Chem.* (submitted).

58. Feroz, M., and M.A.Q. Khan. "Metabolism, Tissue Distribution, and Elimination of [14]C-*cis*-Chlordane in the Tropical Freshwater Fish, *Cichlasoma sp.,*" *J.Agric. Food Chem.* (in press).

59. Feroz, M., and M.A.Q. Khan. "Fate of [14]C-*cis*-chlordane in Goldfish *Carassius auratus* (L.)," *Bull. Environ. Contam. Toxicol.* 23:64-69 (1979).

60. Feroz, M., and M.A.Q. Khan "Metabolic Fate of Photo-*cis*-chlordane in the Rat. I. Excretion, Tissue Distribution, and Preliminary Characterization of Metabolites," *J. Agric. Food Chem.* (in press).

61. Reddy, G., and M.A.Q. Khan. "Metabolism of ^{14}C-photoisodrin in Mice and Houseflies," *Gen. Pharmacol.* 8:285-289 (1977).

62. Barnett, J. R., and H. W. Dorough. "Metabolism of Chlordane in Rats," *J. Agric. Food Chem.* 22:612 (1974).

63. Feroz, M., and M.A.Q. Khan. Metabolism of ^{14}C-heptachlor in Goldfish (*Carassius auratus*)," *Arch. Environ. Contam. Toxicol.* 8:519-531 (1979).

64. Soloway, S. B. "Cyclodiene insecticides," *Adv. Pest. Control. Res.* 6:85-112 (1965).

65. Khan, M.A.Q. "Fate of Cyclodiene Insecticides in Environment," *Proc. Entomol. Soc. Karachi* (Pakistan) 3:25-38 (1973).

66. Martin, L., and M.A.Q. Khan. "Induction of Hepatic Drug Metabolizing Enzymes of Rat with Photo-*cis*-chlordane and Photoheptachlor" (in preparation).

67. Reuber, M. D. "Histopathology of Carcinomas of the Liver in Mice Ingesting Heptachlor or Heptachlor Epoxide," *Exp. Cell Biol.* 45:147 (1977).

68. Hart, L. G., and J. R. Fouts. "Further Studies on the Stimulation of Hepatic Microsomal Drug Metabolizing enzymes by DDT and its

Analogs," *Naun-Schmeidbergs Arch. Exp. Pathol. Pharmakol.* 249: 486 (1965).

69. Hart, L.G., R. W. Shultice and J. R. Fouts. "Stimulatory Effects of Chlordane on Hepatic Microsomal Drug Metabolizing Enzymes in the Rat," *Toxicol. Appl. Pharmacol.* 5:371-386 (1963).

70. Cirelli, D. P. "Patterns of Pnetachlorophenol Usage in the United States of America: An Overview." In: *Pentachlorophenol: Chemistry Pharmacology, and Experimental Toxicology*, K. R. Rao, Ed. (New York: Plenum Publishing Corporation, 1978) pp. 13-18.

71. Rawls, R. L. "Dow Finds Support, Doubt for Dioxin Ideas," *Chem. Eng. News* 57(7):23-29 (1979).

26

N-NITROSO COMPOUND IMPURITIES IN
CONSUMER AND COMMERCIAL PRODUCTS

D. H. Fine, I. S. Krull,
D. P. Rounbehler and G. S. Edwards

New England Institute for Life Sciences
Waltham, Massachusetts 02154

J. G. Fox

Division of Laboratory Animal
   Medicine
Massachusetts Institute of Technology
Cambridge, Massachusetts 02139

INTRODUCTION

If the yield of a chemical process is below about 1%, the reaction is
generally considered of minimal interest. However, in assessing the importance of chemical carcinogens in the environment, the overriding criterion
is the human exposure to the compound of interest. If human exposure
to a particular chemical carcinogen is negligible, then the fact that the
compound can be formed in 90% yield is relatively unimportant. However,
if human exposure is large, then even a reaction which proceeds very
slowly to give only a minimal yield may be of paramount importance.
Thus, the chemist concerned with human exposure to N-nitroso compounds
must be equally familiar with the reaction which gives a 90% yield and
the reaction which gives only a very low yield.

N-nitroso compounds are readily formed from their precursors: amines
and nitrosating agents. The amines can be primary, secondary or tertiary.

The nitrosating species can be derived from nitrite (nitrite salts or nitrous acid), oxides of nitrogen (NO, NO_2, N_2O_3, N_2O_4), certain aliphatic and aromatic $C-NO_2$ compounds or transnitrosation from N-nitroso or C-nitroso compounds. Depending on the reactants and catalysts which are present, N-nitrosation can occur at either acidic, neutral or alkaline pH, or in solely organic media. Because of the wide range of conditions under which N-nitroso compounds can be formed, it is not surprising that N-nitroso compounds are present as contaminants in so many commercial products and processes.

The presence of an N-nitroso compound impurity in a commercial product is a serious matter and can have wide implications. By studying the case histories of known N-nitroso compound impurities, and given some rudimentary knowledge of the chemistry involved, it is often possible to predict the conditions under which N-nitroso compounds are likely to be encountered. In this chapter, a few case histories are discussed.

FISH MEAL

In 1964, Ender et al.[1] isolated and identified N-nitrosodimethylamine (NDMA) as the hepatotoxic factor in fish (herring) meal used for feeding mink. Sen et al.[2] in 1972 again implicated NDMA in fish meal as the causative factor in liver intoxication of mink. Kann et al.[3] in 1977 showed that 80% of the feeds used in Germany for laboratory animals contained NDMA impurity at a level of greater than 1 ppb. Again fish meal was implicated. Some pet and laboratory feeds currently being used in the United States were recently examined. The nitrosamine levels in the commercial dog and cat foods were low (Table 1), with only 1 product containing nitrosamines at a level above 0.1 ppb. By contrast, 8 out of 9 of

Table I. N-nitrosamines in Dog and Cat Food

Pet	Food	Nitrosamine Level μg/kg (ppb)
Dog	Bone biscuit	0.08 NDMA
Dog	Bone biscuit	0.04 NDMA
Cat	Tuna fish	N.D.[a]
Cat	Tuna fish	0.05 NDMA; 0.14 NPYR
Cat	Crab and shrimp	0.01 NDMA
Cat	Seafood	0.01 NDMA
Cat	Fish flavor	0.01 NDEA
Cat	Fish, meat, milk	0.06 NDMA
Cat	Liver	0.02 NDMA

[a] N.D. = none detected ($<$ 0.01 ppb)

feeds used for laboratory animals contained NDMA at levels above 1.0 ppb (Table II). The results are similar to those of the German study.[3]

Table II. N-nitrosamines in the Diet of Laboratory Animals

Type of Diet	NDMA Level μg/kg (ppb)	Contains Fish Meal
Rabbit	N.D.	−
Rabbit	1 NDMA; 2.4 NPYR	−
Dog	0.9	+
Rat and Mouse	3, 0.7	+
Rat and Mouse	0.9	+
Rat and Mouse	1.3, 1.6	+
Rat and Mouse	2.2	+
Rat and Mouse	3.6	+
Rat and Mouse	52	+ (10%)

The highest NDMA level found was 52 ppb in the new National Institutes of Health open formula rat and mouse ration.[4] This carefully controlled product includes 10% fish meal among its 13 major ingredients. The diet is fortified with precise amounts of 13 vitamins and 5 minerals, with routine analyses including crude protein, fat, fiber, ash, 13 amino acids, 11 minerals and 13 vitamins. Even the pellet size is specified to the nearest 1/8 in. Yet, despite these elaborate controls, the diet contains the highest N-nitrosamine level of any food product which we have tested. Almost certainly, fish meal was the source of the contamination.

The presence of appreciable levels of so toxic an animal carcinogen as NDMA in the diets of laboratory animals which are being used for long-term carcinogenesis studies represents an awkward problem,[5] particularly when possible cancer-causing compounds are being tested at low levels. It is recommended that the N-nitrosamine level of the control diet be reported in future carcinogenesis studies.

TERTIARY AMINES

Since the work of Geuther[6] in 1864, it has been known that tertiary amines can yield N-nitrosamines by reaction with nitrous acid. However, fueled by successive college text books, a myth developed that the reaction could not occur. This extraordinary piece of misinformation has been reviewed in the Journal of Chemical Education.[7] The mechanism of tertiary amine nitrosation is discussed by Smith and Loeppky,[8] Lijinsky et al.[9] and Ohshima and Kawabata.[10]

Despite this extensive knowledge, synthetic cutting fluids containing up to 40% triethanolamine and 18% sodium nitrite were widely used. An analysis[11] of the undiluted concentrates showed that they contained N-nitrosodiethanolamine (NDE1A) at concentrations as high as 3%. A contributing factor to the delay in realizing that synthetic cutting fluids contained NDE1A was the fact that the fluids were strongly alkaline. Again, despite evidence to the contrary, the myth persisted that N-nitrosamines could be produced only under acid conditions.[12]

As a direct result of this work, several U.S. manufacturers were able to reformulate their products. "N-nitrosamine-free" cutting fluids are now routinely advertised in trade publications.

The NDE1A in synthetic cutting fluids was formed by nitrosation of triethanolamine (together with a diethanolamine impurity). Since many personal care products, such as hair shampoos, hand and body lotions, and ladies' facial cosmetics, also contain triethanolamine as a major ingredient, it was not altogether surprising to find that these products[13] were comtaminated with NDE1A. In our initial study,[13] NDE1A was reported to be present at levels greater than 10 ppb in 16 out of 29 products, with the highest level being 49,000. A more recent Food and Drug Administration survey[14] shows 19 samples to be positive out of 36, at levels varying from 35 to 130,000 ppb.

The mechanism of formation of NDE1A in cosmetic products has not been demonstrated unequivocally. A likely possiblilty is nitrosation by C-nitro compounds. Fan et al.[15] recently showed that certain bactericides and perfumes, some of which are used cosmetic products, could nitrosate morpholine under certain conditions. This work is continuing, with a view to eliminating the NDE1A impurity in these products.

ROUTE OF MANUFACTURE

In 1977, Ross et al.[16] reported N-nitrosodipropylamine (NDPA) to be present as an impurity in a formulation of the herbicide α,α,α,trifluoro-2,6-dinitro-N,N-dipropyl-p-toluidine (trifluralin) at the 154,000-ppb level. Subsequent studies by other laboratories showed that the nitrosamine contamination was present in all dinitroaniline-based herbicides.[17] The contaminant apparently arose via the synthetic route,[18] wherein the ring was first nitrated with nitric and sulfuric acids, followed by the addition of dipropylamine. Residual nitrosating agent would have N-nitrosated the dipropylamine to NDPA. The manufacturer was able to reduce the NDPA impurity from 154,000 to 18,000.[17] More recent data[19] have shown that the level has now been successfully reduced to below 500 ppb.

It was not surprising to observe that similar problems occur with pharmaceutical products. Eisenbrand *et al.*[20] have reported on the presence of NDMA in all samples of aminopyrine which they analyzed. The most likely source of the contamination was via the synthetic processes used in its manufacture. As a result of their study, sales of the product have been curtailed in Germany.

Schoenhard *et al.* reported on the presence of 1-diphenylmethyl-4-nitrosopiperazine in an antibiotic formulation under development which was never released for general use.[21] Despite the use of a number of different synthetic routes, the final product always contained varying levels of the N-nitroso contaminant. It was eventually determined that the pure drug reacted with singlet oxygen in the air to form the observed N-nitroso impurity. In this case, even the use of a N-nitroso-free synthetic process did not produce a final product devoid of N-nitroso-compound contaminants.

A screen for N-nitroso compounds in 34 prescription drugs and 39 over-the-counter remedies has recently been completed.[22] Comprehensive analytical screening procedures were utilized. Only 3 of the prescription drugs and 2 of the over-the-counter formulations were found to contain suspected N-nitroso compounds. The suspected contaminants have not yet been identified.

AMINE CONTAMINATION

Over the past three years numerous secondary and tertiary amines have been tested for the corresponding N-nitrosamine. Amines which have been tested include dimethylamine, diethylamine, morpholine, triethanolamine and diethanolamine. In each case, most amine samples were contaminated with the corresponding nitrosamine, usually at the 200- to 2000-ppb level. All users of secondary and tertiary amines should be aware that they may contain N-nitrosamine impurities at the high ppb level. Depending on the processes involved, this nitrosamine contamination may carry over into the final product. It is speculated that the parent amine is the source of the very low-level (between 0.03 and 0.1 ppb) NDMA contamination in several household detergents.

ACKNOWLEDGMENTS

The material in this chapter is based upon research supported by the National Science Foundation under Grant No. ENV75 80200 A3. Any opinions, findings and conclusions or recommendations expressed are those

of the authors and do not necessarily reflect the views of the National
Science Foundation.

REFERENCES

1. Ender, F., G. Havre, A. Gelgebosted, N. Koppang, R. Madsen and L.
    Ceh. "Isolation and Identification of an Hepatotoxic Factor in Herring
    Meal Produced from Sodium Nitrite Preserved Herring," *Naturwiss.*
    51:637-638 (1964).
2. Sen, N.P., L. A. Schwinghammer, B. A. Donaldson and W. H. Miles.
    "N-nitrosodimethylamine in Fish Meal," *J. Agric. Food Chem.* 20:
    1280-1281 (1972).
3. Kann, J., B. Spiegelhalder, G. Eisenbrand and R. Preussmann. "Oc-
    currence of Volatile N-nitrosamines in Animal Diets," *Z. Krebsforsch.*
    90:321-323 (1977).
4. Sontag, J. M., and N. P. Page. "NIH Open Formula Rat and Mouse
    Ration." In: *Guidelines for Carcinogen Bioassay in Small Rodents.*
    Division of Cancer Cause and Prevention, National Cancer Institute,
    National Institutes of Health, Bethesda, Maryland, pp. 41-47 (1975).
5. Schmahl, D. "Combination Effects in Chemical Carcinogenesis (Ex-
    perimental Results)," *Oncology*, 33:73-76 (1976).
6. Geuther, A., "Ueber die Einwirkung von Salpetrigsaurem kali auf
    salzaures Trialthylamin," *Archiv Pharmacie*, Weinheim 123:200-
    202 (1864).
7. Hein, G. E. "The Reaction of Tertiary Amines with Nitrous Acid,"
    J. Chem. Educ. 40:181-184 (1963).
8. Smith, P. A. S., and R. N. Loeppky. "Nitrosative Cleavage of Tertiary
    Amines," *J. Am. Chem. Soc.*, 89:1147 1157 (1967).
9. Lijinsky, W., L. Keefer, E. Conrad and R. van de Bogart. "Nitro-
    sation of Tertiary-Amines and some Biologic Implications," *J. Nat.
    Cancer Inst.* 49:1239-1249 (1972).
10. Ohshima, H., and T. Kawabata, "Mechanism of N-nitrosodimethylamine
    Formation from Trimethylamine and Trimethylaminoxide," In: *En-
    vironmental Aspects of N-nitroso Compounds*, E. A. Walker, M.
    Castegnaro, L. Gricuite and R. Lyle Eds., IARC Scientific Publication
    No. 19, Lyon. pp. 143–153 (1978).
11. Fan, T. Y., J. Morrisson, D. P. Rounbehler, R. Ross, D. H. Fine, W.
    Miles and N. P. Sen. "N-nitrosodiethanolamine in Synthetic Cutting
    Fluids: A Part per Hundred Impurity," *Science* 195:70-71 (1977).
12. Mirvish, S. S., "Formation of N-nitroso Compounds: Chemistry,
    Kinetics and *in Vivo* Occurrence," *Toxicol. Appl. Pharmacol.* 31:325-
    351 (1975).
13. Fan, T. Y., U. Goff, L. Song, D. H. Fine, G. P. Arsenault and K.
    Biemann. "N-nitrosodiethanolamine in Consumer Cosmetics, Lotions
    and Shampoos," *Food Cosmet. Toxicol.* 15:423-430 (1977).

14. Wenninger, J. Summary of FDA's Investigations of Nitrosamine Contamination of Cosmetic Products. January 30, 1978. The FDA report lists 6 samples as containing trace (between 10 and 30 ppb) levels. For consistency with the Fan study (Reference 13), these are reported as positive here.

15. Fan, T. Y., R. Vita and D. H. Fine. "C-Nitro Compounds: A New Class of Nitrosating Agents," *Toxicol. Lett.* 2:5-10 (1978).

16. Ross, R., J. Morrison, D. P. Rounbehler, T. Y. Fan and D. H. Fine. "N-nitroso compound impurities in herbicide formulations," *J. Agri. Food Chem.* 25:1416-1418 (1977); Fine, D. H., D. P. Rounbehler, T. Fan and R. Ross. "Human Exposure to Preformed N-nitroso Compounds in the Environment." In: *Cold Spring Harbor Conferences on Cell Proliferation, Vol. 4, Origins of Human Cancer,* H. H. Hiatt, J. D. Watson and J. A. Winsten Eds. (New York: Cold Spring Harbor, 1978).

17. Cohen, S. Z., G. Zweig, M. Law, D. Wright and W. R. Bontoyan. "Analytical Determination of N-nitroso Compounds in Pesticides by United States Environmental Protection Agency-A Preliminary Study." In: *Environmental Aspects of N-nitroso Compounds,* E. A. Walker, M. Castegnaro, L. Griciute and R. Lyle, Eds., IARC Scientific Publication No. 19, Lyon (1978), pp. 333-342.

18. von Rumker, R., E. W. Lawless and A. F. Meiners, Office of Pesticide Programs, Office of Water and Hazardous Materials, U. S. Environmental Protection Agency (1975).

19. Day, E. W., S. D. West, D. G. Saunders and M. J. Bourgeois. "Determination of Volatile Nitrosamine Impurities in Formulated and Technical Products of Dinitroaniline Herbicides," paper PEST 83 presented at the ACS National Meeting, Anahein, CA, March 17, 1978.

20. Eisenbrand, G., B. Spiegelhalder, C. Janzowski, J. Kann and R. Preussmann, "Volatile and Non-volatile N-nitroso Compounds in Foods and Other Environmental Media." In: *Environmental Aspects of N-nitroso Compounds,* E. A. Walker, M. Castegnaro, L. Gricuite and R. Lyle Eds., IARC Scientific Publication No. 19 Lyon (1978), pp. 311-324.

21. Schoenhard, G. L., W. W. Aksamit, R. H. Bible, L. C. Hansen, J. D. Hribar, E. F. LeVon, B. M. Scubeck and H. Wagner. "The Analysis and Source of 1-Diphenylmethyl-4-nitrosopiperazine in 1-Diphenylmethyl-4[(6-methyl-2-pyridyl) Methyleneamino] Piperazine: A Case History." In: *Environmental Aspects of N-nitroso Compounds,* E. A. Walker, M. Castegnaro, L. Griciute and R. Lyle, Eds., IARC Scientific Publication No. 19, Lyon (1978), pp. 75-85.

22. Krull, I. S., U. Goff, A. Silvergleid and D. H. Fine. "N-nitroso Compound Contaminants in Prescription and Nonprescription Drugs," Arzneimittel-Forsch. 29(1) 6:870-874 (1979).

27

TOXICOLOGICAL DYNAMICS

Sheldon D. Murphy

Division of Toxicology
Department of Pharmacology
University of Texas Medical School at Houston
Houston, Texas 77025

INTRODUCTION

Toxicology has recently been defined as "the science which studies the adverse effects of chemicals on living organisms and assesses the probability of their occurrence."[1] Thus, a satisfactory toxicological assessment of a chemical will not only include the identification, quantitation and interpretation of injurious effects of chemicals in living systems, but it will also include a quantitative analysis of the routes and mechanisms by which injurious chemicals reach the sensitive organisms and sensitive cells within the organism. Preceding chapters have focused largely on the physical-chemical, biological and environmental factors that determine to what form and how much of a chemical contaminant of the environment humans and other living organisms will be exposed. This chapter, although not ignoring the role of environmental transport and fate, will be directed more toward the principles and procedures used in assessing the injurious actions of chemicals at target sites of injury in nontarget biological organisms or systems.

The primary objective of toxicological testing is to obtain data on the *dose-response* characteristics of a chemical. These studies provide the primary data base from which estimates of risk to an identified population of organisms may be determined in connection with specific uses or disposal practices for a specific chemicals.

The choice and sequence of toxicity tests will depend on the questions or hypotheses that are developed. The nature and sequence of tests used to satisfy requirements of regulatory agencies may differ markedly from those used in an investigation of basic mechanisms of toxic action. Differences in approach will also depend on whether the investigation in initiated to evaluate the toxicity of a chemical prior to its introduction into use, *i.e., prospective toxicology,* or to confirm in laboratory animals (or under laboratory conditions) epidemiological studies that reveal a statistical association that suggests chemical-induced disease in man, *i.e., retrospective toxicology.* Under ideal conditions, prospective toxicology will eliminate the need for retrospective toxicity evaluation.

The purpose of undertaking research and testing of the potentially injurious effects of chemicals on living organisms is to characterize the nature of the injuries that might be produced and to determine the limiting quantities and/or durations or frequencies of exposure which result in injury. During the decades of the 1940s and 1950s, the Food, Drug and Cosmetic Act of 1938 and its amendments and the Federal Insecticide, Fungicide and Rodenticide Act of 1947 were the major enabling legislation in the United States which formalized requirements for systematic toxicity testing of chemical substances. Even with the toxicological test methods of the 1940s and 1950s, these regulations appear to have provided a sufficient data base to set limits for food additives and pesticide residues adequate to protect the general public against injury from those chemicals under normal use conditions. At least there are no acute or chronic disease states that can be clearly attributed to exposure to these regulated substances in food at the legally permissible residue levels. However, a continuing point of speculation is that the unexplained etiology of a high proportion of birth defects, cancer and some other chronic diseases may be due in part to chemical exposures, and that failure to verify these presumed associations are due to the insensitivity—or total lack—of epidemiological studies. Furthermore, as greater knowledge of biology and highly sensitive methods for measuring biological change have developed, effects have been detected with exposures to chemicals at dosages once thought to be without biological activity. This increased ability to detect effects, the rapid growth of chemical technology and an increasingly informed and concerned public have all been contributing factors which have led to the enactment of numerous laws and regulations which call for an ever increasing quantity and quality of laboratory research and test data concerning the potential of chemicals to produce injury in living organisms. Although several of these laws and regulations acknowledge the potential for adverse effects of chemicals in ecosystems, efforts to develop and validate new toxicological test methods have concentrated heavily on tests intended to evaluate potential for direct effects of exposure on human health.

ASSESSMENT PRIORITIES

Estimates of the numbers of chemicals presently in use which will have to be tested and their hazard to health or environment assessed or reassessed, under laws passed in the 1970s, vary from the tens of thousands to the millions. The number of new chemical substances that will need to be evaluated each year ranges from the hundreds to thousands. In any event the task is formidable, and considerable attention by various committees has been given to methods for assigning priorities for risk assessment. In a recent major study by the National Academy of Sciences,[2] both biological impact and dispersal of the chemical into the environment were scored in order to arrive at a priority classification. Factors considered to contribute to "biological impact" which relate primarily to direct interactions of the chemical with the affected biological system included toxicity, receptor importance and type of effect. Each of these was rated on a scale of one to three with greatest significance rated as one. For example, interference with ecosystem functioning was rated one as a "type of effect," chronic effects at the level of individuals was rated as two, and acute effects on individuals was rated as three. Factors which relate primarily to the role of transport and fate of the chemical in the environment included availability to organism, potential for biomagnification, and stability and persistence. The second rating scheme involved scoring the extent of chemical dispersal by four categories ranked one to four in the following order: widespread-high release; widespread-low release, localized-high release, and localized-low release. The product of the score from the "biological impact" ratings and the "chemical dispersal" ratings can then be taken as representing an overall priority for testing or for regulation, with the lowest products (rated as described above) having the highest priority.

Using this scheme, then, one could readily predict a high priority, for example, for a substance like mercury, which has high toxicity, reacts with important macromolecules in a large number of biological receptors at several trophic levels, and persists in the environment indefinitely in its elemental form, with the potential to biomagnify and incorporate into aquatic and terrestial food chains; and which, because of its diverse properties, is used in numerous and varied applications from which it can be released or made more transportable in the environment.

Of course, the foregoing scheme assumes a considerable knowledge of the biological disposition and effects of a substance in order to assign scores to arrive at the priority rating. It is a larger problem to assign priorities to chemicals for which there are few or no biological data. This must necessarily be done by analogies which place substances into various chemical classes of varied levels of suspicion based on data from one or a few other members of that class.

TOXICITY TESTING

Some knowledge or reasonable basis for suspicion of biological injury is essential for selecting which chemicals are in greatest need of toxicological assessment. Normally, the objective of toxicological testing is to identify the nature of possible adverse effects and to relate response (or injury) to dosage. For any specific adverse effect that may be produced as an organism is exposed to a continuum of doses, various dose-response possibilities can be envisioned as shown in Figure 1. In Case I, either the substance is inert in the test organism or the wrong effect has been chosen for measurement.

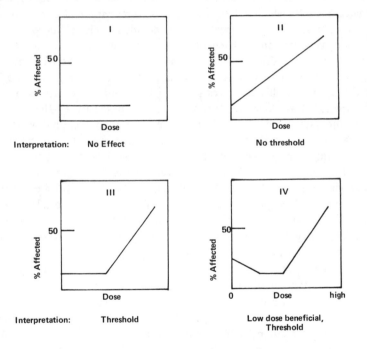

Figure 1. Theoretical dose-response curves.

Case II represents a form of the dose-response curve now most often associated with carcinogenic or mutagenic action. It envisages a continuum of increasing frequency of response at any finite increase in the dose of the chemical. The effects are dose-related, and if the dose-response curves are sufficiently well characterized and understood it may be possible to estimate, with statistical limits, the number of organisms that would be affected at any known or anticipated environmental dosage.

Case III is the most classic form of dose-response curve in toxicology and represents the assumption upon which most of the environmental quality standards and limits in effect today have been based. In short, this form of the dose-response curve says that there is some finite exposure dose below which the rates of the biological protective processes of detoxication, excretion and injury-repair keep pace with or exceed the rates of exposure, absorption and injury-production. The concept of threshhold is no doubt valid for individuals; that is, each individual living organism likely has its own threshold. But, because of individual variation in the rates of the several biological processes cited, application of the threshold concept to populations of organisms is fraught with controversy and misunderstanding.

The last illustration, IV, is one in which adverse effects are associated with either too little or too much of a substance. A common example of such a substance to which humans are exposed (usually in just the right amounts) is oxygen. More realistic examples deal with certain trace nutrients that are toxic at high doses, such as selenium and copper.

CONCLUSIONS FROM TEST DATA

In any experiment intended to provide information for risk assessment, whether the experiment involves acute exposure or chronic, whether it is concerned with an air, food or water contaminant, an industrial chemical, pesticide or food additive, and whether the substance is to be swallowed, inhaled or poured on the skin; the experiment should be designed using a range of exposure doses, some of which will yield results between the 0- and 100%-incidence of injury. With properly designed experiments one can reach decisions concerning hazard assessments which logically flow from experimental data. Descriptive terms have been applied to the conclusions reached on the basis of experimental data; such as "no effect level," "no adverse effect level," or "no observed adverse effect level."[3,4] These terms developed initially out of toxicity assessments of food additives and pesticides, hence "level" refers to level (i.e., concentration) in the diet. No effect level, the oldest of these terms, was taken to mean the highest test concentration in the diet of experimental animals that did not result in biological data that differed from animals fed control diets. Of course it was only as dependable and comprehensive as the test protocol. As an acknowledgement that no assessment could be made beyond the measurements actually conducted, the word "observed" was added. Then, as more and more sensitive biological test methods were applied to toxicity assessments, effects were detected which did not seem to compromise the animals' health. To cover this the term "adverse" was added. This assumes that a given biological effect can, in fact, be interpreted as to its adverseness. This is not always a certainty with effects such as increased liver weights or induction of increased enzyme activities which may occur without detectable cellular pathology.

In spite of the uncertainties in the terminology or the inadequacies of the data base, these experimentally derived estimates of the limiting dosage for effects in animals are used, with appropriate safety factors, to estimate the dosage that humans could ingest (or otherwise be exposed to) *daily* throughout a lifetime without appreciable risk to health. This daily dosage, expressed in terms of mg/kg, has become known as the acceptable daily intake (ADI).[5] This value, coupled with data on what is good manufacturing, agricultural or engineering practice, ultimately leads to standard setting.

TOXICOLOGICAL INFORMATION NECESSARY

What is the nature of the data that must be available to arrive at an assessment of health hazard? The answer to this changes, as it should, with time and the acquisition of new knowledge and development of new test methodology. Furthermore, each chemical substance or proposed use will have characteristics which may require some differences in test protocol from those used to evaluate another substance. Nevertheless, most national or international organizations or agencies involved in chemical hazard assessments as well as producing industries consider certain general types of biological research and testing essential for all substances that may be widely distributed in the environment. These general areas of knowledge about a chemical substance, required for its hazard assessment, include:

1. *Chemical and Physical Properties.* Knowledge of these are essential to the development of suitable analytical methods for monitoring purposes, to enable assessment of purity for standardization of products, to predict potential for biological storage and to suggest environmental distribution. It is necessary to establish chemical identity and to standardize both technical products used in commerce as well as samples subjected to toxicity testing, since extrapolation of toxicity data from test sample to commercial product assumes toxicological (and therefore chemical) equivalence.

2. *Metabolism and Disposition.* Knowledge of these are essential to evaluate the disposition kinetics within an organism as well as in a food chain. It is essential to know if the parent compound or a metabolite is active. This is especially important in ecosystems. For example, are the metabolites in food plants the same as in animals? If not, more extensive tests of plant metabolites may be required. furthermore, knowledge of metabolism helps predict the likelihood of partitioning in different environmental or biological media which will determine the likelihood of biomagnification or accumulation.

3. *Acute Toxicity—LD_{50}*. Although this is the most frequently cited toxicological statistic, it has limited value. For purposes of assessing hazard to human health, it is most useful for the information derived during its determination, *i.e.*, a characterization of signs and symptoms. Comparative acute toxicity data can also be of value in predicting whether at any given dosage a chemical is likely to produce death in one or more organisms in an ecosystem. If a particularly sensitive organism is a member of a food chain, contamination of an ecosystem by a chemical to which the organism is uniquely susceptible may indirectly cause serious health consequences to organisms at a higher trophic level even though direct toxic effects may not be apparent in these organisms. On the other hand, an organism that is uniquely resistant to acute toxicity may provide a medium for subtle bioconcentration of a pollutant, since accumulation in tissues is possible only in surviving organisms.

4. *Repeated Short-Term Studies*. Usually these involve administration of a test substance in the diet, drinking water or inhaled air for a period of several weeks, generally three months in rodents. By thoroughly studying a continuum of doses, a broad biological test program can identify nearly all types of effects that will occur. Obvious exceptions are induction of cancer or heritable mutations.

5. *Long-Term Studies*. Properly designed, these studies can evaluate carcinogenicity, multigeneration reproduction effects, and most other types of delayed chronic injury, *e.g.*, chronic neurologic degenerative diseases. fibrotic processes, etc. These are generally lifetime exposure studies in rodents and for a major fraction of a lifetime in other species.

6. *Special Studies*. Although many tests which were once considered special are now routine for complete toxicity testing, this category has included studies ranging from toxicokinetic studies to studies of carcinogenicity, mutagenicity or teratogenicity, and special tests for effects on reproduction, and tests to determine potentiation by other chemicals. In some cases the potential for a chemical to produce these effects can be assessed in well designed protocols for acute, short-term or long-term exposure studies. In other cases it is necessary to develop protocols to test specifically for an effect, *i.e.*, tests for mutagenic potential. All of these special studies should also be designed to allow dose-response analyses. Unfortunately, a common finding in toxicological assessment or reassessment of many familiar chemicals is that although some test data are available, they were not obtained in experiments designed to give dose-response relationships.

COMPARATIVE TOXICITY ASSESSMENTS

Until recently, systematic assessments of the potential for adverse biological effects of chemicals have been primarily concerned with developing the means to predict and then prevent injury to humans. Common laboratory mammals have served as the experimental models for human biochemical, physiological and pathological response to toxic chemicals. Indeed, there have been, and are, serious concerns as to whether effects observed in common laboratory mammals can accurately predict human response. Amelioration of these concerns will come only through carefully designed studies of comparative metabolism, disposition kinetics, and target site reactivities. These studies are likely to involve measurements in humans and/or their tissues as well as measurements in standard laboratory mammals and their tissues. Ethical considerations preclude many types of experimental studies in humans, but careful epidemiological and clinical studies in persons with high risk of exposure can provide data for correlation with studies in laboratory animals.

Although there are numerous difficulties in extrapolating toxicological data from laboratory mammals to man, these difficulties are small compared with those of evaluating the toxicological hazards of chemicals to the whole environmental biospectrum. The data in Table I illustrate some of the problems of making generalizations regarding the toxicity of one group of environmental contaminants: insecticides.

Table I. Acute Toxicities of Common Insecticides[6-9]

Pesticide	Male Rats LD$_{50}$ (mg/kg) Oral	Mallards LD$_{50}$ (mg/kg) Oral	Bluegills (μg/liter) 96-hr TLM
DDT	113	>2,240	16
Dieldrin	46	381	7.9
Methoxychlor	6,000	>2,000	62
Parathion	13	2.13	95
Methyl Parathion	14	10	1,900
Guthion	13	136	5.2
Malathion	1,375	1,485	90
Carbaryl	850	>2,179	5,300

These data were selected from reports in which the experiments were conducted under standardized conditions. They suggest, first, that one cannot generalize that the organophosphate compounds are more acutely toxic than organochlorine compounds that have been or are being used as insecticides. For the three samples of organochlorines, there does appear to be a measure of predictability of the relative toxicities from one vertebrate class to another, in that the least and most toxic to rats are also the least and most toxic to mallards and bluegills. This is not the case with the organophosphates, however, as parathion, methyl parathion and Guthion (or azinphosmethyl) are all about equally toxic by the oral route in rats, but differ markedly in their relative toxicities to mallards and bluegills.

Several years ago research was undertaken to elucidate the mechanism of the marked differences in toxicity of *parathion* to mammalian, avian and piscine species.[10,11] Following the popular line that species differences in toxicity can generally be explained by differences in rates of metabolic activation or detoxification, the rates of the then known metabolic pathways for parathion in liver homogenates of mice, chickens and bullhead fish were compared. Enzymatic oxidation of parathion to a metabolite, paraoxon, which is a potent inhibitor of cholinesterase, was measured as an index of metabolic activation, and the hydrolysis of paraoxon by the same livers was measured as an index of metabolic detoxication. The in vitro reactivity of the oxygen analog (paraoxon) with acetylcholinesterase in brains from each species was measured to compare sensitivity of target sites. Results showed that comparative rates of biotransformation, both activation and detoxification, failed to correlate with comparative toxicity in intact animals. However, in this case, species differences in sensitivity of a target enzyme, *i.e.,* nerve acetylcholinesterase, to inhibition by paraoxon correlated well with species difference in acute toxicity.

Extensions of these studies[12] indicated that not only are there marked differences in the doses of certain organophosphorus insecticides required to inhibit the acetylcholinesterase of different classes of vertebrates, but there are striking differences in the rates of onset of and recovery from inhibition. Inhibition of brain cholinesterase after i.p. injection of approximately equitoxic doses of azinphosmethyl, parathion and methylparathion occurred much more rapidly in mice than in bluegill sunfish; however, the duration of inhibition was several hours to days longer in fish. Thus, these findings suggest that if one were to estimate probability of accumulation of injury from anticholinesterase insecticides (as opposed to accumulation of the substance) in an ecosystem on the basis of studies in common laboratory mammals, gross underestimates would be made.

But what if toxicity data are collected in one or two representative species of various classes of vertebrates, can toxicity in other members of the class then be predicted? Review of acute toxicity data of pesticides in a large number of species, Tables II and III, suggests that the variance of interspecies response in nonmammalian organisms is probably even larger than the factor of 10-to-12-fold estimated difference in susceptibilities of mice and men. In fish species there was a much greater variation in acute toxicity for organophosphate insecticides than for organochlorines. Similarly,

Table II. Comparative Pesticide Toxicities to Fish[a]

| Species | 96-hr TLM (μg/liter = ppb) | | | | |
	DDT	Dieldrin	Azinphosmethyl	Parathion	Malathion
Bluegills	16	7.9	5.3	95	90
Guppies	43	22.0	110	55	840
Fathead Minnows	32	16.0	93	1300	23,000
Goldfish	27	37.0	1300	2700	450
Largemouth bass	--	----	----	190	50
Ratio, Least:					
Most Sensitive	2.7	4.7	246	49	460

[a]Data selected from References 8 and 9.

Table III. Comparative Pesticide Toxicities to Avian Species[a]

| Species | Oral LD_{50} (mg/kg) | |
	Dieldrin	Parathion
Mallards	381	2.13
Pheasants	79	12.4
Chukars	23.4	24.0
Coturnix	69.7	5.95
Pigeons	26.6	2.52
House Sparrows	47.6	3.36
Fulvous Tree Ducks	100-200	0.125-0.250
Gray Partridges	8.84	16.0
Ratio, least:most sensitive		
Avian Species	45	192

[a]Data selected from Reference 7.

there was as much as 192-fold difference in LD_{50} values for parathion among 9 avian species.

Further complications are apparent when one compares toxicities in a variety of organisms in an ecosystem. For example, Stewart et al.[13] reported on the toxicity of the carbamate insecticide carbaryl and its major hydrolytic degradation product, 1-naphthol. It is of considerable interest that although the parent insecticide was 100 or more times more toxic to mud shrimp and Dungeness crab than was 1-naphthol, the degradation product (1-naphthol) was slightly more toxic to cockle clams and shiner perch.

What do these few examples illustrate? At least, they suggest that the toxicodynamics of an ecosystem is extremely complex. Transport and fate studies are very useful in identifying what compounds and/or metabolites will be available for exposure at different points in the ecosystem and hence may identify the organisms at greatest risk of exposure. However, until we have much more basic information concerning the mechanisms of species differences in susceptibility to toxic effects of chemicals, it will continue to be a most difficult task to predict hazard to a wide spectrum of species in the biosphere. Appropriately, greatest immediate attention must be given to means of assessing hazard of direct exposures of humans, but ultimately human welfare will also be linked with the effects on the ecosystem of which man is only one element. Environmental hazard assessments in the future will be highly dependent on a knowledge of comparative toxicology, and the more mechanistic this knowledge becomes, the more likely it is that an adequate data base for decisions will be attainable from reasonable expenditures on toxicity testing.

REFERENCES

1. Educational Committee of the Society of Toxicology. "Editorial: The Education of a Toxicologist," *Toxicol. Appl. Pharmacol.* 45: 375-376 (1978).
2. NAS. *Principles for Evaluating Chemicals in the Environment.* (Washington, DC: National Academy of Sciences Printing and Publishing Office, 1975), p. 227.
3. NAS/NRC. *Evaluating the Safety of Food Chemicals.* (Washington, DC: National Academy of Sciences Printing and Publishing Office, 1970).
4. "Procedures for Investigating Intentional and Unintentional Food Additives," World Health Organization Technical Report Series *No. 348* (1967).
5. "Toxicological Evaluation of Certain Food Additives with a Review of General Principles and of Specifications," World Health Organization Technical Report Series *No. 539* (1974).
6. Gaines, T. B. "Acute Toxicity of Pesticides," *Toxicol. Appl. Pharmacol.* 14: 515-534 (1969).
7. Tucker, R. K., and D. G. Crabtree. "Handbook of Toxicology of Pesticides to Wildlife," U.S. Department of the Interior Resource Publication No. 84, U.S. Government Printing Office (1972).
8. Pickering, Q. H., C. Henderson and A. E. Lemke. "The Toxicity of Organic Phosphorus Insecticides to Different Species of Warm Water Fishes," *Trans. Am. Fish. Soc.* 91: 175-184 (1962).
9. Henderson, C., Q. H. Pickering and C. M. Tarzwell. Reprinted from "Transactions of Second Seminar on Biological Problems in Water Pollution," Robert A. Taft Sanitary Engineering Center, U.S. Public Health Service, Cincinnati, Ohio, April 20-24, 1959.

10. Murphy, S. D. "Liver Metabolism and Toxicity of Thiophosphate Insecticides in Mammalian, Avian and Piscine Species," *Proc. Soc. Exp. Biol. Med.* 123: 392-398 (1966).

11. Murphy, S. D., R. R. Lauwerys and K. L. Cheever. "Comparative Anticholinesterase Actions of Organophosphorus Insecticides in Vertebrates," *Toxicol. Appl. Pharmacol.* 12: 22-35 (1968).

12. Murphy, S. D., and G. M. Benke. "Anticholinesterase Action of Methyl Parathion, Parathion and Azinphosmethyl in Mice and Fish: Onset and Recovery of Inhibition," *Bull. Environ. Contam. Toxicol.* 12:117-122 (1974).

13. Stewart, N. E., R. E. Millemann and W. P. Breese. "Acute Toxicity of the Insecticide Sevin® and its Hydrolytic Product 1-naphthol to Some Marine Organisms," *Trans. Am. Fish. Soc.* 96:25-30 (1967).

28

TRANSPORT AND FATE OF CHEMICALS IN EVALUATION OF THEIR CARCINOGENICITY

William Lijinsky

Chemical Carcinogenesis Program
Frederick Cancer Research Center
Frederick, Maryland 21701

The assessment of the safety of chemicals has become an important issue since the recognition that many health problems, notably cancer, might be related in part to exposure to chemicals in the environment and diet. The instances of poisoning by kepone in Virginia, methylmercury in Japan, and the many examples of rare types of cancer which can be related to exposure to specific chemicals show the importance of monitoring and examining foreign compounds for adverse effects before large numbers of people are exposed to them. In less enlightened times, society was obliged to take its chances and to wait until danger was suggested by human sickness or even death before taking protective steps, but this is not acceptable today.

Had our knowledge been adequate and our regulations rigorous, we would not have faced the tragedies of thalidomide, kepone, asbestos, aromatic amine bladder cancer, vinyl chloride and diethylstilbestrol, to name only a few. It is now recognized that there is an obligation to examine any substance, to which many people might be exposed for a major part of their lifetime, for long-term toxic effects, the most important of which is carcinogenesis. It was thought at one time that a large single dose of a substance at several dose levels was satisfactory to establish an acute toxic dose, or LD_{50} (the dose which would, on average, kill half of the exposed animals), in one or more experimental animal species. To protect the public it was deemed adequate to ensure that a level of exposure of humans to that substance was set at one hundredth of the LD_{50} for the animals. This, in turn, determined the concentration of such a substance, for example a food additive, in the food product. Little or no consideration was given to the possibility, even then apparent, that an accumulation of miniscule toxic effects of small doses could add up, over a several-decade lifetime exposure of people, to the irreversible production of cancer. There is now not much doubt, from observations in humans and in experimental animals, that this is the route by which most cancers arise and that seldom, if ever, does cancer in humans result from a single exposure to a chemical carcinogen.

Although many compounds are known which are associated with an increased incidence of certain types of cancer in humans (aromatic amines, polynuclear hydrocarbons, vinyl chloride), there is little or no sound information about the doses to which people were exposed—those who developed cancer or those who did not. Consequently, it is not possible to determine which dose produced no observable effect in humans. Even if such information were available it would be only a partial answer to that question, since it would almost certainly not be known which other exposures to carcinogens of other kinds the individuals had. Therefore, there is no present means of assessing the risk to humans of exposure to any particular chemical carcinogen, or to chemical carcinogens in general. Instead we must depend on experiments in animals for information from which to estimate risks to humans of exposure to toxic or carcinogenic chemicals.

It is apparent that very few carcinogens exert their effect on cells as such, and these are mainly direct alkylating agents. Instead, most carcinogens need to be activated enzymically, probably to electrophilic forms[1] which then react with macromolecules to induce the neoplastic change. Whether this activation occurs only in each susceptible organ, or occurs in the liver to produce an intermediate which is then transported in some form to the target cell in which it reacts, is an unresolved question. Our understanding of the mechanism of carcinogenesis depends on the answer, as do our interpretation of the results of carcinogenesis tests in animals and our attempts to correlate bacterial mutagenicity testing and other in vitro tests with carcinogenesis. Also, any attempt we might make to reverse the process of carcinogenesis (assuming that it is reversible) depends on our more or less complete understanding of the mechanism of carcinogenesis. The alternative is to rely entirely on prevention to deal with the cancer problem, which is itself a worthwhile goal.

From these several points of view, the objective of understanding the transport, fate and effect of carcinogens underlies our attempts to make logical decisions about the testing of chemicals for carcinogenicity and the interpretation of the results of those tests. Several assumptions are common sense, such as that compounds which are not absorbed by the body or which cannot enter cells will not be carcinogenic. (This is probably the reason for the noncarcinogenicity of most ionized derivatives of such a potent class of carcinogens as the nitrosamines.) This, in turn, determines which are the effective routes of administration of many carcinogens. For example, polynuclear compounds are highly insoluble in water and often have low solubility in fats, so that they are mainly locally acting carcinogens, inducing skin tumors by skin painting, sarcomas at the site of subcutaneous injection and lung tumors by instillation in the lung. They are largely ineffective by oral administration, although there are a few exceptions.

Many of the best known groups of carcinogens, for example, azo-dyes, aflatoxins and aromatic amines, induce tumors of the liver in rats when administered orally (Figure 1). However, the differences in potency are

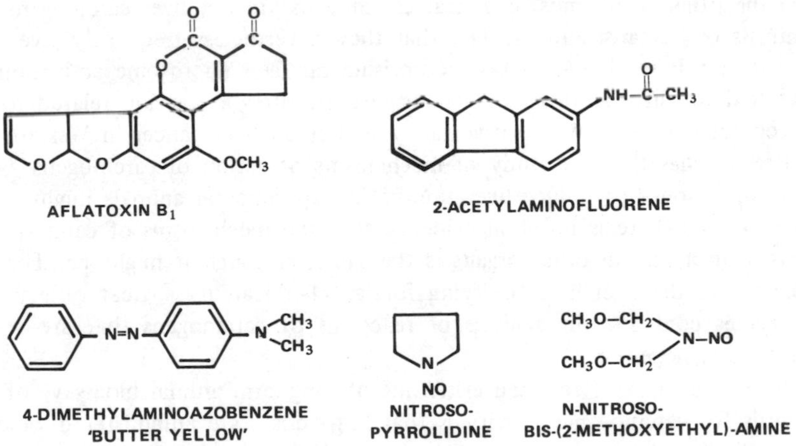
OCH3
AFLATOXIN B1
NH–C–CH3
2-ACETYLAMINOFLUORENE
N=N
CH3
CH3
4-DIMETHYLAMINOAZOBENZENE
'BUTTER YELLOW'
NO
NITROSO-
PYRROLIDINE
CH3O–CH2
CH3O–CH2
N–NO
N-NITROSO-
BIS-(2-METHOXYETHYL)-AMINE

Figure 1. Structures of five carcinogens that induce liver cell tumors (carcinomas) when fed to rats.

enormous, aflatoxin B_1 being several thousand times more effective on a weight basis than the azo dye, the other compounds being intermediate. As little as 1 mg of aflatoxin B_1 has given rise to liver tumors in rats. The outstanding biochemical studies of the Millers during the past 20 years have shown the routes of metabolism in the liver that lead to activation of several of these carcinogens to electrophilic active forms. These active forms seem likely to be those which induce the neoplastic change through inter-action with cell macromolecules, usually assumed to be DNA. While reaction with DNA is a plausible mechanism of action of carcinogens, it is by no means proved, and the Millers recognized the possibility of epigenetic mech-anisms, although what these might be has not been defined. Nevertheless, there is evidence from the work of the Millers and others that formation of a reactive intermediate by particular enzymic reactions is an essential step in carcinogenesis. It is assumed that the reason for the carcinogenic action of the compounds just mentioned, in the liver and not elsewhere, is because the activating enzymes are present in the liver only. The reactive intermediates might have a short life, or might not be transportable else-where in the body, thereby restricting their action to the liver. This might also explain the close correlation of the carcinogenic action of many of these compounds and their active metabolites with mutagenic activity in bacteria (the Ames test), when activated by a rat liver microsomal fraction.[2]

On the other hand, most chemical carcinogens are not liver carcinogens in animals or humans, and the fact that they nevertheless frequently give a positive result in the Ames test comprising rat liver microsome activation makes it doubtful that this mutagenesis test measures a reaction related to carcinogenesis. Indeed, the relative rarity of human liver cancer in Western countries implies that the study of mechanisms of action of carcinogens which appear only to induce liver tumors in experimental animals might be misleading. There is no good evidence that the mechanisms of carcino-genesis in liver and in other organs is the same, although it might be. The major role of the liver in detoxifying foreign chemicals by a great variety of enzymes confuses our analysis of reactions of carcinogens that are re-lated to carcinogenesis.

One of the most often cited criticisms of long-term animal bioassays of chemicals for carcinogenic activity is that large doses are administered which "overload" the enzyme systems in the liver which detoxify these chemicals, thereby leading to accumulation of carcinogenic "insults." This is quite con-trary to logic and to the facts, since most carcinogens require metabolic activation and excessive doses would lead to their being less well metabo-lized and less effective, not more effective. In animal experiments it is usually observed that, by weight, small doses when administered continu-ously are more effective in eliciting tumors than are large doses adminis-tered for a shorter time. This has been demonstrated with polynuclear hydrocarbons, with aflatoxins and with a number of nitrosamines. Single doses of several nitrosamines, close to the acute LD_{50}, have given rise to few, if any, induced tumors in rats, while the same quantity administered in small daily or weekly doses over a 30- to 50-week period will often induce a 100% incidence of tumors. This correlates very well with our studies of the metabolism in rats of three cyclic nitrosamines (Figure 2)

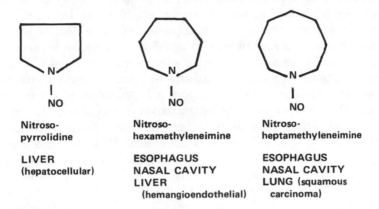

Nitroso-pyrrolidine

LIVER
(hepatocellular)

Nitroso-hexamethyleneimine

ESOPHAGUS
NASAL CAVITY
LIVER
(hemangioendothelial)

Nitroso-heptamethyleneimine

ESOPHAGUS
NASAL CAVITY
LUNG (squamous carcinoma)

Figure 2. Three cyclic nitrosamines labeled with ^{14}C and their target organs. Their metabolism is being compared in rats.

labeled with ^{14}C. When high doses, approaching the LD_{50}, were given, only 2-8% of the dose was converted to $^{14}CO_2$ in 24 hr, whereas at low doses, adequate when continuous to give rise to a large incidence of tumors, there was 40-75% conversion to $^{14}CO_2$.[3] It is probable that the differences in susceptibility of different strains (and sometimes of the two sexes) of rats to a carcinogen are due partly to differences in the content of metabolizing enzymes in their livers (and other organs). Since the mutagenic activity of a compound in a microsomally mediated bacterial test is proportional to the concentration of microsomal enzymes,[4] differences in enzyme content and activity probably explain the inadequate activating ability of liver microsomes from Fischer rats or female Sprague-Dawley rats.

These findings suggest that a small dose of a carcinogen is relatively more effective in adding to the carcinogenic risk than a dose very much larger, some of which is likely to be excreted unchanged. This is probably the situation in man, whose exposures are more likely to be small and frequent than large and rare.

The great differences in potency between one carcinogen and another, even when of similar chemical structure, make it likely that physical and chemical properties determine the effectiveness with which the carcinogen is converted into an active form, and the activity of the latter in inducing the significant lesion. The significant reaction is usually assumed to be with nuclear DNA, and the evidence for this is that such reactions can be demonstrated with simple alkylating agents and that there is a correspondence between carcinogenesis and mutagenesis on the part of many carcinogens. Mutagenesis is assumed to involve direct interaction with DNA and has been given particular attention in the development and application of the Ames *Salmonella* test.[5] Recent attempts to establish a correlation between mutagenic potency in the bacterial systems and carcinogenic potency in animals, including, ultimately, man, seem unlikely to succeed because of the unknown variables of absorption, transport and metabolism of carcinogens in animals. These factors make the interpretation of carcinogenicity tests in animals quite difficult and are the biggest obstacles in extrapolating the results of animal tests to human risk assessment. They also make understanding the mechanisms of carcinogenesis difficult, since carcinogenesis can, at the moment, be considered only a process which occurs in whole animals and not in simpler systems. Similarly, only administration routes which simulate those of man can be considered to be a basis for risk extrapolation to the effects of carcinogens in humans.

All of these points can be illustrated by the enormous number of studies with nitrosamines which, among them, comprise the most broadly acting group of carcinogens with no identified insusceptible species. It is most unlikely that any species, including man, lacks enzymes which can activate nitrosamines to carcinogenic forms. However, there are great differences in response between one species and another to a given nitrosamine, especially in the organs and tissues which are susceptible.[6] This is both confusing and potentially helpful in understanding their mechanism of action.

The nitrosamides seem to be directly acting compounds, both as carcinogens and mutagens, and will act wherever they are placed, so that their actions depend less on their physical properties related to transport than on their instability, which limits their distribution. However, it is noticeable that potency varies considerably even among these compounds, which we assume act in predictable fashion, by direct alkylation of nucleic acids. For example, among the nitrosomethylcarbamate esters, many of which are derivatives of insecticides, there are large differences in both carcinogenicity and mutagenicity which are difficult to reconcile with their chemical similarity.[7,8]

In surveying the more than 120 nitrosamines which have been thus far tested in animals, mostly in rats, several patterns emerge. One is that molecular size, *per se*, does not have a great bearing on whether or not the compound is a carcinogen. However, whether or not it is a neutral molecule or is charged does have a bearing, since almost all nitrosamines which are acidic or basic are weak carcinogens or noncarcinogens; even the exceptions, such as nitrososarcosine and nitrosonornicotine do not induce liver tumors. They are similarly very weak mutagens or nonmutagens in bacteria,[7,9,10] although the possibility exists that they can enter the bacteria, and it is less likely that they can enter mammalian cells.

Among the aliphatic nitrosamines there is a tendency for carcinogenic potency to decrease as the molecular size increases, particularly among symmetrical nitrosamines. Thus, while nitrosodimethylamine and nitrosodiethylamine are equally very potent carcinogens, nitrosodi-n-propylamine is weaker, nitrosodi-n-butylamine is still weaker, nitrosodi-n-amylamine is very weak and nitrosodi-n-octylamine has not induced tumors. Curiously, the mutagenicity in the liver microsomally activated Ames test is in the opposite direction, even nitrosodi-n-octylamine being more active than nitrosodimethylamine.[11] Thus it appears that increasing liposolubility decreases carcinogenic potency but increases mutagenic potency among aliphatic nitrosamines. This seems to be restricted, however, to the symmetrical compounds, since nitrosomethyldodecylamine and nitrosomethylundecylamine, which are both very large molecules and very liposoluble, are potent carcinogens in rats, inducing, respectively, bladder tumors[12] and liver tumors.[13]

Neither is the concept that decreasing lipophilicity (and, therefore, less easy passage across membranes) is favorable to carcinogenicity borne out by study of the carcinogenicity of hydroxylated nitrosamines. Among aliphatic nitrosamines, the presence of oxygen substituents invariably decreases carcinogenic activity. In the case of the 2-hydroxy derivatives of both nitrosodiethylamine and nitrosodi-n-propylamine, they are both orders of magnitude less potent than the unsubstituted nitrosamines. Since nitrosodiethanolamine is a nitrosamine occuring in the environment in relatively high concentrations, it is fortunate that it is a very weak carcinogen in animals. This might be due, partially, to its rapid elimination from the body, a consequence of its very high solubility in water.[14] Most of the hydroxylated nitrosamines tested in the Ames test have been nonmutagenic when activated with rat liver microsomes, although they are liver carcinogens in rats. This suggests that there are more significant factors determining carcinogenicity and mutagenicity than lipid/water partition coefficients.[15]

Among cyclic nitrosamines there is quite a different pattern of carcinogenicity from that seen with the aliphatic nitrosamines. As the ring size increases, and liposolubility increases, there is an increase in carcinogenic potency from the weakly carcinogenic nitrosoazetidine to the extremely potent nitrosoheptamethyleneimine. In this case the mutagenic potency parallels carcinogenic potency quite well,[16] suggesting that the metabolism by liver enzymes to a mutagenic product is dependent to some extent on the liposolubility of the nitrosamine. Again in contrast with the aliphatic nitrosamines, hydroxyl substitution in cyclic nitrosamines does not decrease the carcinogenic potency, even though liposolubility decreases, but the mutagenicity is greatly diminished, almost to the point of insignificance.[6]

In the few studies that have been made of distribution of nitrosamines as a possible factor in the organ-specific effects of these carcinogens, only small differences have been noted in concentration of the compounds or their metabolites between target and nontarget organs. This has been true even when comparisons have been made between species. For example, in our examination of the fate of 2,6-dimethylnitrosomorpholine in rats and hamsters, the most significant difference, and that very small, is in the rate of metabolism of the compound, which appears to be faster in the hamster. Both rats and hamsters were given the same dose (2 mg), and hamsters are considerably smaller than rats. This does not, however, go very far to explain why the compound is a very potent esophageal carcinogen in the rat,[17] but induces tumors of the pancreas in hamsters[18] and, incidentally, only liver tumors in guinea pigs. The structures of some of the metabolites are given in Figure 3. There are some differences, as shown in Figure 4,

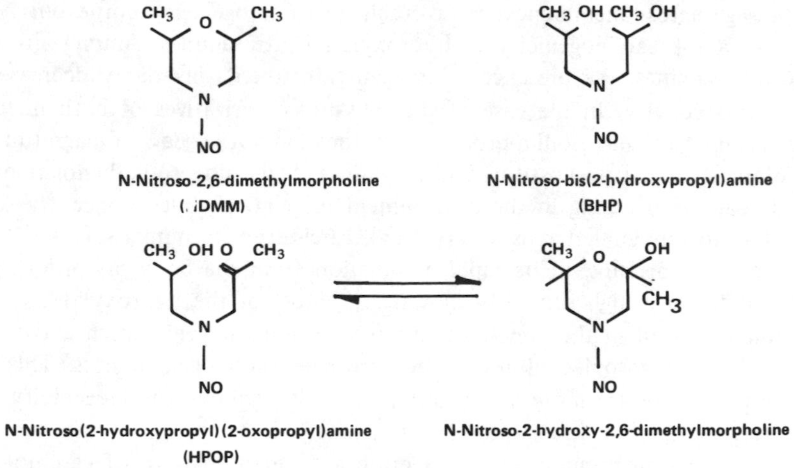

Figure 3. Some metabolites of 2,6-dimethylnitrosomorpholine in rats and Syrian hamsters.

in the pattern of metabolites between the hamster and the rat, but there is no concentration of radioactivity in any particular organ. The problem is further confused in the case of this compound by the fact that the nitrosamine as prepared consists of two isomers (Figure 5) of which the

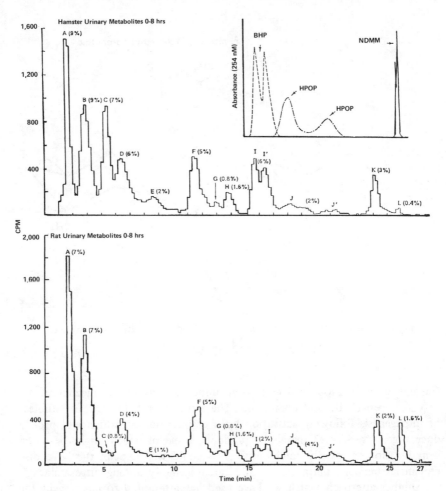

Figure 4. HPLC chromatogram of urine of rats and hamsters given 2 mg of ^{3}H-2,6-dimethylnitrosomorpholine. Urine was collected for 8 hours after administering the compound in corn oil.

minor one, *trans*, is considerably more carcinogenic than the *cis* isomer, at least in rats. Since the gross metabolism of both isomers appears to be very similar, this suggests that most of the metabolism which can be easily observed with this nitrosamine is unrelated to carcinogenesis This might be general for all nitrosamines and it is possibly so for all carcinogens. This will make it unlikely that examining metabolism of carcinogens by our

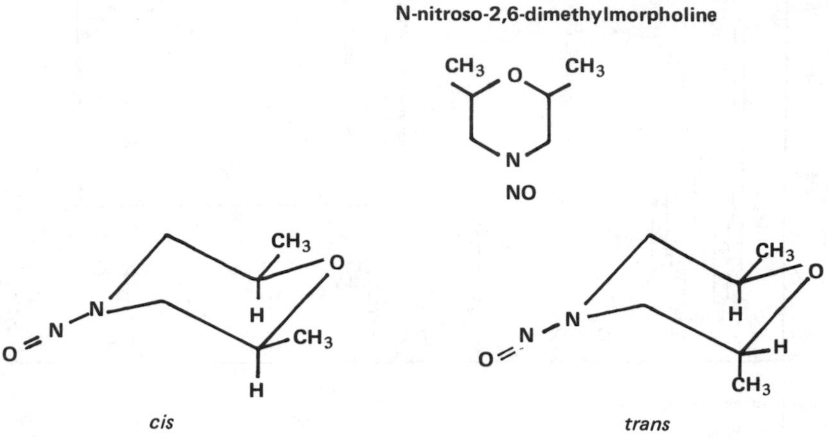

N-nitroso-2,6-dimethylmorpholine

Figure 5. Structures of the two isomers of 2,6-dimethylnitrosomorpholine.

current crude methods will lead to an understanding of the mechanism of carcinogenesis by any compound. The same is probably true of studies of mutagenesis following activation by liver microsomes, which will produce a great variety of metabolic products, some of which might be mutagens but not necessarily carcinogens. This could well explain the large discrepancies between mutagenicity and carcinogenicity among nitrosamines.

Another approach which we have used has attempted to circumvent the complications introduced because of different physical properties of different nitrosamines, which makes comparisons of metabolism difficult. By using nitrosamines specifically labeled with deuterium in certain positions, only the chemical reactivity at those positions is changed and no physical properties. Since it is more difficult to break a carbon-deuterium bond than a carbon-hydrogen bond, metabolism which involves such cleavage will be slowed in the deuterium-labeled nitrosamines. By administering to groups of animals equimolar doses of the labeled and unlabeled compounds, a decrease or increase in the incidence of tumors induced will indicate the effect of deuterium substitution on carcinogenic potency.

Using this system, we were able to show that the effect of deuterium labeling in nitrosodimethylamine or in the alpha positions of nitrosomorpholine was to reduce carcinogenic activity considerably.[19,20] This suggested very strongly that cleavage of an alpha carbon-hydrogen bond was a rate limiting step in carcinogenesis by these compounds. This was quite consistent with the concept that alpha oxidation of nitrosodimethylamine was necessary to convert it to an alkylating agent able to alkylate nucleic acids. This was less satisfactory as an explanation of the action of nitrosomorpholine, since the conversion of this compound to an alkylating agent for nucleic acids has not been satisfactorily demonstrated. In a further experiment of this type with deuterium-labeled nitrosomethylethylamines (Figure 6), it was

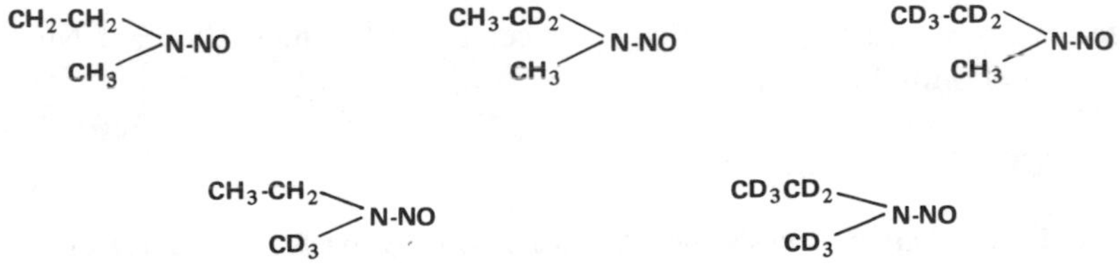

$$CH_2\text{-}CH_2\diagdown N\text{-}NO \qquad CH_3\text{-}CD_2\diagdown N\text{-}NO \qquad CD_3\text{-}CD_2\diagdown N\text{-}NO$$
$$CH_3\diagup \qquad CH_3\diagup \qquad CH_3\diagup$$

$$CH_3\text{-}CH_2\diagdown N\text{-}NO \qquad CD_3CD_2\diagdown N\text{-}NO$$
$$CD_3\diagup \qquad CD_3\diagup$$

Figure 6. Structures of deuterium-labeled nitrosomethylethylamines.

found that deuterium labeling of the methyl group reduced carcinogenic activity, suggesting that oxidation of the methyl group is involved in carcinogenesis. However, deuterium substitution at the alpha carbon of the ethyl group increased carcinogenic activity considerably, suggesting that oxidation at this position is unrelated to carcinogenesis but instead is involved in detoxication by other metabolic routes. Hence, reducing these other reactions increased the relative importance of oxidation of the methyl group, which is related to carcinogenesis. The relative tumor indices observed suggest that the ethyl group oxidation predominates over methyl oxidation by perhaps 4 to 1. These results further indicate, based on previous studies of intact transfer of alkyl groups in nucleic acid alkylation, that methylation of nucleic acids by nitrosomethylethylamine cannot be involved in carcinogenesis, although ethylation could be. In general, as with nitrosomethyl-*n*-butyl-amine, methylation is observed with unsymmetrical nitrosamines of this structure, and not alkylation by the large alkyl group. This might be another instance of the ease of being misled by examining the most obvious, which is probably unrelated to carcinogenesis; the latter appears to involve a pathway of metabolism of most carcinogens which is very minor, and therefore difficult to discover. Also methyl deuterium-labeled nitrosomethyl-n-butylamine is much more toxic than either the unlabeled compound or alpha-n-butyl-labeled nitrosomethyl-n-butylamine; carcinogenicity might very well be in the opposite direction.

These observations emphasize the need for studies of metabolism and distribution of carcinogens as part of any examination of mechanisms of carcinogenesis to avoid being misled by preconceived and simplistic ideas into observations of the most obvious, which is likely to be unrelated to carcinogenesis.

ACKNOWLEDGMENT

Research sponsored by the National Cancer Institute under Contract No. N01-CO-75380 with Litton Bionetics

REFERENCES

1. Miller, J. A. "Carcinogenesis by Chemicals—An Overview," *Cancer Res.* 30:559-576 (1970).
2. McCann, J., E. Choi, E. Yamasaki and B. N. Ames. "Detection of Carcinogens as Mutagens in the *Salmonella*/Microsome Test: Assay of 300 Chemicals," *Proc. Nat. Acad. Sci. (USA)* 72:5135-5139 (1975).
3. Snyder, C. M., J. G. Farrelly and W. Lijinsky. "Metabolism of Three Cyclic Nitrosamines in Sprague-Dawley Rats," *Cancer Res.* 37:3530-3532 (1977).
4. Elespuru, R. K., and W. Lijinsky. "Mutagenicity of Cyclic Nitrosamines in *E. coli* Following Activation with Rat Liver Microsomes," *Cancer Res.* 36:4099-4101 (1976).
5. Ames, B. N., W. E. Durston, E. Yamasaki and F. D. Lee. "Carcinogens Are Mutagens: a Simple Test System Combining Liver Homogenates for Activation and Bacteria for Detection," *Proc. Nat. Acad. Sci (USA)* 70: 2281-2285 (1973).
6. Magee, P. N., R. Montesano and R. Preussman, "N-Nitroso Compounds and Related Carcinogens," ACS Monograph No. 173 (1976), pp. 491-625.
7. Lijinsky, W., and R. K. Elespuru. "Mutagenicity and Carcinogenicity of N-nitroso Derivatives of Carbamate Insecticides," IARC Scientific Publications, No. 14 (1976), pp. 425-428.
8. Lijinsky, W., and D. Schmahl, "Carcinogenesis by Nitroso Derivatives of Methylcarbamate Insecticides and Other Nitrosamides in Rats and Mice," IARC Scientific Publications, No. 19 (1978), pp. 495-501.

9. Rao, T. K., A. A. Hardigree, J. A. Young, W. Lijinsky and J. L. Epler. "Mutagenicity of N-nitrosopiperidines with *Salmonella Typhimurium*/Microsomal Activation System," *Mutation Res.* 56:131-145 (1977).

10. Rao, T. K., J. A. Young, W. Lijinsky and J. L. Epler. "Mutagenicity of N-nitrosopiperazine Derivatives in *Salmonella Typhimurium*," *Mutation Res.* 57:127-134 (1978).

11. Rao, T. K., J. A. Young, W. Lijinsky and J. L. Epler. "Mutagenicity of Aliphatic Nitrosamines in *Salmonella Typhimurium*," *Mutation Res.* 66:1-7 (1979).

12. Lijinsky, W., and H. W. Taylor. "Carcinogenicity of Nitrosomethyldodecylamine in Sprague-Dawley and Fischer Rats Bearing Transplanted Bladder Tissue," *Cancer Lett.* 5:215-218 (1978).

13. Lijinsky, W., M. Mangino, H. W. Taylor and G. M. Singer. "Carcinogenicity of Nitrosomethylundecylamine in Fischer Rats," *Cancer Lett.* 5:209-213 (1978).

14. Preussman, R., G. Würtele, G. Eisenbrand and B. Spieglehalder. "Urinary Excretion of N-nitrosodiethanolamine Administered Orally to Rats," *Cancer Lett.* 4:207-209 (1978).

15. Singer, G. M., H. W. Taylor and W. Lijinsky. "Liposolubility as an Aspect of Nitrosamine Carcinogenicity: Quantitative Correlations and Qualitative Observations," *Chem-Biol. Interact.* 19:133-142 (1977).

16. Rao, T. K., D. W. Ramey, W. Lijinsky and J. L. Epler. "Mutagenicity of Cyclic Nitrosamines in *Salmonella Typhimurium*: Effect of Ring size. *Mutation Res.* 67:21-26 (1979).

17. Lijinsky, W., and H. W. Taylor. "Increased Carcinogenicity of 2,6-dimethylnitrosomorpholine Compared with Nitrosomorpholine in Rats," *Cancer Res.* 35:2123-2125 (1975).

18. Reznik, G., U. Mohr and W. Lijinsky. "Carcinogenic Effect of N-nitroso-2,6-dimethylmorpholine in Syrian Golden Hamsters," *J. Nat. Cancer Inst.* 60:371-378 (1978).

19. Keefer, L. K., W. Lijinsky and H. Garcia. "Deuterium Isotope Effect on the Carcinogenicity of Dimethylnitrosamine in Rat Liver. *J. Nat. Cancer Inst.* 51:299-302 (1973).

20. Lijinsky, W., H. W. Taylor and L. K. Keefer. "Reduction of Rat Liver Carcinogenicity of 4-nitrosomorpholine by Alpha Deuterium Substitution," *J. Nat. Cancer Inst.* 57:1311-1313 (1976).

29

AQUATIC HAZARD EVALUATION
STATE OF THE ART

Richard A. Kimerle

Monsanto Company
St. Louis, Missouri 63166

INTRODUCTION

During the past five years there has been a conscious effort on the part of scientists from government, industries and universities to establish a framework for appropriate evaluation of the safety of chemicals in the environment in general, and in particular aquatic organisms in aquatic ecosystems. This has happened partly in response to proposed regulations under the Toxic Substances Control Act and partly from the general recognition of the need to conduct more extensive in-depth safety testing of chemicals. Because of the cooperation which has existed, much progress has been made. The purpose of this chapter is to review the current state-of-the-art of aquatic hazard evaluation, while recognizing that it is still evolving.

At Monsanto an effort was made some five years ago to develop a system to study the environmental safety of a new high-volume chemical to partially replace phosphate in detergents. In sharing that safety program in meetings and through publication[1] it became obvious that the topic of how to perform aquatic hazard assessments was an important new subject. A subsequent publication by Monsanto presented more details on this subject.[2]

HAZARD EVALUATION PRINCIPLES

There now exist a number of approaches and procedures to evaluate aquatic safety/hazard of chemicals. However, a few consistent principles or facts of hazard evaluation have emerged.

1. Hazard assessment can only be performed by comparison of the toxicity of a chemical with the exposure concentration. A very toxic material with no exposure potential is not hazardous. Conversely, a chemical of low toxicity but high exposure could be much more hazardous.
2. Aquatic toxicity can be determined only by actually testing the chemical on the appropriate representatives of aquatic environments. Methodologies do exist to perform a number of tests from simple acute lethality through chronic tests on growth, reproduction, physiology and behavior using both freshwater and marine organisms. Field tests on effects are not yet developed to the same degree as clean-water laboratory studies.
3. Methodologies also exist to estimate and/or measure the exposure concentrations of chemicals in various compartments of the aquatic environment. Much research is currently underway to improve this area of environmental science.
4. Acquisition of data on toxic effects and environmental fate can most appropriately be done in a stepwise sequential manner or in tiers. This principle recognizes that not all chemicals require, or should be expected to undergo, the same amount of testing. It is also valuable for an industry to develop data in a sequential manner to facilitate necessary business decisions to stop or continue toward commercializing a new product.
5. Decision criteria must be built into the tiers to give guidance for when (1) the risk is acceptable and no more data are needed, (2) the risk is unacceptable and commercialization should be stopped or risk management practices must be developed, and (c) the risk is marginally acceptable and can only be resolved with additional data.
6. It is generally recognized that the closer the assessment is made to real-world condition, the more confidence we can have in our estimate of hazard. A chemical that receives only simple laboratory testing must demonstrate a much greater margin of safety between the effect and exposure concentration than a chemical with a narrow margin of safety. This means that marginally acceptable assessments of hazard may have to be ultimately resolved with actual field studies under use conditions.
7. No subjective system of hazard assessment should be expected to replace good scientific judgement weighing the risks with the societal benefits of a chemical.

HAZARD EVALUATION PROCESS

Aquatic hazard assessment dictates that a chemical be tested using the currently available array of toxicity and environmental fate tests. Because of the limited resources and time available, and the fact that there are so many chemicals which need testing under the new vigorous testing schemes, decisions must be made to test chemicals only to the point that a confident decision can be reached on the hazard of the material. At Monsanto we have found that use of the tier approach facilitates this decision process. Four tiers are used: (1) screening tests of short duration and minimum expense, which help eliminate obvious potential problem materials, (2) predictive tests of greater utility for estimating hazard but with a greater investment of time and resources, (3) confirmative tests which take us into the field to confirm some of the earlier laboratory data, and (4) monitoring studies which are conducted after commercialization to validate the safety of a material under actual use conditions.

Figure 1 summarizes the concepts and details of an aquatic hazard assessment procedure which is in the process of being developed at Monsanto. It incorporates most of the principles of other procedures. Data are acquired sequentially in the tiers of screening, predictive, confirmative and monitoring. Hazard is assessed by comparison of exposure concentration in environmental fate studies to toxic effects. Three types of decisions emerge as a result of this comparison:

1. Risk is unacceptable—stop testing and development of the chemical or develop risk management practices. This happens when there is no safety margin because the exposure concentration (either estimated, predicted, measured or validated) exceeds the toxic effect concentration.
2. Risk is acceptable—no further testing is needed because the margin of safety is judged more than adequate.
3. Risk is acceptable but the margin of safety is not as large as would be desired— acquire additional data in order to increase the confidence in the hazard assessment. This third case frequently means performing real-world studies of the confirmative type and/or monitoring the impact of the chemical on aquatic ecosystems under actual use conditions.

It should be noted that these "safety margins" have been presented only as a point of discussion. Adoption of any rigid guidelines at this time in the emerging science of aquatic hazard assessment would be inappropriate. However, there is a definite need to establish an understanding of what the acceptable and unacceptable criteria are for the protection of aquatic ecosystems. It is precisely beausec we don't understand all there is to know about assessing real aquatic hazard that we must bring the subject up front and discuss it.

Figures 2 and 3 graphically demonstrate in a simple way the concept of acceptable and unacceptable risks. As data are collected in the tiers of screening, predictive and confirmative, the biological effect concentration is greater than the expected and measured exposure concentration (Figure 2); then the risk is not as great as when the exposure concentration exceeds the biological effect concentration (Figure 3). Obviously, in this latter case the chemical in question probably would not have been developed beyond the

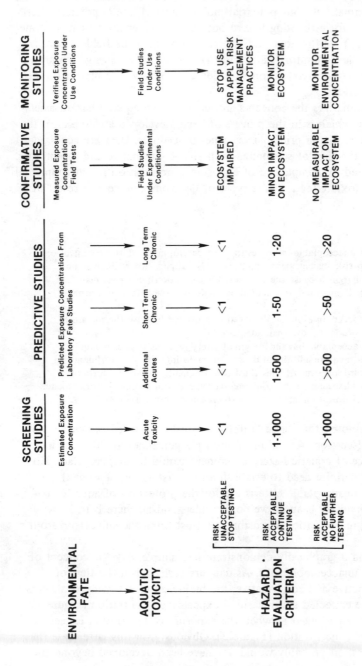

Figure 1. Incorporation of aquatic toxicity and environmental fate data into the aquatic hazard evaluation process.

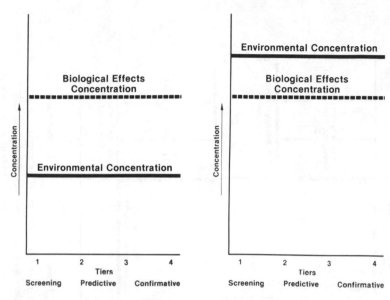

Figure 2. Comparison of biological effects concentration to environmental concentration when there is a large margin of safety.

Figure 3. Comparison of biological effects concentration to environmental concentration when there is no margin of safety.

early tiers. Both these cases are an oversimplification of a very complicated matter which needs to be brought ot the attention of the scientific community and resolved.

HAZARD EVALUATION NEEDS

As a result of this newly emerged understanding of aquatic hazard assessment, a dilemma has arisen. Simply stated, we, as aquatic toxicologists, have failed to demonstrate what the relationship really is between our clean-water laboratory toxicity data and the toxicity of the same chemical under real-world conditions. For those chemicals which fall in the marginal range of acceptability it may be crucial to know the effect of the real world on toxicity. It could either reduce toxicity through mitigating effects to an acceptable level or synergistically enhance toxicity to an unacceptable level.

Figure 4 and 5 present these two scenarios. Figure 4 demonstrates a hypothetical case where the heavily depended on clean-water laboratory studies of the screening and predictive tiers result in a fairly close estimate of biological effect and exposure concentration. However, when the confirmatory studies are conducted in natural waters of the real world, a different

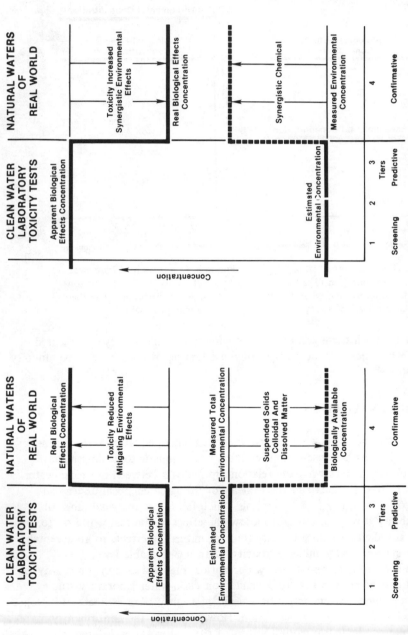

Figure 5. Hypothetical situation of an apparent large margin of safety from clean-water laboratory toxicity data actually being much smaller because of synergistic effects of natural waters.

Figure 4. Hypothetical situation of an apparent small margin of safety from clean-water laboratory toxicity data actually being much greater because of mitigating effects of natural waters.

understanding of hazard results. Although the total concentraton of a chemical may be confirmed as predicted, factors such as suspended solids, colloidal material, and dissolved matter make only a fraction of the total chemical available to the aquatic organisms. The effect is that these factors "mitigate" the toxicity so that it really takes much more total exposure to obtain the same significant biological effect. Thus, the margin of safety is really much greater than was perceived from the clean-water laboratory data.

On the other hand, Figure 5 demonstrates the reverse case, where the perceived wide margin of safety is significantly reduced because of some synergistic effect present in the natural water.

Although both these cases are hypothetical, it is important to get a better understanding of how the factors of the real world can influence our estimates of hazard. It seems quite likely that we have overlooked the role of mitigating effects while emphasizing the difficult to document cases of significant synergistic effects. Both no doubt operate. The challenge is to conduct more quality field studies and find out what the real utility is of clean-water laboratory studies. After the data have been obtained, perhaps we will have much more confidence in our laboratory data, or we may realize that field studies must play a larger role in aquatic hazard assessment.

CONCLUSION

This chapter presented some of the more-or-less accepted principles and facts of aquatic hazard assessment and indicated where the challenges are for the future. It is hoped that the cooperation that has existed over the past few years will continue and will permit forward progress based on the foundations which have been laid down.

REFERENCES

1. Kimerle, R. A., G. J. Levinskas, J. S. Metcalf and L. G. Scharpf. "An Industrial Approach to Evaluating Environmental Safety of New Products," in Proceedings of First Annual Symposium on Aquatic Toxicology, *Aquatic Toxicology and Hazard Evaluation*, F. L. Mayer and J. L. Hamelink, Eds., American Society for Testing and Materials, STP 634 (1977).
2. Kimerle, R. A., W. E. Gledhill and G. J. Levinskas. "Environmental Safety Assessment of New Materials," in *Estimating the Hazard of Chemical Substances To Aquatic Life*, John Cairns, Jr., K. L. Dickson and A. W. Maki, Eds., American Society for Testing and Materials, STP 657 (1978).

30

ENVIRONMENTAL TESTING OF TOXIC SUBSTANCES

Arthur M. Stern

Office of Toxic Substances
Environmental Protection Agency
Washington, DC 20460

INTRODUCTION

The Toxic Substances Control Act (TSCA) was passed in 1976 and probably represents the most far-reaching piece of legislation ever enacted with regard to its effect on the chemical industry. The act covers all chemicals except radionuclides, explosives, etc., and those covered by the Pesticide and Food and Drug Acts. Its language clearly defines its intent, which is the protection of both health and the environment. To meet these commitments, the act requires that the EPA develop test protocols for existing chemicals and, because of their importance, provide testing guidelines for new chemicals. It has been estimated that between 200 and 1000 new chemicals were introduced into commerce annually. Section 5 of the act requires manufacturers of new chemicals to notify EPA prior to beginning production. The agency requires a brief period of time in which to regulate those chemicals which present unreasonable risks. To implement the notice review process, the EPA must identify the information that will enable it to evaluate the risk these chemicals pose. Section 4 of the act addresses the approximately 70,000 chemicals already in commercial use and will cover only those assigned a high priority for consideration.

Section 5 guidelines are in the process of being developed. They are designed to identify the information that EPA considers necessary in a risk assessment and include a set of preliminary reference tests that industry can follow in preparing premanufacturing notifications. It is up to the submitter to determine whether additional testing should be carried out, and this decision would be based on information acquired from the base set of tests. It should be emphasized that these guidelines are intended to be flexible, in that the test protocols described in the base set are offered as reference methodologies. That is, certain exemptions from some testing and departures from prescribed protocols are possible, provided substantive reasons are presented for such omissions and deviations.

Section 4 test rules will be requirements that some existing commercial chemicals will have to satisfy. Since they will cover all testing needs for risk assessment rather than preliminary tests for general Section 5 screening purposes), they will be more extensive than the Section 5 guidelines. It is expected that Section 4 test rules will be arranged in tiers going from short-term and relatively inexpensive testing to longer-term testing. Figure 1 is a schematic of the relationship between Section 4 protocols and

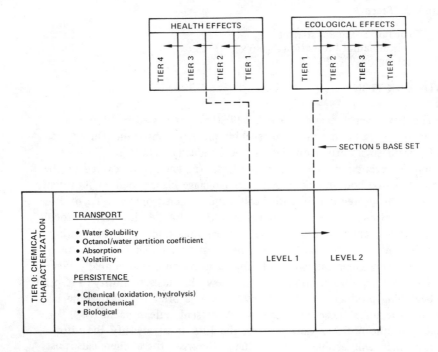

Figure 1. Relationship of section 4 test protocols to section 5 test guidelines.

Section 5 guidelines. It is expected that, as the higher tiers of Section 4 evolve, additional guidance will be provided to industry with regard to the types of tests that could be run beyond the base set where necessary. The basic goal here is for the agency to acquire enough information to make an assessment of a chemical's actual or potential risk to health and the environment. One further general fact to keep in mind is that neither the Section 5 guidelines nor the Section 4 rules are going to be static documents. The law requires that protocols be evaluated annually to take advantage of new and evolving technologies. Therefore, change with time is inevitable. In fact, it is the subject of an aggressive R&D program that the Office of Toxic Substances (OTS) is initiating.

This chapter will discuss the approaches the agency has taken to develop guidelines and protocols related to protection of the environment. It will also cover some major issues that must ultimately be resolved and present examples of protocols that have been tentatively selected.

APPROACHES TO DEVELOPMENT OF GUIDELINES AND PROTOCOLS

Figure 2 depicts the way test protocols (rules) are being developed for Section 4. With minor modifications, the same division of effort has been followed in developing Section 5 guidelines. The major organization responsible for developing test rules is a work group which has at least one representative from each office in the agency. For convenience of operation, the work group has been divided into several subgroups. The subgroup

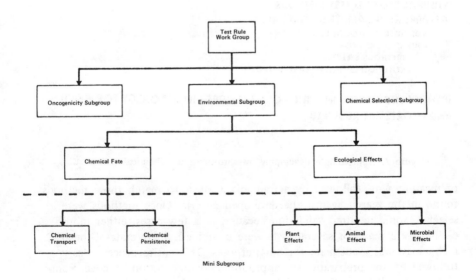

Figure 2. Test rule work breakdown structure.

of interest to this book is the Environmental Subgroup, which has been further subdivided into the Chemical Transport, Chemical Persistence, Plant Effects, Animal Effects and Microbial Effects Minisubgroups. The first two cover chemical fate, and the last three cover ecological effects. Although this book is concerned with chemical fate, the application of chemical fate testing results to the evaluation of ecological effects cannot be ignored; therefore, some of the approaches and issues that concern the latter will also be covered.

Before test protocols could be selected, key parameters for making assessments had to be identified, and these will be discussed later. Following the identification of key parameters, criteria for the selection of tests to measure these parameters had to be developed. Some of these criteria are summarized in Figure 3. Next came a state-of-the-art search for methods

DEVELOP CRITERIA FOR SELECTION OF BEST TEST METHODS
- APPLICABLE TO WIDEST SPECTRUM OF CHEMICALS/CHEMICAL CLASSES
- AMENABLE TO STANDARDIZATION
- PROVIDES CLEAR CUT AND REPRODUCIBLE RESULTS
- IS CONSIDERED AS STATE - OF - THE - ART TECHNOLOGY
- REQUIRES MINIMUM FACILITIES, EQUIPMENT, TIME AND TRAINED PERSONNEL

CONDUCT STATE - OF - THE - ART METHODOLOGY SEARCH
CONDUCT TRADE-OFF ANALYSIS
VALIDATE SELECTED TEST METHODS
PREPARE APPROPRIATE DOCUMENTS
- DETAILED PROTOCOL DESCRIPTIONS
- JUSTIFICATION FOR
  - METHODS SELECTED
  - REJECTION OF ALTERNATIVE METHODS

IDENTIFY WEAK AREAS IN STATE - OF - THE - ART METHODOLOGY FOR WHICH R&D EFFORTS ARE REQUIRED

Figure 3. Approach to developing test protocols for chemical transport.

of measurement, followed by a trade-off analysis to match those methods found in the search against the developed criteria. Once methods were selected, some required validation because, in a few cases, either modifications were made or several tests were combined where state-of-the-art technology was inadequate for our purposes. This entire process was followed by the preparation of appropriate documentation to cover complete protocol descriptions and justification for their selection. One spin-off of this procedure was the identification of soft spots in the state-of-the art, which made it possible to pinpoint those research and development areas needing support.

Figure 4 depicts the slightly different approach that was taken to develop test rules for both persistence and ecological effects. The major difference is the use of workshops to assist in the process. The second workshop involves participation of individuals both in and out of government. However, in both processes, the participation, advice and comments of experts in the disciplines covered are genuinely appreciated and sought.

CHEMICAL TRANSPORT

Having chemical transport information (1) identifies the possible environmental sites of primary chemical deposition and (2) provides an important element in estimating environmental exposure. More will be said about these applications later. Figure 5 lists those parameters considered to be significant in developing a concept of a chemical's mobility into and through the environment following its release. Thus, water solubility, volatility, adsorption, desorption and partition coefficient characteristics have been identified as key transport parameters.

There are a number of issues that have to be kept in mind when selecting test protocols to serve as predictors of chemical transport. Unfortunately, resolution of some of these problems must be deferred until the yield from research, development and experience produces new methodologies or modifies existing ones. Examples of such problems are:

- the extrapolation of laboratory results to events that occur in nature;

- the identification of environmental variables than can be appropriately incorporated into laboratory experimental designs. Of difficulty here is the cost effectiveness of pursuing this in an absolute sense. A good deal of effort has to be expended to determine where to draw the line.

- the judicious selection of reference materials (sometimes called benchmark chemicals). It is important that reference compounds be incorporated into test programs without appreciably raising the cost of testing beyond practical limits. The advantages of reference compounds is self-evident.

In spite of obvious gaps in our knowledge, a number of methods have been tentatively selected to measure the parameters felt to be critical in predicting chemical transport. A few examples are:

- *Water Solubility*. Four methods have been selected that are based on the physical nature of the test compound and the range of expected solubilities. Thus, there are tests for highly soluble (greater than 0.5 mg/l) solids, hydrophobic (ppb) solids and hydrophobic (ppb) liquids. It is important to know the solubility of materials fairly accurately because this information is important in the design of ecological and health effects testing experiments.

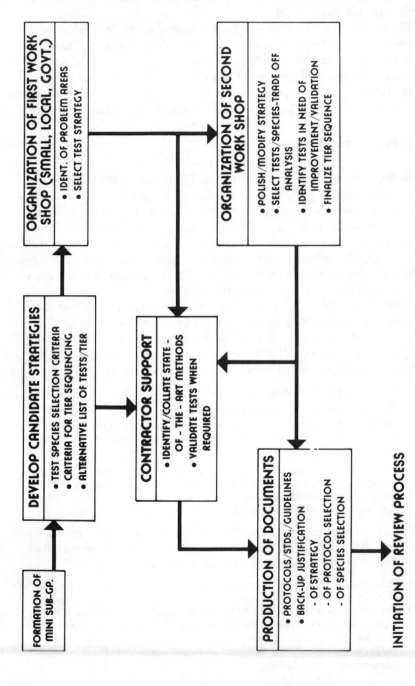

Figure 4. Approach to development of ecological effects test rules.

- SOLUBILITY
- VOLATILITY
- ADSORPTION
- DESORPTION
- PARTITION COEFFICIENT

Figure 5. Selected parameters for predicting chemical transport.

- *Volatility.* Volatility is to be determined by measurement of vapor pressure using an isotheniscope and a gas saturation method for determining values of from 1 to 760 mm Hg and 0.1 to 10 mm Hg, respectively. Only limited information can be derived from these tests because they don't allow for predictions to be made with regard to volatility from solids or perturbed bodies of water.

As has already been indicated, information regarding chemical transport can be utilized in a number of ways. An example is to be found in Figure 6, in which approximations can be made of probable sites of chemical distribution based on some physical/chemical characteristics of the test compound. This, in turn, could reduce the amount of ecological effects testing that would have to be carried out, because test species found in ecological compartments where disposition is unlikely to occur could be exempt from such testing. As stated before, chemical distribution information coupled with other data such as production volume, usage and disposal could allow an order-of-magnitude estimate to be made with regard to environmental concentrations or exposure. This could then be factored into the design of both health and ecological effects testing programs. Finally, chemical transport data could provide some direction to current or projected monitoring programs.

CHEMICAL PERSISTENCE

Chemical persistence can be defined as the property of a compound to retain its salient physical, chemical and functional characteristics in the environment through which it is transported and/or distributed for a finite period following its release. The determination of persistence of a chemical contaminant is an important aspect of the overall evaluation of a compound's potential environmental hazard. As long as the pollutant remains in existence, it may enter the food chain where it can affect man directly.

The three mechanisms by which an organic chemical may be degraded are biological, chemical and photochemical. Although there are specific

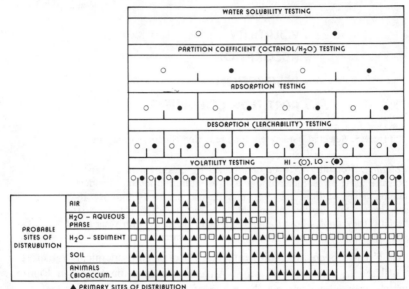

Figure 6. Testing for environmental mobility.

problems associated with each, there are also certain issues that are applicable to all mechanisms. Some of these can be summarized as follows:

- The identification of degradation products can present enormous analytical problems that are dependent on the nature of the parent compound. However, this can be an important exercise because there have been many cases where degradation products pose more serious health and ecological problems than the parent compounds from which they were derived. The decision with regard to what one considers "important" degradation products is also difficult because the risk these compounds present is not always a function of their relative concentration.

- The term "significant degradation" is difficult to define because its meaning is intimately associated with the effect of accumulated degradation products.

- Where multiple routes of degradation are operative, it is difficult to ascertain whether their effects are additive or mutually exclusive in the real world. Associated with this is the usual problem of extrapolating results of isolated laboratory tests to multiple events taking place *in situ*.

- There is the usual problem of designing into an experiment the pertinent environmental conditions found in nature. Adjusting pH and temperature is relatively easy, but designing an experiment to account for the protective and/or concentrating influence of particulates is much more difficult. Again, testing procedures have to be cost effective—*e.g.*, the effort and cost of generating information must be commensurate with its importance and usefulness.

In addition to the above, there are issues that have to be addressed for each mechanism of degradation. For example, in biodegradation there have been many arguments concerning the following problems:

- the use of mixed or pure cultures;
- the sources of mixed and/or pure cultures;
- the use of adapted or unadapted inocula;
- selection of chemical concentrations to use in a test and, in particular, the ratio of biomass to substrate concentration;
- the influence of cometabolites on an organism's response to chemical challenges;
- the role of microbial interactions in nature, such as synergism, symbiosis, antagonism, commensalism, etc.;
- differentiating chemical from biological degradation if natural materials are utilized as inocula.

Specific methods currently under consideration include measurement of changes in dissolved organic carbon, carbon dioxide evolution and oxygen uptake. They involve the use of relatively high and low substrate concentrations, high (sludge) and low biomass densities, and aerobic and anaerobic methods of culture. These methods have been worked out for Section 5 and, as such, represent only a base set of reference tests. They have quite a few deficiencies, many of which will be corrected before the Section 4 protocols are subjected to review and comment.

Some issues that are pertinent to chemical degradation are similar to those described for biodegradation. However, the following appear to be unique for chemical degradation:

- What are the most important natural reactants of concern?
- If the rate of reaction in laboratory testing is artificially accelerated in order to derive measurable results in a reasonable period of time (as has been suggested), how does this influence the validity of our conclusion regarding expected events *in situ*?
- In view of the diversity of structures and compositions of chemicals under study, is it feasible to pinpoint specific tests that have to be performed, or will general guidelines have to suffice?

Methods now under consideration for assessing the potential suceptibility of chemicals to undergo oxidation include (1) simulation of atmospheric oxidation with hydroxy radicals and ozone and (2) the use of peroxy-generating radicals to similate oxidation in water. At the present time, improved methods for evaluating oxidation are under investigation and should be utilized in our testing scheme once they are validated.

Issues related to photochemical degradation are also in need of solution. Some of these are:

- selecting wavelengths/energies to utilize in laboratory testing;

- dealing with indirect photochemical degradation via activators; and
- accounting for seasonal and geographical variations in developing experimental designs.

Right now, a photochemical test is under consideration for the Section 5 guidelines. This, again, is a part of a base set of reference tests and utilizes natural sunlight. Although there is still concern about seasonal and geographical variations, it is hoped that the selection and use of reference compounds will at least partially resolve this particular problem.

Figure 7 presents one scheme for handling persistence data. Level 1 is designed to provide "presumptive evidence" of degradation, while Level 2 testing is directed toward establishing the degree of significance of observed degradation. Obviously, both "presumptive evidence of degradation" and "significant degradation" require meaningful definitions. The third level of testing requires the qualitative and quantitative assessment of degradation products where significant degradation has taken place. The extent to which this type of analysis has to be carried must also be carefully defined. There is a break between levels 2 and 3. It is our thinking that, for Section 4, degradation mixtures should be subjected to biological activity (health and ecological effects) screening and that work on identification of reaction products should be carried out only if positive results are obtained from either screen. This has the advantage of allowing observations of the synergistic effects of multiple components of degradation mixtures and offers the possibility of exemptions from arduous work on the identification of degradation products.

SUMMATION

The Toxic Substances Control Act was designed for our protection and, in so doing, imposes a burden on all of us. It requires that (1) the EPA mobilize an appreciable number of well-trained individuals to carry out the mandates of the act, (2) the chemical industry apply its resources to conform to the act , and (3) the general public be prepared to share part of the costs associated with industry conformance. Above all, it makes mandatory the participation of all segments of the public sector in assisting the EPA in implementing the act.

The initial guidelines and protocols which will be issued by the agency should reflect this level of general participation. As far as those guidelines and protocols related to chemical fate testing are concerned, while they are deficient in many respects, it is felt that these deficiencies reflect weaknesses in current methodologies and that, as research and development are directed toward these areas, the quality of the testing package will improve accordingly.

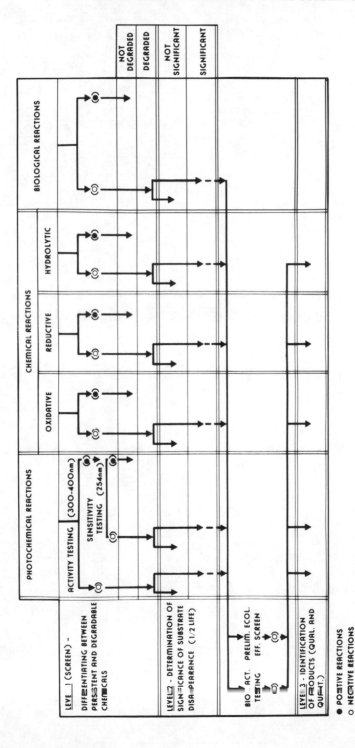

Figure 7. Persistence testing.

31

METHODOLOGY FOR ENVIRONMENTAL
HUMAN EXPOSURE AND HEALTH
RISK ASSESSMENT

A. A. Moghissi and R. E. Marland

Office of Research and Development
U. S. Environmental Protection Agency
Washington, DC 20460

F. J. Congel and K. F. Eckerman*

Office of Nuclear Reactor Regulations
Nuclear Regulatory Commission
Washington, DC 20555

INTRODUCTION

One of the major problems facing the environmental community is the assessment of human health risk subsequent to the release of an environmental pollutant. There appears to be considerable uncertainty as to what constitutes a health risk and how human exposure to a pollutant is established and quantified. One area in which experience in risk assessment has developed to a certain maturity is radiation effects of radionuclides released by nuclear power reactors. Because of controversy associated with this field, it has withstood repeated court tests. Many national and international organizations, notably the International Commission for Radiological Protection (ICRP), the International Commission on Radiation Units and Measurements (ICRU), the National Council on Radiation Protection and Measurements (NCRP) and the National Academy of Sciences (Washington) have published reports dealing with this subject. For example, ICRP and NCRP have developed many reports on biokinetics, exposure assessment and biological effects of radiation.[1] The National Academy of Sciences

---
*Present address: Oak Ridge National Laboratory, Oak Ridge, TN.

has prepared a report on the subject of biological effects of low levels of radiation, the so-called BEIR report.[5] These valuable contributions define a number of sequences in evaluating public health implications of radiation exposure. These principles can be used to compare risks associated with alternative actions.[6]

This chapter is intended as an application of these principles to toxicants other than radionuclides. Appropriate modifications are made to accommodate a generalized hazard-assessment methodology.

Figure 1 shows a simplified model for environmental exposure.[7] In order to simplify the applicability of the model, all boxes except the box containing source are termed environmental elements. The modes of transport

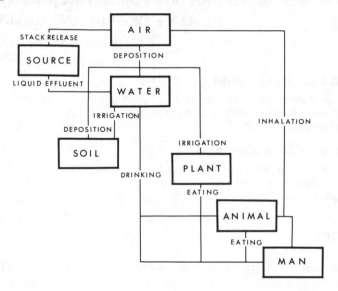

Figure 1. Environmental model.

of pollutants from one element to others are also shown in the figure. Figure 1 must be subdivided for each specific application. In particular, transfer of toxicants through the human food chain must be considered in more detail. The following description contains such a detailed discussion of human exposure as a result of release of toxic pollutants.

KINETICS OF ENVIRONMENTAL TOXICANTS

The movement of toxicants in air, water, biota and their ultimate sink in land or ocean is usually complex. In certain rare cases exact equations for their behavior either are known or are readily available. For example, the concentration of krypton-85 in the atmosphere is reasonably well known[8] and can be predicted on the basis of an assumed rate of energy generation. Similarly, the concentration of halogenated hydrocarbon can be measured and its distribution in the atmosphere can be predicted.[9]

In most cases, however, one is forced to make simplifying assumptions to predict the behavior of toxicants. As can be seen from the following, these cases can be treated advantageously by a first-order reaction. The dilution of toxicants in air and water follows diffusion processes. Dilution equations resemble first-order reaction equations, and therefore their validity does not need to be further elaborated here. However, the case of second-order reactions, which under special conditions of the environment can be treated as first-order reactions, will be discussed.

If a toxicant reacts with a compound, E, in the environment to yield a compound, X

$$T + E \; \rightleftharpoons \; X + \ldots \tag{1}$$

The classical kinetics equation for the reaction, 1, is as follows:

$$\frac{dX}{dt} = K(C_T - C_X) \; (C_E - C_X) \tag{2}$$

where C is the concentration and t is the time. If $E \gg T$, the section $(C_E - C_X)$ can be regarded as constant, and thus Equation 2 is reduced to

$$\frac{dX}{dt} = K_m(C_T - C_X) \tag{3}$$

where K_m is the modified reaction rate. Equation 3 is a first-order reaction, and its validity is more than generally appreciated. This validity does not depend upon the absolute value of C_E. The requirement is that C_E be considerably larger than C_X at the time t under consideration.

In cases where C_E is not considerably larger than C_X (or C_T), a series of several pseudo-first-order reactions can be used. For any given concentration range of T, the value of K can be regarded independent of T, and thus a first-order reaction can be assumed.

The following transport and exposure model assumes first-order reactions. This model has been used for many years by the Nuclear Regulatory Commission to predict radiation dose from nuclear power plants. It has been confirmed by experimental data and has been shown to be generally conservative. The model has been designed for atmospheric transport,[10] aquatic dispersion,[11] food chain contamination, exposure, biokinetics in the human body[12] and biological effects.[5] The model has been appropriately modified to accommodate the general nature of all toxicants.

KINETICS OF TOXICANTS IN THE HUMAN BODY

Environmental toxicology is largely based on the assumption that the toxicity of pollutants occurring in the environment at low levels is directly proportional to their concentration. It also assumes that the longer a toxicant is in contact with the tissue, the more probability of toxic effects. Therefore, in order to establish the toxicity of an environmental pollutant, the concentration of the toxicant ingested, inhaled, or otherwise taken into the human body along with its residence time must be known.

Environmental literature contains a number of definitions for the word "dose". Also words such as "exposure", "absorbed dose", "integrated dose", "toxic dose" and similar terms have been used to relate the intake of a toxicant to its toxicity. It is therefore necessary to define expressions used in this chapter (Figure 2).

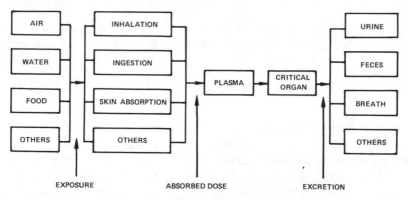

Figure 2. Intake and excretion of toxicants.

Exposure is defined as that fraction of toxicant which is in the immediate vicinity of various ports of entry such as lung, GI tract and skin. The word "others" in the exposure column deals with absorption through wounds or other potential intakes. The term "absorbed dose" is the quantity of toxicant which has passed appropriate membranes and is about to enter the body water pool.

Subsequent to the exposure and absorption, the toxicant is distributed into various organs of the body. A distinction is being made between the critical organ (or tissue) and the target organ (or tissue). Whereas target organ constitutes that organ which has the highest concentration of the toxicant, the critical organ is the one mostly affected by it regardless of whether it contains the highest concentration. Although in general critical organs will coincide with target organs, there may be cases where organs with lesser concentrations will exhibit higher degrees of risk because of their higher biosensitivity to the toxicant.

The kinetics of toxicants in the human body often follows an exponential function as follows:

$$A_t = A_o e^{-\lambda t} \tag{4}$$

where A constitutes the concentration of the toxicant at times t or o in an organ or in total body and λ is the excretion constant relating to the half-life $T_{1/2}$ by the relationship

$$\lambda = 0.693/T_{1/2} \tag{5}$$

Figure 3 shows retention of a toxicant following Equation 4.

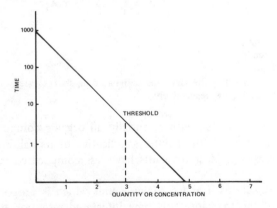

Figure 3. Retention of a toxicant in the human body, single exponent.

Frequently retention of toxicants in the human body follows a more complicated pattern (Figure 4). For example, tritium retention[13] follows Equation 6.

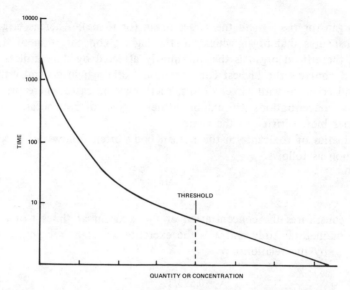

Figure 4. Retention of a toxicant in the human body, multiple exponent.

$$A = A_s e^{-0.693t/T_s} + A_w e^{-0.693t/T_w} + A_e e^{-0.693t/T_e} + A_1 e^{-0.693t/T_1} \quad (6)$$

where

A = the pool size,
t = the elapsed time,
T = the half-life,
subscripts s, w, i and 1 = the fractions designated as short, water,
    intermediate and long, respectively.

Because hydrogen is associated with essentially all organic compounds, the validity of this equation for tritium is indicative of its validity for other organic compounds. Most toxicants have less complicated retention patterns.

Integration of Equations 4 and 6 yields a unit which is proportional to the toxic effects of the toxicant. This time integrated retention dose is called retention dose (D_r) and is identical to Fridberg's target dose.[14]

In general, D_r will be calculated for a certain length of time, such as one year. The lifelong retention dose is further defined as dose commitment D_c. D_r and D_c are expressed in units such as gram days (g-day) or ng-day. Excretion of toxicants occur through urine, feces, breath and other routes such as perspiration.

REFERENCE HUMAN

The impact of population exposure to a pollutant depends not only on the nature and the concentration of the toxicant but also on a variety of other factors, including age, food habits and many physiological parameters. In rare cases all these parameters are known and can be applied for risk assessment. More often they are unknown and must be assumed or estimated. Recognizing this problem, ICRP has proposed a "reference man", a "reference woman" and "reference children" of various ages.[15] Although ICRP specifies that caution should be used in application of reference human data for areas other than radiation protection, a great deal of the information gathered in preparing reference human data is applicable to other environmental toxicants. Tables I and II contain some of these data.

Table I. Certain Data for Reference Human[15]

Description	M	F
Length of the Body (cm)	170	160
Surface Area of the Body (cm^2)	70	60
Specific Gravity of the Total Body	1.026	1.022
Mass of the Body (kg)	70	60

Table II. Tissue and Organ Weights for Reference Human According to ICRP[15]

Tissue/Organ	Weight Male (g)	Weight Female (g)	Remarks
Adult Human	70,000	58,000	
Skeleton—Wet	10,000	6,000	Ash = 28%
Skeleton—Dry	5,000	3,400	Ash = 56%
Cartilage	1,100	900	
Tendons, Fascia and Periarticular Tissues	1,500	1,200	900 for ♂, 700 for ♀ included with skeleton
Bone Marrow (total)	3,000	2,600	50% red, 50% yellow
Lymphocytes (total)	1,500	1,200	
"Fixed" Lymphatic Tissue	700	580	0.1% body weight
Spleen	180	150	
Thymus	20	20	
Skeletal Muscle	2,800	1,700	
Heart—Without blood	330	240	
Blood	5,200 ml	3,900 ml	
Tongue	70	60	
Salivary Glands	85	70	

Table II. continued

| Tissue/Organ | Weight | | Remarks |
	Male (g)	Female (g)	
Esophagus	40	34	
Stomach	150	140	
Intestine (total)	1,000	950	
Small	(640)	(600)	
Duodenum	(60)	(60)	
Jejunum	(280)	(250)	
Ileum	(300)	(290)	
Large	(370)	(360)	
Upper Ascending	(90)	(90)	
Upper Transverse	(120)	(110)	
Lower Descending	(90)	(90)	
Sigmoid Colon and Rectum	(70)	(70)	
Liver	1,800	1,400	
Gall Bladder	10	8	
Pancreas	100	85	
Nasal Mucosa	32	27	
Larynx	28	19	
Trachea	10	8	
Lungs	440	360	Includes associated lymph nodes, but not blood or bronchial tree.
Bronchial Tree	30	25	
Pulmonary Blood	530	430	
Arterial	(200)	(160)	
Venous	(230)	(190)	
Capillary	(100)	(80)	
Breast	26	360	Pair
Kidneys	310	275	Pair
Ureter	16	15	Pair
Bladder	45	45	Empty
Urethra	10	3	
Testes	35	–	Pair
Prostate	17	–	
Ovaries	–	11	Pair
Uterus	–	80	
Thyroid	20	17	
Parathyroid	0.12	0.14	Total for four
Adrenal	14	14	Pair
Pineal	0.13	0.15	
Pituitary	0.6	0.7	
Brain	1,400	1,200	
Spinal Cord	30	28	
Cerebrospinal Fluid	120 ml	100 ml	Volume measurement
Meninges	65	65	
Eyes	15	15	Pair

In addition to these "static" data, ICRP has also established certain data dealing with dynamics of pollutants. For example, Table III contains data dealing with the inhalation of particles containing toxic materials.[10]

Table III. Retention of Particulates in the Lung in %

	Readily Soluble	Others
Exhaled	25	25
Dep. upp. res. tract. (swallowed)	50	50
Dep. low res. tract	25[a]	25[b]

[a]Taken up into the body.
[b]About 12.5% swallowed in 24 hr and 12.5% taken into the body with a half-life of 120 days.

THE MODEL

A toxicant released to the environment becomes involved in the physical, chemical and biological processes occurring in the environment. Estimation of the retention dose implied by a toxicant's release can be derived from a mathematical model formulated to characterize the toxicant's dispersion in the dispersion media (air or water), the movement through various exposure pathways leading to humans, and human intake and metabolism of the toxicant. Rather sophisticated predictive mathematical models of atmospheric and hydrologic disperison and ecological movement of radionuclides have been developed and are routinely employed in the assessment of potential radiological impact of effluents of the nuclear fuel cycle. This established methodology served as the basis in proposing models for the evalution of nonradioactive pollutants.

General Retention Dose Expression

A generalized equation for estimating the retention dose can be written as

$$Rapj = Cp \ Uap \ Dapj$$

where Cp =   the concentration of a toxicant in the pth pathway medium, in $\mu g/m^3$ (inhalation) or $\mu g/kg$ (ingestion).

       Uap =   the intake rate (usage) of the pth pathway medium for individuals in the ath age group, in units of m^3/yr (inhalation) or kg/yr (ingestion).

       $Dapj$ =   the retention dose factor specific to age group a, pathway medium

p, and organ j; it represents the retention dose due to a unit intake of the toxicant, in μg-day/g-organ per μg inhaled or ingested

Rapj = the annual retention dose to organ j of an individual of age group
a through the pth exposure pathway in μg-days/g-organ

Presented below are mathematical models for estimating each of the factors noted above.

Atmospheric and Aquatic Dispersion

Material released to the atmosphere or hydrospere are transported and diluted as characterized by the direction and speed of the flow of the dispersing media, the intensity of the turbulent mixing, the topography of the mixing layer and the characteristics of the material being transported. Detailed description[5] of the models for aquatic and atmospheric dispersion have been published.[10,11]

Atmospheric Dispersion

The basic input data of atmospheric dispersion models include the wind speed, atmospheric stability and airflow patterns in the region of interest. In the commonly employed Gaussian straight-line trajectory model, the wind speed and atmospheric stability conditions at the release point are assumed to determine the dispersion characteristics throughout the region of interest in the direction of the wind. The equation for this model[10] is

$$X/Q = 2.032 \ \Sigma \ f \, i \, j \ [xui \ zj(x)]^{-1} \ exp \ [-he^2/2 \ \sqrt{} \ zj^2(x)] \tag{8}$$

where X/Q = the atmospheric dispersion factor representing the effluent con-
centration, x(units/m^3) normalized by source strength Q (units/sec), at the location a distance x (meters) downwind of the release

f_{ij} = the fraction of the time weather conditions are observed to be at the given wind direction, windspeed class i, and atmospheric stability class j

u_i = the midpoint of wind speed class i (m/sec)

x = the distance downwind of the release (m)

$\sqrt{}zj(x)$ = the vertical plume spread for stability class j (m)

he = the effective release height which reflects the physical height and considers the momentum and buoyancy of the effluent

Adjustments to the above equations may be necessary to reflect considerations of entrapment of the effluent in the building wake and mechanisms which remove the pollutant from the dispersing plume. These considerations are discussed in Reference 11.

Application of Equation 8 with knowledge of the local meteorological data permits the calculation of atmospheric dispersion factors as a function of distance and direction from the point of release. Once these factors are obtained, the airborne concentration of a toxicant can then be estimated at any location in the environment as

$$X = Q.X/Q \tag{9}$$

where Q = the toxicant release rate (μg/sec)
  X/Q = the atmospheric dispersion factor (sec/m^3)
  X = the atmospheric concentration (μg/m^3)

It is this airborne concentration that largely determines human exposure to the toxicant.

The aquatic environment includes a variety of surface water bodies, *e.g.*, nontidal river, lakes, reservoirs, estuaries and open coastal waters. In addition to the turbulent and flow character of the dispersing media, the size, geometry and bottom topography of the receiving water body influence the dispersion patterns and dilution. Models of hydrological dispersion employed in radiological impact assessments have been published.[11] The mathematical form of the models are somewhat distinct for each water body type and thus are not presented here.

Exposure Pathways

A toxicant released to the environment can reach man through a number of pathways, *e.g.*, the grass-cow-milk pathway. Figure 5 shows a generalized exposure pathway diagram delineating possible environmental transfers and resulting pathways to human beings.

Mathematical models of the environmental transfer of the toxicant can be developed through two basic approaches: the concentration-factor approach and the system analysis approach. The concentration-factor approach relates the media concentration of the toxicant at each point in the pathway to that of the previous medium, the ratio of the i to the i-l medium being the concentration factor. The initial medium (i = 1) is the waterborne or airborne concentration. The concentration factors are estimated from experimental observations, presumably at steady-state conditions, or can be inferred from the knowledge of the distribution of elements in the environment. The major limitation of this approach is the lack of any dynamic characterization of the pathway concentration.

The system-analysis approach is mathematically a more elegant model which considers the dynamic behavior of the toxicant in the environment. In this approach the environment is represented by a number of compartments which interact with each other through transfer coefficients. For each of the compartments, a differential equation describing the change in the inventory or concentration is written based on the input and output flux of the compartment. The solution to this system of differential equations

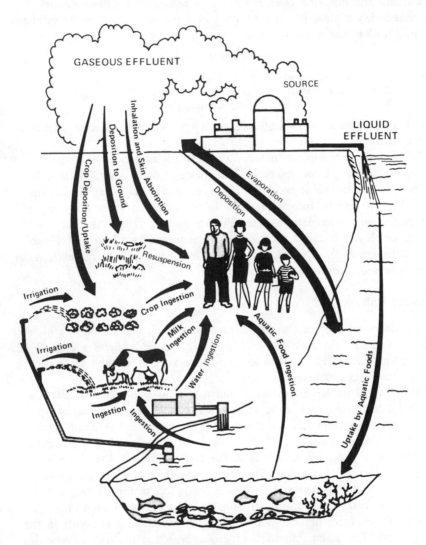

Figure 5. Generalized exposure pathway diagram.

is obtained through numerical methods with a computer. Unlike the concentration-factor approach, the system-analysis model can characterize the time-dependent behavior of the toxicant in the environment. The major limitation of this approach lies with the paucity of data upon which one can base the derivation of the compartment transfer coefficients. The two modeling approaches will yield comparable results under comparable conditions, *e.g.*, steady-state conditions which are usually readily achieved in chronic release considerations. The models presented below are based on the concentration-factor approach.

Concentrations in Liquid Pathway Media

The exposure modes associated with toxicants released to the aquatic environment are primarily the result of ingestion of drinking water and aquatic foods. The first step in estimating pathway media concentrations is the determination of the dispersed waterborne concentrations.

The dispersed waterborne concentration at a location downstream of a discharge can be related to the discharge concentration by a mixing ratio, M. This ratio can be estimated through use of an appropriate hydrological model, *e.g.*, in the case of a river where the location is sufficiently remote from the discharge that such uniform mixing in the river can be assumed, the mixing ratio is simply the volumetric ratio of the discharge flow to the river flow rates. The water concentration, C_w, is then given as

$$C_w = MQ/F \qquad (10)$$

where Q = the toxicant release rate ($\mu g/yr$)
M = the mixing ratio
F = the discharge flow rate (kg/yr)

Drinking Water Pathway

The consumption of drinking water represents a pathway of direct use of the dispersing media. Thus the dispersing water concentration as estimated from the mixing ratio may be used directly in the evaluation of this pathway. For some toxicants the drinking water concentrations may be altered due to the chemical and physical processes occurring within the water purification system of the public water supply. In these instances use of a concentration factor between the raw water and the drinking water should be considered. If this concentration factor is less than unity, it is actually a decontamination factor. The concentration, C_d, of the toxicant in the drinking water is thus

$$C_d = DF \cdot C_w \qquad (11)$$

where C_w = the dispersing water concentration of the toxicant, $\mu g/kg$
DF = the concentration factor
C_d = the concentration of the toxicant in drinking water ($\mu g/kg$)

In the absence of information on the concentration factor between the dispersing and drinking water, a value of unity may be assumed.

Concentration in Aquatic Foods

Aquatic organisms may take up a toxicant through various processes, *i.e.*, absorption and adsorption, metabolic processes, etc. The details of such processes are complex, involving feeding habits, osmoregulation and metabolic considerations, and are dependent on external environmental factors, such as water temperature and salinity.

The concentration-factor model approach avoids the nearly impossible task of assessing the details of the geological, environmental and ecological factors governing the uptake of the toxicant. The concentration of the toxicant in the aquatic organisms is related to the waterborne concentration by a concentration factor generally referred to as a bioaccumulation factor. Estimates of the bioaccumulation factor are obtained from experimental observations or inferred from other distribution factors in the aquatic environment. The toxicant's concentration, ca, in aquatic orgainsms is thus given as

$$Ca = BAF.Cw \tag{12}$$

where BAF is the toxicant's bioaccumulation factor specific for the toxicant and the organism of interest, and Cw is the dispersed waterborne concentration ($\mu g/kg$).

Concentrations in Atmospheric Pathway Media

The exposure modes associated with a toxicant's release to the atmosphere are primarily inhalation and ingestion of terrestrial food items. The toxicant's airborne concentration at any location in the environment can be estimated from knowledge of the magnitude of release and the atmospheric dispersion factor, X/Q. The airborne concentration is thus

$$X = Q(X/Q) \tag{13}$$

where X/Q is the atmospheric dispersion factor applicable to the location of interest, and Q is the toxicant's release rate. The airborne concentration is directly applicable to estimating the inhalation retention base and serves as the basis for estimating additional pathway media concentrations.

Terrestrial Food Pathways

The input of a toxicant into the terrestrial biosphere occurs as a result of deposition of the dispersed airborne toxicant onto the vegetative cover.

The flux of material crossing the atmosphere-ground plane interface ($\mu g/m^2$-sec) can be determined as the product of the airborne pollutant concentration, $X(\mu g/m^3)$, and the deposition velocity, $v(m/sec)$, *i.e.*, the flux is given as $X \cdot v$.

This material transfer may be the result of "settling out" of the airborne toxicant and/or arises as a result of physical-chemical interactions occurring at the atmosphere-vegetation interface. Estimates of the magnitude of this transfer, as represented by the deposition velocity for particulate matter, range from about 0.002 to 0.05 m/sec with a value of 0.01 generally considered as applicable. For chemically inert substances, *e.g.*, noble gases, this transfer can be neglected.

If the deposition velocity represents the total flux through the atmosphere-ground plane interface, then a fraction of the material transfer can be considered as captured by the vegetative cover. The deposited material will be subject to weathering and plant growth processes which reduce the concentration. The plant concentration of the toxicant can then be expressed as

$$CU = \frac{rXv\ [1-\exp(-\lambda_w t)]}{Y\lambda_w} \tag{14}$$

where
Cu = the toxicant concentration on vegetation ($\mu g/kg$)
r = the fraction of the flux captured by the vegetation
X = the airborne concentration of the toxicant ($\mu g/m^3$)
v = the deposition velocity (m/sec)
Y = the vegetative density (kg/m^2)
λ_w = the weathering loss rate coefficient ($time^{-1}$)
t = the plant exposure period (time)

In addition to the foliar deposition discussed above, vegetation can also be contaminated by the plant's uptake of the toxicant from the soil. The toxicant's concentration in the soil under conditions of chronic release is given as

$$Cs = \frac{Xv\ [1-\exp(-\lambda_s t)]}{P\lambda_s} \tag{15}$$

where
P = the density thickness of the root zone (kg/m^2)
λ_s = a loss-rate coefficient describing the toxicant's movement through the root zone
t = the duration of the chronic exposure

All other terms are defined above.

The toxicant's concentration in vegetation via the uptake from the soil is then estimated using a concentration factor, Bv, thus

$$C_u = B_v C_s \tag{16}$$

where Bv is the soil-to-plant concentration factor. Estimates of Bv are derivable from λ_s, the "weathering" loss rate coefficient, and t, the plant exposure period (growth period).

The retention of the deposited material was assumed above to be characterized by a first-order process represented by a loss coefficient,

λw. Experimental observations have suggested a rather broadly applicable loss coefficient corresponding to a half-time of about 14 days, *i.e.*, X_w = $1n2/14$ days half-time. For chronic exposure situations, the vegetative growth period is long composed to the romoval half-time, *e.g.*, t = 60 days, the factor containing the exponential will approach unity, and the above equation can be written as

$$Cu = \frac{rXv}{Y\lambda w} \qquad (17)$$

The total vegetation concentration including both foliar and soil uptake can then be written as

$$Cu = Xv \frac{r}{Y\lambda w} + \frac{Bv[1-\exp(-\lambda st)]}{P\lambda s} \qquad (18)$$

Concentration in Animal Products—meat and milk

The toxicant's concentration in milk and meat products can be related to the amount of material ingested by the animal, as follows:

$$C_m = F_m C_v Q_f$$

where   Fm   =   the fraction of the animal's daily intake of the toxicant which appears in each kg of milk or meat (d/kg)

         QF   -   the amount of feed consumed by the animal per day (kg/d)

         Cv   =   the toxicant's concentration in the animals feed (μg/kg)

The toxicant's concentration in the animal's feed should be estimated considering the diet of the animal, *e.g.*, fraction obtained from pasture. Typical value for the density of pasture grass, the parameter, Y, above, is 0.7 kg/m^2, while a value of 2.0 kg/m^2 is representative of forage crops. Thus, if fp is the fraction of the year animals are on pasture and fz denotes the fraction of their daily intake obtained from pasture grass while the animal is in pasture, Cv can be written as

$$Cu = \frac{rXv[\frac{fpfs}{Yp} + \frac{(1-fpfs)}{Ys}]}{\lambda w} \qquad (20)$$

where   Yp   =   the density of pasture grass (0.7 kg/m^2) and

         Ys   =   the density of forage (2.0 kg/m^2). In lieu of specific information as to the agricultural practice of the specific region it is often assumed, conservatively, that fp and fs are unity. Specific values for the other parameters are outlined in section

Pathway Usage Ratios

The retention dose is in part determined by the pathway concentrations discussed above and man's usage of the pathway medium. The usage is generally expressed as a consumption rate (kg/yr or m^3/yr). Numerical values have been developed for radiological assessments as appropriate for the various pathways and age group under consideration. In these evaluations, it is typical to consider an individual whose pathway usage is in excess of the normal usage, *i.e.*, the maximum individual, as well as a so-called "average" individual. Table IV presents values recommended for radiological assessments.

Table IV. Recommended Values for Uap[a]

Pathway	Infant	Child	Teen	Adult
Fruits, Vegetables[b] and Grain (kg/yr)	-	520 (200) [c]	630 (240)	520 (190)
Milk (kg/yr)	300	330 (170)	400 (200)	310 (110)
Meat (kg/yr)	-	41 (37)	65 (59)	110 (95)
Fish (kg/yr)	-	6.9 (2.2)	16 (5.2)	21 (6.9)
Other Seafood (kg/yr)	-	1.7 (0.33)	3.8 (0.15)	5 (1.0)
Drinking Water (hg/yr)	330	510 (260)	510 (260)	330 (370)
Inhalation (m^3/yr)	1400	3700	8000	8000

[a]Based on Table Ef and E-5 of Reference 12.
[b]Consists of the following (on a mass basis): 22% fruit, 54% vegetables (including leafy vegetables) and 24% grain.
[c]The values tabulated are for the maximum individual; the values in parenthesis are the average individual values.

ASSESSMENT OF BIOLOGICAL EFFECTS

The objective of this chapter is to illustrate how retention dose D_r and dose commitment D_c can be estimated. However, because D_r and D_c must be related to the established animal and human data, the relationship between D_r (and D_c) and health effects must be discussed. Although environmental toxicants cause a number of diseases, there appears to be a preoccupation with cancer both in the general public and in the scientific community. Therefore, this chapter will discuss the methodology of carcinogenic rask, although the method may be applied to other diseases with appropriate modifications.

One of the required modifications relates to the presence of a threshold in the induction of a disease. The correction for "threshold effects" is advantageously done during the calculation of D_r (and D_c). In these cases the integration of the concentration-time curve is carried out using the appropriate limit (Figures 3 and 4). Most carcinogenic effects are, however, considered linear and nonthreshold (LNT) in their dose-response curve. For example, it is generally agreed that ionizing radiation follows a LNT dose response. Similarly it is expected that most chemical carcinogens will follow LNT effect, proveded they reach the critical organ.

Hoel et al.[16] have described various techniques for estimation of carcinogenic risk. A recent symposium contains excellent additional information on this subject.[17] Briefly, the cancer risk, is presumed to follow the following equation

$$R = B + aD_r + b\ D_r^2 \tag{21}$$

where B is the spontaneous (background) risk and a and b are constants specific for each toxicant. At low levels, D_r^2 becomes small as compared to D_r and thus Equation 21 is reduced to

$$R = B + D_r\ Q \tag{22}$$

The relationship between Q and R can be illustrated by rearranging Equation 22 as follows

$$Q = (R-B)/D_r \tag{23}$$

Q is therefore the cancer risk resulting from the toxicant per unit of retention dose. The determination of Q is thus the key factor in establishing the carcinogenic rish by using data from feeding experiments using rodents. Typically, rats or mice are fed a few mg of the toxicant per day (or a few ppm per daily feed). This intake is subsequently related to the body weight of the animal. This mg toxicant/g body weight is assumed to be constant for all species. Hoel et al.[16] recommend the application of the body surface area, which is 2/3 power of body weight, in place of body weight. This extrapolation method is based on a number of assumptions including the following:

1. The GI tract of the rodents and humans absorb toxicants at the same rate.
2. Chemical changes occurring during the absorption are similar, resulting in identical products.
3. Biological half-lives of the toxicant in the two species are relatable by body mass or body area.
4. The metabolism of the toxicants, subsequent to the absorption, are similar, leading to the same critical organ in both species.
5. Biological impacts on the tissues of both species are similar.

If any of the above or other factors are known for specific toxicants, they are taken into account in the cancer risk estimation. Whereas in the field of radiation these and other factors are routinely considered, for many chemical toxicants they are unavailable. It appears that the application of procedures used for environmental health risk assessment of ionizing radiation to other environmental toxicant's would simplify and improve the accuracy of present risk-assessment methods.

REFERENCES

1. International Commission on Radiological Protection. *Report of Committee II on Permissible Dose for Internal Radiation* (Oxford: Pergamon Press, Inc., 1959).
2. International Commission on Radiological Protection. *Report of Committee IV on Evaluation of Radiation Doses to Body Tissues from Internal Contamination Due to Occupational Exposure,* (Oxford: Pergamon Press, Inc., 1967).
3. International Commission on Radiological Protection. *Radiation Protection,* ICRP 26 (Oxford: Pergamon Press, Inc., 1977).
4. "Basic Radiation Protection Criteria," NCRP 39 (Washington, D C: National Council on Radiation Protection and Measurements, 1971).
5. "The Effect on Populations of Exposure to Low Levels of Ionizing Radiation," BEIR Report (Washington, D C: National Academy of Sciences, 1972).
6. Moghissi, A. A., and M. W. Carter. "Public Health Aspects of Radioluminous Materials" FDA 768001, (Washington, D C: Food and Drug Administration, 1976).
7. Moghissi, A. A., R. G. Patzer and M. W. Carter. In: *Tritium,* A. A. Moghissi and M. W., Carter, Eds. (Phoenix, AZ: Messenger Graphics, 1972) p. 314.
8. Stanley, R. E., and A. A. Moghissi Eds. *Noble Gases,* U.S. Government Printing Office (1976).
9. "Nonflourinated Halomethanes in the Environment," (Washington, D C: National Academy of Sciences, 1978).
10. "Methods for Estimating Atmospheric Transport and Dispersion of Gaseous Effluents in Routine Releases from Light Water Cooled Reactors," U.S. Nuclear Regulatory Commission, Regulatory Guide 1.111 (1977).
11. "Estimating Aquatic Dispersion of Effluents from Accidental and Routine Reactor Releases," U.S. Nuclear Regulatory Commission, Regulatory Guide 1.113 (1977).
12. "Calculation of Annual Doses to Man from Routine Releases of Reactor Effluents for the Purpose of Evaluating Compliance with 10 CFR Part 50 Appendix I," U.S. Nuclear Regulatory Commission Regulatory Guide 1.109 (1977).
13. Moghissi, A. A., M. W. Carter and R. Liberman. *Health Physics* 21: 57-60 (1971).
14. *Environ. Health Pers.* 22:1-12 (1978).
15. "Report of the Task Group on Reference Man," International Commission on Radiological Protection 23 (Oxford: Pergamon Press, 1975).
16. Hoel, D. G., D. W. Gaylor, R. L. Kirschstein, U. Saffiotti and M. A. J. Schneiderman. *Toxicol. Environ. Health* 1: 133-151 (1975).
17. "Air Pollution and Cancer-Risk Assessment Methodology and Epidemiological Evidence," *Environ. Health Pers.* 22: 1-181 (1978).